Jayle R Matthews
Vancouver.
August 09, 2010

Springer Series on Environmental Management

Series Editors
Bruce N. Anderson
Planreal Australasia, Keilor, Victoria, Australia

Robert W. Howarth
Cornell University, Ithaca, NY, USA

Lawrence R. Walker
University of Nevada, Las Vegas, NV, 89154

For further volumes:
http.//www.springer.com/series/412

Derek Armitage · Ryan Plummer
Editors

Adaptive Capacity and Environmental Governance

Editors
Prof. Dr. Derek Armitage
Department of Geography and
Environmental Studies
Wilfrid Laurier University
N2L 3C5 Waterloo, Ontario
Canada
darmitag@wlu.ca

Prof. Dr. Ryan Plummer
Department of Tourism and Environment
Brock University
500 Glenridge Avenue
L2S 3A1 St. Catharines, Ontario
Canada

Stockholm Resilience Centre
Stockholm University
Stockholm, SE-106 91
Sweden
rplummer@brocku.ca

ISSN 0172-6161
ISBN 978-3-642-12193-7   e-ISBN 978-3-642-12194-4
DOI 10.1007/978-3-642-12194-4
Springer Heidelberg Dordrecht London New York

Library of Congress Control Number: 2010930468

© Springer-Verlag Berlin Heidelberg 2010
This work is subject to copyright. All rights are reserved, whether the whole or part of the material is concerned, specifically the rights of translation, reprinting, reuse of illustrations, recitation, broadcasting, reproduction on microfilm or in any other way, and storage in data banks. Duplication of this publication or parts thereof is permitted only under the provisions of the German Copyright Law of September 9, 1965, in its current version, and permission for use must always be obtained from Springer. Violations are liable to prosecution under the German Copyright Law.
The use of general descriptive names, registered names, trademarks, etc. in this publication does not imply, even in the absence of a specific statement, that such names are exempt from the relevant protective laws and regulations and therefore free for general use.

Cover design: WMXDesign GmbH, Heidelberg, Germany

Cover illustrations: D. Armitage

Printed on acid-free paper

Springer is part of Springer Science+Business Media (www.springer.com)

Preface and Acknowledgments

How communities, governments, and other social actors respond to environmental issues and nonlinear change is a pressing challenge. The implications for human societies and their sustainable economic development are profound. In response, a growing number of researchers and practitioners are examining governance dilemmas using the concept of adaptive capacity, defined here as an ability to learn, experiment, and foster novel solutions in complex social–ecological circumstances. With the rapidly growing interest in environmental governance, there is both a need and a tremendous opportunity to examine the concept of adaptive capacity from different perspectives, explore opportunities for conceptual development, and debate its role within an emerging scholarship on governance in complex social–ecological systems.

This volume emerged from our initial work in exploring adaptive capacity in community-based resource management contexts, and a subsequent interest in extending this analysis into the realm of governance. This is the first integrated volume we are aware of that seeks to connect the theoretical foundations of adaptive capacity with the identification of governance attributes and practices that build adaptive capacity.

The volume is a result of many valuable contributions. Foremost, we thank the contributing authors for their participation. Individually and collectively, the chapters in the volume provide a wide range of insights about adaptive capacity and governance. A striking feature of the volume is the geographic scope and varied resource contexts in which insights, ideas, and lessons are offered – from the boreal forests of Sweden, to fisheries in Brazil, and agricultural systems in South Africa. The case studies presented in this volume are offered by applied researchers who have long engagements with their respective research sites. However, the cases in this book are set alongside a number of contributions devoted to the exploration of cross-cutting themes and conceptual innovations on the frontier of research. This diversity of cases and conceptual exploration reveals the multifaceted nature of linkages between adaptive capacity and governance, and the context-specific strategies that must be employed to build governance processes that will adapt to change and steer societies towards sustainable pathways.

Each of the contributions in this volume has been peer-reviewed by at least two individuals. We are grateful for the valuable insights and constructive feedback provided by anonymous reviewers and the following individuals: Fikret Berkes, Joshua Cinner, Beatrice Crona, Alan Diduck, Frank Duerden, Georgina Cundill, Ioan Fazey, George Francis, Jennifer Fresque, Thomas Hahn, Kevin Hanna, Marco Janssen, Carina Keskitalo, Gita Laidler, Patrick McConney, Melissa Marschke, Dan McCarthy, Melanie Muro, Don Nelson, Mark Pelling, Lisen Schultz, Scott Slocombe, Robert Summers, and Sonia Wesche.

Several individuals and organizations have provided essential support for this initiative. We thank especially Chris Luckhart, Jennifer Fresque, and Kate Cave for the editorial and logistical support they have provided. We would also like to thank the editorial team at Springer, Lawrence Walker and Andrea Schlitzberger in particular, for their assistance with this project. The volume has been made possible with support of our research programs from the Social Sciences and Humanities Research Council of Canada, Wilfrid Laurier University, and Brock University through the Chancellor's Chair for Research Excellence.

April 2010   Derek Armitage and Ryan Plummer

Contents

1 **Integrating Perspectives on Adaptive Capacity and Environmental Governance** 1
Ryan Plummer and Derek Armitage

Section I Adaptive Capacity in Theory and Practice

2 **Adaptive Capacity in Theory and Reality: Implications for Governance in the Great Barrier Reef Region** 23
Erin Bohensky, Samantha Stone-Jovicich, Silva Larson, and Nadine Marshall

3 **Building Adaptive Capacity in Systems Beyond the Threshold: The Story of Macubeni, South Africa** 43
Christo Fabricius and Georgina Cundill

4 **Learning and Adaptation: The Role of Fisheries Comanagement in Building Resilient Social–Ecological Systems** 69
Daniela C. Kalikoski and Edward H. Allison

5 **Adaptive Capacity and Adaptation in Swedish Multi-Use Boreal Forests: Sites of Interaction Between Different Land Uses** 89
E. Carina H. Keskitalo

6 **From the Inside Out: A Multi-scale Analysis of Adaptive Capacity in a Northern Community and the Governance Implications** 107
Sonia Wesche and Derek Armitage

7 **Vulnerability and Adaptive Capacity in Arctic Communities** 133
Robin Sydneysmith, Mark Andrachuk, Barry Smit, and Grete K. Hovelsrud

8 **Climate Change, Adaptive Capacity, and Governance for Drinking Water in Canada** 157
Rob de Loë and Ryan Plummer

9 **Institutional Fit and Interplay in a Dryland Agricultural Social–Ecological System in Alberta, Canada** 179
Johanna Wandel and Gregory Marchildon

Section II Frontiers in Adaptive Capacity

10 **The Learning Dimension of Adaptive Capacity: Untangling the Multi-level Connections** 199
Alan Diduck

11 **Adaptive Capacity as a Dynamic Institutional Process: Conceptual Perspectives and Their Application** 223
Ralph Matthews and Robin Sydneysmith

12 **Sociobiology and Adaptive Capacity: Evolving Adaptive Strategies to Build Environmental Governance** 243
David A. Fennell and Ryan Plummer

13 **Building Transformative Capacity for Ecosystem Stewardship in Social–Ecological Systems** 263
Per Olsson, Örjan Bodin, and Carl Folke

14 **Adapting and Transforming: Governance for Navigating Change** 287
Derek Armitage and Ryan Plummer

Index 303

Contributors

Edward H. Allison Policy, Economics and Social Sciences, The World-Fish Center, Penang, Malaysia

Mark Andrachuk Department of Geography, University of Guelph, Guelph, ON, Canada

Derek Armitage Department of Geography and Environmental Studies, Wilfrid Laurier University, N2L 3C5 Waterloo, ON, Canada

Örjan Bodin Stockholm Resilience Centre, Stockholm University, Stockholm, SE-106 91, Sweden

Erin Bohensky CSIRO Sustainable Ecosystems, Davies Laboratory, PMB Aitkenvale, QLD 4814, Australia

Georgina Cundill Rhodes University, P.O. Box 94, Grahamstown, 6140, South Africa

Rob de Loë Environment and Resource Studies, Faculty of Environment, University of Waterloo, N2L 3G1 Waterloo, ON, Canada

Alan Diduck Environmental Studies Program, The University of Winnipeg, 515 Portage Avenue, R3B 2E9 Winnipeg, Canada

Christo Fabricius Rhodes University, P.O. Box 94, Grahamstown, 6140, South Africa; Nelson Mandela Metropolitan University, P/Bag X 6531, George 6530, South Africa

David A. Fennell Department of Tourism and Environment, Brock University, 500 Glenridge Avenue, L2S 3A1 St., Catharines, ON, Canada

Carl Folke Stockholm Resilience Centre, Stockholm University, SE-106 91, Stockholm, Sweden; The Beijer Institute, The Royal Swedish Academy of Sciences, Stockholm, 104 05, Sweden

Grete K. Hovelsrud Center for International Climate and Environmental Research, Oslo, Norway

Daniela C. Kalikoski Institute of Humanities and Information Science, Federal University of Rio Grande (FURG), Rio Grande, Brazil

E. Carina H. Keskitalo Department of Social and Economic Geography, Umeå University, Umeå 901 87, Sweden

Silva Larson CSIRO Sustainable Ecosystems, Davies Laboratory, PMB Aitkenvale, QLD 4814, Australia

Gregory Marchildon Johnson Shoyama Graduate School of Public Policy, University of Regina, Regina, SK, Canada

Nadine Marshall CSIRO Sustainable Ecosystems, Davies Laboratory, PMB Aitkenvale, QLD 4814, Australia

Ralph Matthews The University of British Columbia, 6303 North West Marine Drive, V6T 1Z1 Vancouver, BC, Canada

Per Olsson Stockholm Resilience Centre, Stockholm University, SE-106 91, Stockholm, Sweden

Ryan Plummer Department of Tourism and Environment, Brock University, 500 Glenridge Avenue, L2S 3A1 St. Catharines, ON, Canada; Stockholm Resilience Centre, Stockholm University, Stockholm, SE-106 91, Sweden

Barry Smit Department of Geography, University of Guelph, Guelph, ON, Canada

Samantha Stone-Jovicich CSIRO Sustainable Ecosystems, Davies Laboratory, PMB Aitkenvale, QLD 4814, Australia

Robin Sydneysmith Department of Sociology, The University of British Columbia, 6303 North West Marine Drive, V6T 1Z1 Vancouver, BC, Canada

Johanna Wandel Department of Geography, University of Waterloo, N2L 3G1 Waterloo, ON, Canada

Sonia Wesche Geography and Environmental Studies, Wilfrid Laurier University, Waterloo, Canada; National Aboriginal Health Organization - Métis Centre, 220 Laurier Ave. West, Suite 1200, Ottawa ON K1P 5Z9, Canada V2N 4Z9

Chapter 1
Integrating Perspectives on Adaptive Capacity and Environmental Governance

Ryan Plummer and Derek Armitage

1.1 Introduction

Promising governance strategies for complex social–ecological conditions are emerging. Although these strategies build upon well established ideas from environmental management, emphasis has increasingly been directed at recognizing feedback processes, nonlinearity, and the problem of "fit" between institutions and biophysical systems (Young 2002; Galaz et al. 2008). Strategies better suited to embrace resource uncertainty and environmental change via collaborative processes and systematic learning highlight the importance of multilevel interactions among social actors with conflicting objectives (Adger et al. 2005; Lebel et al. 2005), innovative ways of producing and sharing knowledge (van Kerkhoff and Lebel 2006; Berkes 2009), and linking science and policy through communities of practice (Reid et al. 2006; Davidson-Hunt and O'Flaherty 2007) to cope with uncertainty and adapt to change. Building adaptive capacity is at the core of these promising governance strategies.

Adaptive capacity is broadly defined as the ability of a social–ecological system (or the components of that system) to be robust to disturbance and capable of responding to change (Walker and Salt 2006; Carpenter and Brock 2008). As Folke et al. (2003) identified, four general factors may foster adaptive capacity in social–ecological systems, particularly during periods of crisis: (1) learning to live with

---

R. Plummer
Department of Tourism and Environment, Brock University, 500 Glenridge Avenue, L2S 3A1 St., Catharines, ON, Canada
Stockholm Resilience Centre, Stockholm University, Stockholm, SE-106 91, Sweden
e-mail: rplummer@brocku.ca

D. Armitage
Department of Geography and Environmental Studies, Wilfrid Laurier University, N2L 3C5 Waterloo, ON, Canada
e-mail: darmitag@wlu.ca

change and uncertainty; (2) nurturing diversity for resilience; (3) combining different types of knowledge for learning; and (4) creating opportunity for self-organization toward social–ecological sustainability. Adaptive capacity, therefore, provides a valuable analytic construct around which managers, scientists, resource users, and policy makers can come together in theoretically engaged but decidedly applied ways to address the challenges of governance. Yet, several important questions need to be addressed to realize the potential of adaptive capacity to enhance environmental governance: What is the theoretical basis of adaptive capacity? How can adaptive capacity be effectively applied through policy and cultivated in practice? What issues require future consideration?

The goal of this volume is to address the need for a consolidated, interdisciplinary approach to the theoretical advances and practical implications of adaptive capacity in the context of an emerging environmental governance discourse that emphasizes local–global interactions and linkages, adaptiveness, and learning. Specifically, the volume seeks to (1) synthesize current knowledge and understanding of adaptive capacity in the context of environment and natural resource governance (e.g., climate change, fisheries, water resources, forests); (2) build theory from synthesis of experiences with adaptive capacity by identifying principles and critically examining key features of the concept; (3) highlight the implications of theory and experience for practice; (4) foster policy innovation; and (5) encourage efforts to build capacity for novel governance approaches to address complex environment and natural resource challenges.

This introductory chapter sets the stage for the book. Contemporary environmental challenges and the emerging governance agenda provide a backdrop and rationale for the volume. Adaptive capacity is then defined in relation to its interdisciplinary heritage and situated in the context of previous research. Areas of scholarship that inform adaptive capacity are highlighted as key themes, upon which the interdisciplinary contributors to this volume advance concepts, tools and knowledge about the critical relationships among global environmental change, multilevel environmental governance, and adaptive capacity.

1.2 Contemporary Environmental Challenges: A Synopsis

Kofi Annan called for the Millennium Ecosystem Assessment (MA) in his report to the United Nations General Assembly in the year 2000 (Millennium Ecosystem Assessment 2005). From 2001 to 2005, the MA was conducted to "... assess the consequences of ecosystem change for human well-being and to establish the scientific basis for actions needed to enhance the conservation and sustainable use of ecosystems and their contributions to human well-being" (Millennium Ecosystem Assessment 2005, p. ii). The conceptual framework developed to orient the assessment concentrates on human well-being, assumes dynamic interactions between people and parts of ecosystems, acknowledges other factors that also influence humans and ecosystems, and recognizes the need for a multiscale approach (Millennium Ecosystem Assessment 2003). The multiscale assessment involved the compilation, judgment, and interpretation of existing knowledge by some 1,360 experts from 95 countries. It yielded four main findings: (1) over the past half century, humans have changed ecosystems more quickly and extensively than in the past due primarily to the growing demands for ecosystem services by humans and have caused a "substantial and irreversible loss in diversity of life on Earth"; (2) while changes have furthered "substantial net gains in human well-being and economic development", they have also come with increasing costs that will diminish the benefits possible to future generations; (3) degradation of ecosystem services is an impediment to accomplishing the Millennium Development Goals and could worsen substantially in the next 50 years; and (4) options exist to reduce negative trade-offs and bolster positive synergies with ecosystem services, but these will involve new and substantial changes by humans (Millennium Ecosystem Assessment 2005).

Since conveying the detailed findings for all systems are beyond the scope of this book, we draw attention to the contemporary challenges of global–local environmental change captured by the MA (2005) and more recent evidence and assessments that support these findings (Stern 2007). Most substantial has been the transformation of one quarter of the land surface area from terrestrial ecosystems (e.g., forests and grasslands) to cultivated systems. Forest cover in particular has declined at 0.2% per year, with Africa as well as Latin America and the Caribbean losing forests at higher rates (FAO 2007). Marine and freshwater ecosystems have also been altered considerably. Fishing pressure in marine systems has been intensified dramatically in response to the world demand for food to the point where trophic levels harvested are declining and biomass is generally reduced to less than one-tenth of preindustrial fishing levels. Modification of freshwater systems is pervasive throughout the world with flows in more than half of the large river systems worldwide being moderately or strongly affected by capture structures and withdrawals. Substantial change is also evident in polar systems, where average temperatures are warmer than the past 400 years, precipitating widespread reductions in sea ice and thawing of permafrost.

The MA (2005) also focused on the critical factors responsible for ecosystem changes. The term "driver" is used to describe natural or human-induced factors that cause changes; direct drivers unequivocally influence ecosystem processes while indirect drivers cause ecosystem change by influencing one or more direct drivers. Direct drivers include habitat change, overexploitation, invasive alien species, pollution and climate change. Indirect drivers identified in the MA (2005) include human population change, change in economic activity, sociopolitical factors, cultural factors, and technological change. The scenarios developed by the MA to explore potential future situations suggest that these drivers will continue to cause changes and the relative importance of particular drivers will increase; climate change and associated impacts may be the dominant direct drivers of change in ecosystems by the end of the century (Millennium Ecosystem Assessment 2005). The Intergovernmental Panel on Climate Change (IPCC) corroborated the importance of this driver as "there is *high agreement* and *much evidence* that with current climate change mitigation policies and related sustainable development practices, global GHG emissions will continue to grow over the next few decades" (2007, p. 44).

This synopsis of global–local environmental change signals the scope and severity of contemporary environmental challenges within which the chapters in this book are located. The interrelationships among current and future drivers clearly illustrate the complicated nature of the causes and outcomes of these environmental challenges. The attendant issues of how to approach these challenges is the quintessential question for those concerned with environmental governance.

1.3 Environmental Governance

The limitations of conventional command and control approaches to environmental problems that will dominate this century are now recognized. These limitations include substantial economic costs associated with compliance and enforcement of regulations; extensive litigation that is often associated with regulatory approaches and the management decisions of regulatory agencies; political conflicts and polarization of stakeholders; limited gains with respect to initial problems and often unforeseen (and undesirable) outcomes (Holling and Meffe 1996; Cortner 2000; Kettle 2002; Durant et al. 2004); and more generally a "pathology of natural resource management" (Holling and Meffe 1996; Cortner 2000; Briggs 2003). The nature of social–ecological problems thus combines with limitations of conventional approaches and pervasive uncertainties to demand broader consideration of governance systems forged to address recurring and emerging environmental challenges.

Governance is distinct from government. Governance refers to "... the whole of public as well as private interactions taken to solve societal problems and create societal opportunities. It includes the formulation and application of principles guiding those interactions and care for institutions that enable them (Kooiman and Bavinck 2005, p. 17)". More recently, Biermann et al. (2009, p. 3) defined governance in the context of an earth systems perspective as, "... the interrelated and increasingly integrated system of formal and informal rules, rule-making systems, and actor-networks at all levels of human society (from local to global) that are set up to steer societies towards preventing, mitigating, and adapting to global and local environmental change". This definition has a number of advantages, including a stronger emphasis on scale, integration, and the normative emphasis on seeking ways to "steer" societies away from harm and toward sustainability. Finally, the definition emphasizes the importance of adapting to change. Flexibility to respond quickly and proactively to uncertain circumstances is recognized as a focal point of governance.

Both these definitions and the issues they raise point to the varied manner in which environmental governance is conceived. Indeed, environmental governance as a specific form or subcategory of a broader governance concerns has been conceptualized from a variety of perspectives, determined largely by the disciplinary orientation of the scholar or analyst. One common typology is to distinguish among several models or agents, including regulatory control through bureaucracies of the state and international regimes, market-based approaches, and a range of more civil society-oriented approaches. These forms and the related actors often combine or hybridize in practice in the form of governance networks with different degrees of formality (e.g., private–social partnerships, public–private partnerships, or cooperative arrangements between states and communities), to address critical environmental challenges (see Glasbergen 1998; Lemos and Agrawal 2006).

Recognizing the importance of interplay between structure and agency in different governance forms is necessary. However, we argue here that in light of global environmental change, an overly static and structural perspective of governance is limiting, as is the tendency to consider actors through a lens of those that govern as opposed to those being governed. Instead, the crucial task as summarized by Galaz et al. (2008, p. 169) is how to "... create governance that is able to 'navigate' the dynamic nature of multilevel and interconnected socio-ecological systems..." This view draws attention to several crucial aspects of governance, such as (1) embracing integrative science of sustainability and interconnections between social and ecological systems (Berkes et al. 2003; Folke 2007); (2) fostering attributes of institutions that are devolved, participatory, deliberative, accountable, just, multilayered and polycentric (e.g., Kettle 2002; Durant et al. 2004; Lebel et al. 2006; Huitema et al. 2009); and (3) framing governance in the context of a complex adaptive systems approach and dynamics of cross-scale and cross-level interactions (Dietz et al. 2003; Lemos and Agrawal 2006; Cash et al. 2006; Folke 2007; Ostrom 2007).

Although governance strategies are emerging with slightly different names (e.g., adaptive governance, adaptive comanagement, cogovernance) and points of emphasis (e.g., consequences during periods of change or transformation, the intricacies of co-management and learning processes, partnerships extending across state-market-society), they share the sentiment that making environmental governance operational requires collaboration among heterogeneous actors with diverse interests, institutions that are flexible and nested across scales and levels, and analytic deliberation that develops understanding through multiple knowledge systems; builds trust through repeated interactions; and fosters learning and adaptive responses through continuous feedback (Dietz et al. 2003; Folke et al. 2003, 2005; Armitage et al. 2009). However, there are many unanswered questions regarding the institutional systems that facilitate or constrain the capacity of actors to respond to change and learn through uncertainty. Similarly, greater clarity regarding the specific attributes and relational factors (e.g., collaborative processes, social capital, and social networks) that foster efforts to navigate multilevel and interconnected socio-ecological systems is required. Adaptive capacity provides a valuable analytical entrée into those strategies and requirements paramount to the making of environment governance.

1.4 Adaptive Capacity

Adaptive capacity has a myriad of meanings and a diverse intellectual ancestry from which various perspectives emerge. For example, adaptive capacity is traceable to the natural sciences and evolutionary biology, in which features that permit adaptation and success (i.e., fitness) have received attention from individual to ecosystem scales (see Smit and Wandel 2006; Gallopín 2006). In the social sciences, anthropologist Julian Steward is credited with transferring the biological idea of adaptation to human systems by introducing the theory of cultural ecology, defined as "the study of the processes by which a society adapts to its environment" (1968, p. 337). This general concept, like other biological terms (e.g., carrying capacity, homeostasis) was taken up by social scientists in human ecology, and the related fields of anthropology and geography, in an effort to understand cultural practices in light of environmental changes (Smit and Wandel 2006; O'Brien and Holland 1992). Building on these antecedents, ideas of adaptation and adaptive capacity have been a central theme in a variety of environment and resource literatures and applied areas including risks and hazards, political ecology, climate change studies, resilience thinking and social–ecological systems. The various meanings associated with adaptive capacity from the above perspectives are summarized in Table 1.1 (see also Smit and Wandel 2006; Füssel 2007). It is important to acknowledge that these meanings did not develop in complete isolation and to varying degrees have been shaped by multiple perspectives.

In the domain of governance, the more recent use of adaptive capacity has been relatively generic, thus highlighting a rationale for this volume. Adaptive capacity is generally referred to as the capability of a social–ecological system to be robust to disturbance, and to adapt to actual or anticipated changes (whether exogenous or endogenous). From a social systems vantage point, adaptive capacity is determined by the suite of resources (technical, financial, social, institutional, political) held, and the social processes and structures through which they are employed and mediated (i.e., governance). This definition thus frames ideas of adaptive capacity within the body of scholarship on institutional dynamics and environmental governance (Folke et al. 2002a, b, 2005; Ostrom et al. 2002; Brunner et al. 2005; Nelson et al. 2007; Folke 2007; Armitage 2008). Positioning adaptive capacity in this way recognizes the contemporary context of environmental challenges, which are characterized by complexity, discontinuity, surprise, and change (social and ecological). It also builds upon the assumption that strategies to build resilience will be found in the domain of governance and require collaborative and adaptive interactions among a diverse set of actors (e.g., scientists, resource users, policy makers) at local to global scales.

The approach taken to adaptive capacity in this volume is integrative. It brings together many strands of scholarship on adaptive capacity (Table 1.1) and seeks to bridge the bodies of work that inform adaptive capacity. In so doing, this volume (1) acknowledges the importance of the evolutionary dimension of adaptive capacity and recognizes that key determinants must be better understood (e.g., institution building, trust building, and social learning for adaptive capacity are long-term concerns); (2) underscores that adaptive capacity in environment and resource governance is a social process, and must confront a diversity of social actors (e.g., government agencies, "communities", industry) with different roles and relationships and a diversity of interests; and (3) cultivates the connection between adaptive capacity and complex adaptive systems thinking, and the subsequent need

Table 1.1 Meanings of adaptive capacity

Perspective	Meaning	Key references
Natural sciences (evolutionary biology)	Adaptations generally refer to developments or changes in an organism or system that promote its fitness. Adaptive traits are viewed in the historical sense and are adaptations if they have been arrived at by natural selection. Adaptedness, on the other hand, refers to the fit of an organism and their environment or features that make the organism better adapted and enhance their fitness. Differentiating traits that occur through adaptation and other processes are the subject of the contested field of adaptionism.	Hamilton (1964), Dawkins (1976), Dobzhansky et al. (1977), Burian (1983), O'Brien and Holland (1992), Dennett (1995), Shanahan (2004)
Social sciences (e.g., anthropology, geography, sociology)	Adaptation concerns how culture and institutions arbitrate humans and the environment to explain societal evolution. Emphasis has been placed on broadening cultural repertoire and improved coping methods that foster cultural endurance. Approaches such as human behavioral ecology connect to evolutionary theory and focus on adaptive design (e.g., traits, behaviors) while dual inheritance theory directs attention to both genes and culture.	Steward (1968), Braun (1990), O'Brien and Holland (1992), Denevan (1983), Cronk et al. (2000), Stone (2008)
Environment and resource studies and application — Political ecology	Analysis concentrates on the vulnerability of people: the ability of an entity (person through community) to cope with and adapt to stresses on their livelihoods. Vulnerability can be understood in terms of exposure, capacity (or response ability), and potential Understanding who is vulnerable and why they are vulnerable in terms of political, economic and social	Chambers and Conway (1992), Sen (1992), Watts and Bohle (1993), Adger and Kelly (1999)

(*continued*)

Table 1.1 (continued)

Perspective	Meaning	Key references
	factors is the main focus. The ability to adapt is thus influenced by endowments, capability, and entitlements.	
Risks and hazards	Analysis focuses on assessing the possibility of loss (risk) to a system from exposure to a hazard. Risk to a particular system is determined by characteristics of the hazard (e.g., intensity, location, frequency) and the vulnerability (i.e., hazard severity and amount of damage). Most simply it is schematized $R = H \times V$. Vulnerability is countered by the capacities of individual protection and collective action (e.g., resist, avoid, adapt) and employment of abilities to create security. Several models have emerged (e.g., pressure and release (PAR), access, hazard, impact, risk and vulnerability (HIRV)) to investigate the intersection of forces, process of impacts, and responses to disasters and hazards.	Burton et al. (1978), Blaikie et al. (1994), Cutter (1996, 2003), Brooks (2003), Dayton-Johnson (2004), Wisner et al. (2004), Haque (2005)
Climate change studies	Adaptive capacity is defined as "the ability of a system to adjust to *climate change* (including *climate variability* and extremes) to moderate potential damages, to take advantage of opportunities, or to cope with the consequences" (IPCC 2007, p 869). Adaptive capacity is synonymous with adaptability and manifests in adaptations that reduce vulnerabilities. Drivers of adaptive capacity include the interconnected forces (local to global) that act upon capability (reactive and proactive) of the system to adapt. More	Smithers and Smit (1997), Kelly and Adger (2000), Burton et al. (2002), Smith et al. (2003), O'Brien et al. (2004), Brooks and Adger (2005), Luers (2005), Smit and Wandel (2006), IPCC (2007), Füssel (2007)

Resilience thinking and social–ecological systems	specific understanding of adaptive capacity hinges on competing interpretations of vulnerability; in end-point analysis the focus is on the current capacity for technological adaptations while in starting-point analysis the focus is on adjustments to changing conditions. Knowledge of adaptive capacity has not moved beyond broad categories. The resilience approach is concerned with the dynamics of integrated systems of humans and nature (a social–ecological system), which interact across scales. Resilience (ecological and social) is the capacity of a system (i.e., robustness) to experience perturbations or disturbances while retaining its general properties (e.g., functions, structure, etc.) as well as taking advantage of opportunities (i.e., new system trajectories). Adaptive capacity entails a dynamic balance between sustaining and developing and is paramount to sustainability. Adaptive capacity: entails resources and their employment that are prerequisites to adaptation; is a multidimensional construct, consisting of learning to live with uncertainty, nurturing diversity, bringing together multiple knowledge systems and ways of learning, and fostering opportunities for self-organization; it is influenced by operational (technical, financial, political and strategic issues (e.g., power, scale, culture).	Berkes and Folke (1998), Adger (2000), Gunderson (2000), Gunderson and Holling (2002), Folke et al. (2002a, b), Berkes et al. (2003), Armitage (2005), Folke (2006), Gallopín (2006), Smit and Wandel (2006), Nelson et al. (2007), Fazey et al. (2007)

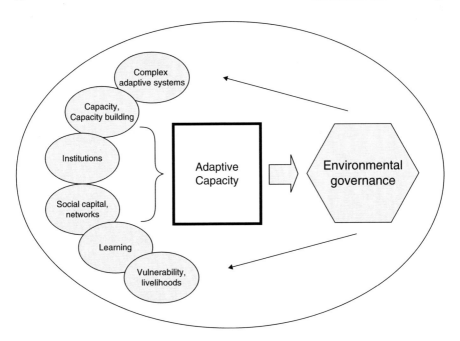

Fig. 1.1 Adaptive capacity and environmental governance

to deal with issues of scale, multiple perspectives and epistemologies, uncertainty and nonlinearity, self-organization and emergence.

The main proposition put forward in this volume is that adaptive capacity is a poorly understood yet critical enabling factor in efforts to build multilevel governance systems for complex social–ecological systems (Fig. 1.1). Although a growing body of literature examines adaptive capacity in specific places and resource systems, we, here, seek to offer theoretical and applied insights into the still evolving connections between adaptive capacity and environmental governance. Indeed, regardless of the scale or particular form of governance, understanding the basis of collaborative and adaptive interactions and the ability of governance actors to respond to uncertainty and change depends on interdisciplinary, theoretical and applied advances in the bodies of work that inform adaptive capacity: complex systems thinking, capacity and capacity building, institutions, social capital and networks, learning, vulnerability and livelihood studies. The particular relevance of these bodies of work is briefly outlined as they serve as cross-cutting themes in the book.

1.4.1 Complex Adaptive Systems

Complex systems thinking offers a way of examining, describing, interpreting, and cognitively structuring not only ecological systems but also increasingly linked social–ecological systems as well (Berkes et al. 2003). Specifically, complex systems thinking highlights the dynamic, nonlinear relations among coupled social and ecological phenomenon that result in discontinuities, surprises, system flips, and the potential for multiple equilibrium states. Complex systems thinking provides valuable heuristics for understanding environmental governance and natural resource management and emphasizes relationships, networks, and feedback processes. Complex systems thinking thus indicates the importance of institutional diversity and flexibility to improve the fit between ecological and social systems. Adaptive capacity has been identified as a central concern and a foundational dimension of environmental governance that is adaptive in the context of change and uncertainty (Folke et al. 2005).

1.4.2 Capacity and Capacity Building

Capacity and capacity building have received considerable attention in a wide range of disciplinary settings (e.g., public administration and development, organizations and management). Ivey et al. explain that "...capacity is typically conceptualized from a *functionalist* perspective that focuses on the ability of individuals, organizations to perform efficiently, effectively and on an ongoing basis, a set of externally defined goals. However, capacity for self-determination, conversely, is grounded on a *relational* perspective that focuses on the ability of individuals, organizations, communities and governments to establish and achieve their own goals and agendas" (2006, p. 946). Issues associated with capacity building thus draw attention to factors that facilitate or constrain these ends. These factors may encompass technical, financial, human resource, institutional, and social components (e.g., Ivey et al. 2006; Timmer et al. 2007), as well as the issues that form a "critical subtext" to such processes, such as power, scale, and culture (Armitage 2005). Unilateral actions are particularly challenging in turbulent problem domains and flexible, multiparty collaborations are important to constructively manage differences (Gray 1989). Evidence suggests that similar to ecological systems, these multiparty collaborations move through patterns of transformation that are generally predictable (Scheffer et al. 2002).

1.4.3 Institutions

Adaptive capacity involves human choice. Choices, the conditions (opportunities and constraints) surrounding them, and the arrangements that shape our decisions are the domain of institutions and institutional choice theorists (Ostrom 2005; Vatn 2005). Vatn explains that "institutions are the conventions, norms and formally sanctioned rules of a society. They provide expectations, stability and meaning essential to human existence and coordination. Institutions regularize life, support values and produce and protect interest" (2005, p. 60). Understanding institutions necessitates considering social theories about human nature, interactions among individual choices, and interconnections between social and ecological systems. Adaptive capacity has an important social dimension and understanding the roles of institutions (formal and informal) influences governance outcomes. Interconnections between social and ecological systems highlight the need to consider the relationship between institutions (and governance systems) and biophysical system dynamics in terms of fit, interplay, and scale (Galaz et al. 2008; Young 2008).

1.4.4 Social Capital and Networks

Relationships, trust, and networks among individuals are important considerations in collective action dilemmas and small (community) scale efforts to manage natural resources (Ostrom and Ahn 2003; Crona and Bodin 2006; Plummer and FitzGibbon 2006; Bodin and Crona 2009). Although the concept of social capital has been variously defined, it is generally understood to involve "networks together with shared norms, values and understanding that facilitate cooperation within or among groups" (OECD 2001, p. 41). Social capital is related to adaptive capacity through its various forms and functions. Bonding enables strong ties between close associates or friends, bridging connects individuals of greater social distances, and vertical linkages leverage political and financial advantages (see Côté 2001; Woolcock 2001; Plummer and FitzGibbon 2006). Social capital may be the outcome of social interactions (and other processes such as social learning), and it may also influence collaborative endeavors (Plummer and FitzGibbon 2007). Although networks are identified to contribute to building resilience and increasing adaptive capacity (Tompkins and Adger 2004), they are complicated multidimensional constructs shaped by their structure and types of linkages, distribution of nodes, and connectivity or density of linkages (Newman and Dale 2005; Bodin et al. 2006).

1.4.5 Learning

A key feature of innovative governance approaches is an explicit focus on linking collaborative efforts with systematic learning. Learning involves the collaborative or mutual development and sharing of knowledge by multiple stakeholders, and feeds directly into the development of capacity for adaptation by individuals and social collectives (Keen et al. 2005; Fazey et al. 2007). Much learning is directed at modifying management strategies or actions (e.g., harvest rates, techniques) without challenging the assumptions upon which those strategies are based. This type of learning is sometimes referred to as "single-loop" learning. In contrast, "double-loop" or transformative learning involves resolving fundamental conflict over values and norms, and promoting change in the face of significant uncertainty (Argyris and Schön 1978). However, effort to foster double-loop learning requires a commitment to valuing different knowledge sources and epistemologies. Double-loop learning is also linked to social capital, or the social norms, networks of reciprocity and exchange, and relationships of trust that enable people to act collectively – conditions and traits of individuals and groups with fundamental importance to adaptive capacity and innovative governance efforts.

1.4.6 Vulnerability and Livelihoods

Vulnerability involves the extent to which individuals and communities (e.g., resource users) are susceptible to conditions and situations that indirectly or directly affect their well-being and prospects for sustainability (Smit and Wandel 2006). This vulnerability includes the sensitivity of the ecological systems within which social systems are embedded. A productive lens through which to understand vulnerability is the sustainable livelihoods framework, which helps to conceptualize the cross-scale and complex economic, social, ecological and behavioral choices confronting predominately rural, agricultural producers (Chambers and Conway 1992). Ideas of complexity and resilience, already embedded in the livelihood approach, are being emphasized (Adger 2003; Barrett and Swallow 2006; Marschke and Berkes 2006). There is increasing emphasis in understanding the connections between vulnerability and the resilience or adaptive capacity of livelihoods which can: (1) cope with and are able to recover from shocks and stresses; (2) maintain or enhance existing capabilities and assets despite uncertainty; and (3) ensure the provision of sustainable livelihood opportunities for future generations. These insights help to put issues of adaptive capacity and environmental governance in the context of real, everyday challenges and opportunities facing individuals and communities.

1.5 A Roadmap to This Volume

Contributions to this volume reflect an integrative approach to adaptive capacity. Each contributor has a wealth of experience with collaborative and adaptive approaches involving diverse actors at multiple levels and an ability to these areas of scholarship to inform the concept and practice of adaptive capacity as a response to pressing environmental problems. The international composition of the contributors spans a wide geographic area, and accordingly there is case study material covering many different environmental and resource contexts and themes.

Advancing knowledge of how adaptive capacity builds environmental governance in complex social–ecological systems requires bringing together conceptual understandings with applied experiences. It also involves thinking about adaptive capacity in new ways. This volume is therefore divided into the following two sections.

Section I (Adaptive Capacity in Theory and Practice) consists of eight chapters that draw inspiration from the perspectives and bodies of scholarship related to adaptive capacity (outlined in this chapter) and emphasize empirical results or applied experiences. The first three contributions in this section explore influential aspects on adaptive capacity in complex systems under pressure. Erin Bohensky, Samantha Stone-Jovicich, Silva Larson, and Nadine Marshall (Chap. 2) investigate theory and practice of adaptive capacity in the Great Barrier Reef; Christo Fabricius and Georgina Cundill (Chap. 3) explore the issue of building adaptive capacity in social–ecological systems that have been pushed past their limits using the case of Macubeni South Africa; and Daniela Kalikoski and Edward Allison (Chap. 4) discuss factors affecting the capacity of co-management to build robust and resilient organizations by critically examining case studies of fisheries from Africa and South America.

The next series of chapters examine adaptive capacity and the implications for building governance in the context of the North. Carina Keskitalo (Chap. 5) asks where vulnerability lies and probes this question in the forest lands of Gällivare municipality in northern Sweden. Sonia Wesche and Derek Armitage (Chap. 6) undertake a multiscale analysis of adaptive capacity in Fort Resolution (Northwest Territories, Canada) and draw implications to support adaptation by focusing on the determinants of adaptive capacity. Robin Sydneysmith, Mark Andrachuck, Barry Smit, and Grete Hovelsrud (Chap. 7) outline the CAVIAR project and draw upon its framework to analyze vulnerability and adaptive capacity in Arctic communities.

The first section closes with a pair of chapters that address the implications from climatic and other changes for institutions and governance. Rob de Loë and Ryan Plummer (Chap. 8) argue that a broad and integrative perspective is required in response to greater uncertainty and complexity relating to drinking water. In applying such a perspective to very different settings in Canada (urban water supply and drinking water quality in Aboriginal communities), they gain insights into appropriate strategies for addressing these challenges. In Chap. 9, Johanna Wandel and Gregory Marchildon examine the social–ecological systems that have emerged in response to climatic stimuli, political changes and macro-economic conditions in the Special Areas of Alberta, Canada. Lessons from their analysis of the Special Areas case highlight broad public policy implications concerning institutional fit, interplay, and adaptive capacity.

Section II (Frontiers in Adaptive Capacity) consists of four chapters that push the boundaries of adaptive capacity scholarship in new directions. These chapters direct readers to future concerns and issues, reveal the potential to gain insights from controversial theories, and highlight the necessity of moving from adaptive capacity to holistic notions of adaptive governance. In Chap. 10, Alan Diduck develops the idea of multilevel learning to overcome current uncertainties and gaps associated in the resource and environmental governance literature. In Chap. 11, Ralph Matthews and Robin Sydneysmith depart from conventional perspectives on environmental and climatic change by stressing the need to understand adaptive capacity as a dynamic institutional process. The relationship between evolutionary biology and adaptation in the natural sciences is used as a departure point by Fennell and Plummer in Chap. 12. They bring together ecological and socio-institutional perspectives to argue that evolutionary biology is foundational to adaptive capacity and has important implications for behavioral and governance changes. In Chap. 13, Per Olsson, Örjan Bodin, and Carl Folke aim to increase understanding about transformative capacity in social–ecological systems. They draw attention to three key dimensions that serve as an important starting point for researchers to move forward on the imperative issue of understanding how transformations take place.

The volume concludes with a final synthesis chapter where the main themes, ideas and points made by the various contributors are brought together in order to develop key lessons and implications for policy and practice. The chapter raises consciousness of important challenges anticipated for adaptive capacity in building environmental governance. Directions for further research are also identified.

The overarching goal of this effort from the start has been to advance innovative solutions to meet contemporary environmental challenges. In drawing upon the various perspectives and the bodies of knowledge associated with adaptive capacity and environmental governance, the contributors of this volume weave an integrative approach to elucidate its dynamism (contributing to robustness and realizing opportunities for changes) and functions (resources, the social process through which they are employed, the institutions by which they operate). Ultimately, this volume enriches knowledge of the multifaceted ways in which adaptive capacity builds environmental governance.

Acknowledgements We thank Fikret Berkes and Rob de Loë for their valuable comments and feedback on an earlier version of this chapter.

References

Adger WN (2000) Social and ecological resilience: are they related? Prog Hum Geogr 24(3):347–364
Adger WN (2003) Social aspects of adaptive capacity. In: Smith JB, Klein RJT, Huq S (eds) Climate change, adaptive capacity and development. Imperial College, London
Adger WN, Kelly PM (1999) Social vulnerability to climate change and the architecture of entitlements. Mitig Adapt Strat Glob Change 4:253–266
Adger WN, Brown K, Tompkins EL (2005) The political economy of cross-scale networks in resource co-management. Ecol Soc 10(2):9, https://www.ecologyandsociety.org/vol10/iss2/art9/. Accessed 8 Apr 2009
Argyris C, Schön D (1978) Organizational learning: a theory of action perspective. Addison-Wesley, Reading
Armitage D (2005) Adaptive capacity and community-based natural resources management. Environ Manage 35(6):703–715
Armitage D (2008) Governance and the commons in a multi-level world. Int J Commons 2(1):7–32
Armitage DR, Plummer R, Berkes F, Arthur RI, Charles AT, Davidson-Hunt IJ, Diduck AP, Doubleday N, Johnson DS, Marschke M, McConney P, Pinkerton E, Wollenberg E (2009) Adaptive co-management for social–ecological complexity. Front Ecol Environ 7(2):95–102

Barrett C, Swallow B (2006) Fractal poverty traps. World Dev 34(1):1–15
Berkes F (2009) Evolution of co-management: role of knowledge generation, bridging organizations and social learning. J Environ Manage 90:1692–1702
Berkes F, Folke C (1998) Linking social and ecological systems: management practices and social mechanisms for building resilience. Cambridge University Press, Cambridge
Berkes F, Colding J, Folke C (2003) Navigating social–ecological systems: building resilience for complexity and change. Cambridge University Press, Cambridge
Biermann F, Betsill MM, Gupta J, Kanie N, Lebel L, Liverman D, Schroeder H, Siebenhüner B, Conca K, da Costa Ferreira L, Desai B, Tay S, Zondervan R (2009). Earth system governance: people, places and the planet. Science and implementation plan of the Earth system governance project. ESG Report No. 1. The Earth system governance project, Bonn, IHDP
Blaikie P, Cannon T, Davis I, Wisner B (1994) At risk. Routledge, London
Bodin Ö, Crona B (2009) The role of social networks in natural resource governance: what relational patters make a difference? Glob Environ Change 19:366–374
Bodin Ö, Crona B, Ernstson H (2006) Social networks in natural resource management: what is there to learn from a structural perspective? Ecol Soc 11(2):2, http://www.ecologyandsociety.org/vol11/iss2/resp2/. Accessed 30 Mar 2009
Braun DP (1990) Selection and evolution in nonhierarchical organization. In: Upham S (ed) The evolution of political systems: sociopolitics in small-scale sedentary societies. Cambridge University Press, New York, pp 62–86
Briggs S (2003) Command and control in natural resource management: revisiting Holling and Meffe. Ecol Manage Restor 4(3):161–162
Brook N (2003) Vulnerability, risk and adaptation: a conceptual framework. Tyndall Centre Working Paper No. 38. Tyndall Centre for Climate Change Research and Centre for Social and Economic Research on the Global Environment, University of East Anglia, Norwich
Brooks N, Adger WN (2005) Assessing and enhancing adaptive capacity. In: Chopra K, Leemans R, Kumar P, Simons H (eds) Adaptation policy frameworks for climate change: developing strategies policies and measures. Cambridge University Press, Cambridge, pp 165–181
Brunner R, Steelman T, Coe-Juell L, Cromley C, Edwards C, Tucker D (2005) Adaptive governance: integrating science policy and decision making. Columbia University Press, New York
Burian R (1983) Adaptation. In: Grene M (ed) Dimensions of darwinism. Cambridge University Press, New York, pp 287–314
Burton I, Kates RW, White GF (1978) The environment as hazard. Oxford University Press, New York
Burton I, Huq S, Lim B, Pilifosova O, Schipper EL (2002) From impacts assessment to adaptation priorities: the shaping of adaptation policy. Clim Policy 2:145–159
Carpenter SR, Brock WA (2008) Adaptive capacity and traps. Ecol Soc 13(2):40, http://www.ecologyandsociety.org/vol13/iss2/art40. Accessed 22 Mar 2009
Cash DW, Adger W, Berkes F, Garden P, Lebel L, Olsson P, Pritchard L, Young O (2006) Scale and cross-scale dynamics: governance and information in a multilevel world. Ecol Soc 11(2):8
Chambers R, Conway GR (1992) Sustainable rural livelihoods: practical concepts for the 21st century. IDS Discussion Paper 296
Intergovernmental Panel on Climate Change (2007) Climate change 2007: synthesis Report
Cortner HJ (2000) Making science relevant to environmental policy. Environ Sci Policy 3:21–30
Côté S (2001) The contribution of human and social capital. ISUMA 2(1):29–36
Crona B, Bodin Ö (2006) WHAT you know is WHO you know? – Communication patterns among resource extractors as a prerequisite for co-management. Ecol Soc 11(2):7, http://www.ecologyandsociety.org/vol11/iss2/art7/. Accessed Apr 14 2009
Cronk L, Chagnon N, Irons W (eds) (2000) Adaptation and human behavior: an anthropological perspective. Aldine de Gruyter, New York
Cutter SL (1996) Vulnerability to environmental hazards. Prog Hum Geogr 20(4):529–539
Cutter SL (2003) The vulnerability of science and the science of vulnerability. Ann Assoc Am Geogr 93(1):1–12

Davidson-Hunt I, O'Flaherty M (2007) Researchers, indigenous peoples, and place-based learning communities. Soc Nat Resour 20:1–15

Dawkins R (1976) The selfish gene. Oxford University Press, Oxford

Dayton-Johnson J (2004) Natural disasters and adaptive capacity. Working Paper No. 237. OECD Development Centre, Issy-les-Moulineaux, France

Denevan WM (1983) Adaptation, variation and cultural geography. Prof Geogr 35(4):399–406

Dennett DC (1995) Darwin's dangerous idea. Touchstone, New York

Dietz T, Ostrom E, Stern PC (2003) The struggle to govern the commons. Science 302:1907–1912

Dobzhansky T, Ayala FJ, Stebbins GL, Valentine JW (1977) Evolution. Freeman, San Francisco

Durant RF, O'Leary R, Fiorino DJ (2004) Introduction. In: Durant RF, Fiorino DJ, O'Leary R (eds) Environmental governance reconsidered. MIT, Cambridge, pp 1–29

Fazey I, Fazey JA, Fischer J, Sherren K, Warren J, Noss RF, Dovers SR (2007) Adaptive capacity and learning to learn as leverage for social–ecological resilience. Front Ecol Environ 5(7):375–380

Food and Agriculture Organization of the United Nations (2007) State of the World's Forests, Rome

Folke C (2006) Resilience: the emergence of a perspective for social–ecological systems analysis. Glob Environ Change 16:253–267

Folke C (2007) Social–ecological systems and adaptive governance of the commons. Ecol Res 22:14–15

Folke C, Carpenter S, Elmqvist T, Gunderson L, Holling CS, Walker B, Bengtsson J, Berkes F, Colding J, Danell K, Falkenmark M, Moberg M, Gordon L, Kaspersson R, Kautsky N, Kinzig A, Levin SA, Mäler K-G, Ohlsson L, Olsson P, Ostrom E, Reid W, Rockström J, Savenije S, Svedin U (2002a) Resilience and sustainable development: building adaptive capacity in a world of transformations. The Environmental Advisory Council to the Swedish Government Scientific Background Paper. Ministry of the Environment, Stockholm

Folke C, Carpenter S, Elmqvist T, Gunderson L, Holling CS, Walker B (2002b) Resilience and sustainable development: building adaptive capacity in a world of transformations. Ambio 31(5):437–440

Folke C, Colding J, Berkes F (2003) Synthesis: building resilience and adaptive capacity in social–ecological systems. In: Berkes F, Colding J, Folke C (eds) Navigating social–ecological systems. Cambridge University Press, Cambridge, pp 252–387

Folke C, Hahn T, Olsson P, Norberg J (2005) Adaptive governance of social–ecological systems. Annu Rev Environ Resour 30:441–473

Füssel H-M (2007) Vulnerability: a generally applicable conceptual framework for climate change research. Glob Environ Change 17:155–167

Galaz V, Olsson P, Hahn T, Folke C, Svedin U (2008) The problem of fit among biophysical systems, environmental and resource regimes, and broader governance systems: insights and emerging challenges. In: Young OR, King LA, Schroeder H (eds) Institutions and environmental change. MIT, Cambridge, pp 147–186

Gallopín GC (2006) Linkages between vulnerability, resilience, and adaptive capacity. Glob Environ Change 16:293–303

Glasbergen P (1998) The question of environmental governance. In: Glasbergen P (ed) Co-operative environmental governance. Kluwer, Dordrecht, pp 1–20

Gray B (1989) Collaborating. Jossey-Bass, San Francisco

Gunderson LH (2000) Ecological resilience – in theory and application. Annu Rev Ecol Syst 31:425–439

Gunderson LH, Holling CS (eds) (2002) Panarchy: understanding transformations in human and natural systems. Island, Washington

Hamilton WD (1964) The genetical evolution of social behavior (I and II). J Theor Biol 7:1–52

Haque CE (ed) (2005) Mitigation of natural Hazards and disasters: international perspectives. Springer, Dordrecht

Holling CS, Meffe GK (1996) Command and control and the pathology of natural resource management. Conserv Biol 10(2):328–337
Holling CS, Gunderson LH, Peterson GD (2002) Sustainability and panarchies. In: Gunderson LH, Holling CS (eds) Panarchy. Island, Washington, pp 63–102
Huitema D, Mostert E, Egas W, Moellenkamp S, Pahl-Wostl C, Yalcin R (2009) Adaptive water governance: assessing the institutional prescriptions of adaptive (co-)management from a governance perspective and defining a research agenda. Ecol Soc 14(1):26, http://www.ecologyandsociety.org/vol14/iss1/art26/. Accessed 3 Apr 2009
Ivey JL, de Loë R, Kreutzwiser RD (2006) Planning for source water protection in Ontario. Appl Geogr 26:192–209
Keen M, Brown VA, Dyball R (2005) Social learning: a new approach to environmental management. In: Keen M, Brown VA, Dyball R (eds) Social learning in environmental management. Earthscan, London, pp 3–21
Kelly PM, Adger WN (2000) Theory and practice in assessing vulnerability to climate change and facilitating adaptation. Clim Change 47:325–352
Kettle DF (ed) (2002) Environmental governance. A Report on the next generation of environmental policy. Brookings Institution, Washington
Kooiman J, Bavinck M (2005) The governance perspective. In: Kooiman J, Jentoft S, Pullin R, Bavinck M (eds) Fish for life: interactive governance for fisheries. Amsterdam University Press, Amsterdam
Lebel L, Garden P, Imamura M (2005) The politics of scale, position, and place in the governance of water resources in the Mekong region. Ecol Soc 10(2):18, http://www.ecologyandsociety.org/vol10/iss2/art18/. Accessed 29 Mar 2009
Lebel L, Anderies JM, Campbell B, Folke C, Hatfield-Dodds S, Hughes TP, Wilson J (2006) Governance and the capacity to manage resilience in regional social–ecological systems. Ecol Soc 11(1):19, http://www.ecologyandsociety.org/vol11/iss1/art19/. Accessed 29 Mar 2009
Lemos MC, Agrawal A (2006) Environmental governance. Annu Rev Environ Nat Resour 36:297–325
Luers AL (2005) The surface of vulnerability: an analytical framework for examining environmental change. Environ Change 15:214–223
Marschke M, Berkes F (2006) Exploring strategies that build livelihood resilience: a case from Cambodia. Ecol Soc 11(1):42, http://www.ecologyandsociety.org/vol11/iss1/art42/. Accessed 3 Apr 2009
Millennium Ecosystem Assessment (2003) Ecosystems and human well-being: a framework for assessment. Island, Washington, pp 1–15
Millennium Ecosystem Assessment (2005) Ecosystems and human well-being: synthesis. Island, Washington
Nelson DR, Adger WN, Brown K (2007) Adaptation to environmental change: contributions of a resilience framework. Annu Rev Environ Resour 32:395–419
Newman L, Dale A (2005) Network structure, diversity, and proactive resilience building: a response to Tompkins and Adger. Ecol Soc 10(1):2, http://www.ecologyandsociety.org/vol10/iss1/resp2/. Accessed 26 Mar 2009
O'Brien MJ, Holland TD (1992) The role of adaptation in archeological explanation. Am Antiq 57(1):36–59
O'Brien MJ, Eriksen KS, Schjolden A, Nygaard L (2004) What's in a word? Conflicting interpretations of vulnerability in climate change research. CICERO Working Paper 2004:04. Center for International Climate and Environmental Research, Oslo, NO
Organisation for Economic Cooperation and Development (OECD) (2001) The well-being of nations: the role of human and social capital. Centre for educational research and innovation, organisation for economic cooperation and development, Paris
Ostrom E (2005) Understanding institutional diversity. Princeton University Press, Princeton
Ostrom E (2007) A diagnostic approach for going beyond panaceas. Proc Natl Acad Sci USA 104 (39):15181–15187

Ostrom E, Ahn TK (2003) Introduction. In: Ostrom E, Ahn TK (eds) Foundations of social capital. Edward Elgar, Cheltenham, pp Xi–xxxix
Ostrom E, Dietz T, Dolsak N, Stern PC, Stonich S, Weber EU (eds) (2002) The drama of the commons. National Academy, Washington
Plummer R, FitzGibbon JE (2006) People matter: the importance of social capital in the co-management of natural resources. Nat Resour Forum 30:51–62
Plummer R, FitzGibbon JE (2007) Connecting adaptive co-management, social learning and social capital through theory and practice. In: Armitage D, Berkes F, Doubleday N (eds) Adaptive co-management: learning. Collaboration and multi-level governance. University of British Columbia Press, Vancouver, pp 38–61
Reid WV, Berkes F, Wilbanks T, Capistrano D (eds) (2006) Bridging scales and knowledge systems: linking global science and local knowledge in assessments. Millennium Ecosystem Assessment and Island, Washington
Scheffer M, Westley F, Brock WA, Holmgren M (2002) Dynamic interactions of societies and ecosystems – linking theories from ecology, economy and sociology. In: Gunderson LH, Holling CS (eds) Panarchy. Island, Washington, pp 195–240
Sen A (1992) Inequality reexamined. Oxford University Press, Oxford, p 207
Shanahan T (2004) The evolution of Darwinism: selection, adaptation, and progress in evolutionary biology. Cambridge University Press, Cambridge
Smit B, Wandel J (2006) Adaptation, adaptive capacity and vulnerability. Glob Environ Change 16:282–292
Smith J, Klein R, Huq S (2003) Climate change adaptive capacity and development. Imperial College, London
Smithers J, Smit B (1997) Human adaptation to climatic variability and change. Glob Environ Change 7(2):129–146
Stern N (2007) The economics of climate change. Cambridge University Press, Cambridge
Steward JH (1968) Cultural ecology. International encyclopedia of the social sciences, vol 4. Macmillan, New York, pp 337–344
Stone BL (2008) The evolution of culture and sociology. Am Soc 39:68–85
Timmer DK, de Loë R, Kreutzwiser R.D (2007) Source water protection in the Annapolis Valley, Nova Scotia: lessons for building local capacity. Land use policy, pp 187–198
Tompkins EL, Adger WN (2004) Does adaptive management of natural resources enhance resilience to climate change? Ecol Soc 9:10, http://ecologyandsociety/org/vol9/iss2/art10
van Kerkhoff L, Lebel L (2006) Linking knowledge and action for sustainable development. Annu Rev Environ Resour 31:445–477
Vatn A (2005) Institutions and the environment. Edward Elgar, Cheltenham
Walker BH, Salt D (2006) Resilience thinking. Island, Washington
Watts MJ, Bohle HG (1993) The space of vulnerability: the causal structure of hunger and famine. Prog Hum Geogr 17:43–67
Wisner B, Blaikie P, Cannon T, Davis I (2004) At risk, 2nd edn. Routledge, London
Woolcock M (2001) Microenterprise and social capital: a framework for theory, research, and policy. J Socio Econ 30(2):193–198
Young OR (2002) The institutional dimensions of environmental change: fit, interplay, and scale. MIT Press, Cambridge Massachusetts
Young OR (2008) Institutions and environmental change: the scientific legacy of a decade of IDGE research. In: Young OR, King LA, Schroeder H (eds) Institutions and environmental change. MIT, Cambridge, pp 3–46

Section I
Adaptive Capacity in Theory and Practice

Chapter 2
Adaptive Capacity in Theory and Reality: Implications for Governance in the Great Barrier Reef Region

Erin Bohensky, Samantha Stone-Jovicich, Silva Larson, and Nadine Marshall

2.1 The Great Barrier Reef Region: A Complex Governance Challenge

The Great Barrier Reef (GBR) is among the world's iconic ecosystems. Extending some 2,300 km along the coast of Queensland, Australia, the GBR was listed as a World Heritage Area in 1981 in recognition of its outstanding universal ecological and cultural values. The direct economic value of the Great Barrier Reef Marine Park to the Australian economy, in terms of marine tourism, commercial fishing and recreational use, was estimated at $5.4 billion Australian dollars in 2006–2007 (Access Economics 2008).

As in many marine ecosystems, three major processes pose threats to the GBR: overharvesting of marine resources, water quality decline from land use in the adjacent catchment, and climate change (Hughes et al. 2007). Each threat operates at a particular scale, but is itself a manifestation of cross-scale processes. For example, overharvesting reflects pressures not only from local fishers but also international demand for marine resources (Hughes et al. 2003). Changes in water quality are influenced by national and international demand for agricultural and mineral commodities produced in the catchment, national and state environmental policy, the regional economy and farm-scale management. The impacts of climate change are the result of actions, including mitigation and adaptation, from global to local scales. The multiple-scale nature of the processes influencing the GBR argues for the importance of incorporating scale into any analyses, monitoring systems, and management strategies to address these processes.

Although governance of the GBR clearly needs to span multiple scales, this presents a highly complex challenge. First, adopting a multiple-scale approach to governance has implications for multiple actors who have different objectives and

---

E. Bohensky, S. Stone-Jovicich, S. Larson, and N. Marshall
CSIRO Sustainable Ecosystems, Davies Laboratory, PMB Aitkenvale QLD 4814, Australia
e-mail: erin.bohensky@csiro.au

values associated with the GBR. Furthermore, each of the three major threats is managed by different organizations and addressed by largely separate policies, some of them relatively new. For example, land and natural resource management (NRM), including fisheries, are managed at state level, while national marine waters and reef tourism fall under the jurisdiction of the Great Barrier Reef Marine Park Authority (GBRMPA), a federal government agency responsible for the management of the reef. Climate change affecting the marine park is also managed by the GBRMPA, as well as the Australian Department of Climate Change, created in 2007. Water quality is addressed by the Reef Water Quality Protection Plan (Reef Plan), an initiative started in 2003 involving all levels of government, industry organizations, community, and Indigenous groups and scientists.

The ways in which the various changes noted above will play out depends in part on the region's capacity to adapt. Adaptive capacity, by most definitions and measures, is considered relatively high for this region (Nelson et al. 2007) and for Australia as a whole (Haddad 2005). Yet, these definitions and measures may have limited utility for management in reality because a greater understanding is needed of the sources and determinants of adaptive capacity that appreciates the complex dynamics of the region. Our aim in this chapter is to draw on a combination of theory and empirical data to examine where regional adaptive capacity for environmental governance in the GBR region lies. Our main questions are: (1) how is adaptive capacity defined in theory? (2) how is adaptive capacity defined in "reality," as illustrated by empirical research on perceptions held by resource users, managers, and other decision-makers in the GBR? and (3) to what extent do the theoretically-derived and empirically-derived definitions differ? We first review theoretical definitions of adaptive capacity, and compare and contrast these to individual and organizational perceptions of adaptive capacity elicited through four separate research efforts that were recently carried out in the GBR region. We then discuss key messages emerging from this comparison of definitions and implications for the future governance of this region.

2.2 Adaptive Capacity in Theory

2.2.1 Review of Definitions

Below we identify major contributions in the last 10 years to the theory of adaptive capacity as it relates to environmental governance. Significant contributions have been made prior to this, but as many of these are acknowledged in the current literature as the basis for contemporary thinking on adaptive capacity, we did not include these earlier papers in our analysis. We also note that these "theoretical" definitions may in fact be derived from or supported with empirical data to varying degrees.

The literature we discuss falls into one of two broad and often overlapping domains: vulnerability and adaptation, and resilience (Janssen et al. 2006).

We focus on these because they deal with adaptive capacity at multiple scales and multiple aspects of environmental change, and as such are most relevant to the GBR. We acknowledge, but do not review, a growing body of literature dealing with specific aspects of adaptive capacity at specific scales, such as community-level vulnerability to climate change, natural disasters, or other disturbances (see Day and Dwyer 2003; Norris et al. 2008; Cinner et al. 2009; Ford et al. 2009).

2.2.1.1 Vulnerability and Adaptation

The concept of vulnerability is often discussed in relation to natural hazards and the ability of individuals or social groups to cope with these hazards (Adger and Vincent 2005). The study of adaptation of humans to environmental variability has its roots in anthropology (Janssen et al. 2006), but in recent decades has been applied to issues such as global climatic change and its impacts (IPCC 2001; Adger et al. 2005). Within the vulnerability and adaptation domain, adaptive capacity has been defined in several ways. The Intergovernmental Panel on Climate Change (IPCC) defines adaptive capacity as: "the general ability of institutions, systems, and individuals to adjust to potential damage, to take advantage of opportunities, or to cope with the consequences" (IPCC 2001). This definition has been adopted widely by other scholars and scientific assessments such as the Millennium Ecosystem Assessment (MA 2005).

Adaptive capacity, along with exposure and sensitivity, is considered a determinant of vulnerability (Adger and Vincent 2005). A region is thus more vulnerable if its adaptive capacity is low, but having high adaptive capacity in itself does not render it immune from disturbance; the nature of the disturbance and its impact also matter. However, adaptive capacity does represent "a vector of resources and assets that represents the asset base from which adaptation actions and investments can be made" (Adger and Vincent 2005). Adaptive capacity may be latent, realized only when sectors or systems are exposed to the actual or expected stimuli (Adger et al. 2005), and can only be observed when realized through some form of concrete adaptation (Lemos et al. 2007).

Adaptive capacity can be created by: "(1) investing in information and knowledge, both in their production and in the means of distributing and communicating them; (2) encouraging appropriate institutions that permit evolutionary change and learning to be incorporated; and (3) increasing the level of resources such as income and education to those in which they are presently lacking" (Lemos et al. 2007). Institutional arrangements are also important; adaptive capacity depends on the structure of institutions, the ability of decision-makers to manage information (Yohe and Tol 2002), and the potential of institutions to reduce impacts of risks (Smit et al. 2000).

Although vulnerability and adaptation are often discussed together, some authors make a distinction between the vulnerability and adaptation literature (Janssen et al. 2006). On the one hand, vulnerability studies may give more attention to the hazard itself, or to the risk of being detrimentally affected by the hazard

(Adger et al. 2004). Adaptation, on the other hand, may focus on actual management of the impact of and response to the hazard, and to do so successfully, by one account, requires heterogeneity of adaptive capacity across different stakeholders (Adger and Vincent 2005). Vulnerability may reflect "stocks" of adaptive capacity that are determined by a range of factors, whereas adaptation transfers adaptive capacity into action; we can assume that if one is adapting one has sufficient capacity to do so (although the converse is not necessarily true, having adaptive capacity does not necessarily imply adaptation).

2.2.1.2 Resilience

Resilience in ecological and social systems is the ability to undergo change and still retain the same controls on function and structure, the capability to self-organize, and the ability to build and increase the capacity for learning and adaptation (Gunderson and Holling 2002). The resilience literature discusses adaptive capacity in various contexts, including ecosystems (Carpenter et al. 2001), prehistoric societies (Redman and Kinzig 2003), organizations (Olsson et al. 2004), and governance (Lebel et al. 2006). A distinction is made between adaptive capacity in ecosystems and social systems. The former is thought to be related to genetic diversity, biological diversity, and the heterogeneity of landscape mosaics (Peterson et al. 1998; Carpenter et al. 2001; Bengtsson et al. 2003). In social systems, adaptive capacity is enhanced by institutions and networks that learn and store knowledge and experience, create flexibility in problem solving and balance power among interest groups (Scheffer et al. 2000; Berkes et al. 2002).

Much of the resilience literature is concerned with social–ecological systems (SES) (Walker et al. 2002); that is, coupled systems of humans and the environment (Westley et al. 2002) connected through a complex array of linkages and feedbacks (MA 2005). Central to social–ecological resilience theory is the concept of alternative regimes maintained by a small number of slow variables. Disturbance and change can result in abrupt, nonlinear shifts that move the system past a threshold and into a new regime. Such regimes manifest as "basins of attraction" that can be difficult to enter or escape when desired. The adaptive capacity of the SES is the collective ability of human actors in the system to manage these basins such that the system is kept within critical thresholds (Walker et al. 2004).

Views differ regarding the relationship between adaptive capacity and resilience (Gallopín 2006). Folke et al. (2002) maintain that resilience is key to enhancing adaptive capacity, though Walker et al. (2002) remark that "adaptive capacity is an aspect of resilience," which together suggest a mutually reinforcing relationship between these concepts. One view holds that resilience is about negotiating vulnerability and adaptation under different conditions; "true resilience will lie in knowing when to change course and when to forge ahead" (Redman and Kinzig 2003). In this vein, Folke et al. (2005) identify four dimensions of adaptive capacity in SES undergoing change and reorganization: (1) Learning to live with change and uncertainty; (2) Combining different types of knowledge for learning; (3) Creating opportunity for self-organization toward social–ecological resilience; and (4) Nurturing sources of resilience for renewal and reorganization.

2.2.2 A Conceptual Lens for Assessing Adaptive Capacity

We note three themes that are common to the vulnerability and adaptation and resilience literatures, and together offer a lens through which our empirical research on adaptive capacity can be viewed. First, the context of adaptive capacity matters. Carpenter et al. (2001) argue for the importance of defining "of what" and "to what" in studies of resilience and adaptive capacity; that is, whom or what is resilient or adaptive, and to what change or disturbance is this individual, community, or system resilient or adaptive? Walker et al. (2009) distinguish between specified resilience, when the "to what" can be defined, and general resilience, when the threat is unknown. In the vulnerability literature, a similar concept is expressed in the terms of sensitivity and exposure (e.g., Adger and Vincent 2005), where sensitivity reflects the characteristics of a particular individual, community, or society, and exposure reflects the interaction of the individual, community, or society with a particular change or disturbance. Thus, adaptive capacity may vary depending on the specific change processes, and may also vary depending on the ecological characteristics of the system.

Second, the scale of adaptive capacity matters. Folke et al. (2005) observe that adaptive capacity results from critical factors that interact across spatial and temporal scales. Although the adaptive capacity of individuals may be linked to community, regional or even global adaptive capacity (Smit and Wandel 2006), specific attributes and determinants of adaptive capacity may be scale-dependent (Adger and Vincent 2005). They may also be culture- and place-specific, such that scaling up is not possible (Adger 2003). Furthermore, enhancing adaptive capacity at one scale may undermine adaptive capacity at other scales: a sector may benefit at the expense of a region (Allison and Hobbs 2004) or individual at the expense of a community (Pelling and High 2005). Similarly, short-term adaptive capacity may differ from long-term adaptive capacity (Folke et al. 2002; Pelling and High 2005), implying trade-offs between present and future outcomes. Thus, adaptive capacity is "the ability of a socio-ecological system to cope with novelty without losing options for the future" (Folke et al. 2002).

Third, information and knowledge to support adaptive capacity matter, as do the processes by which they are created and transmitted. Adaptive capacity requires communication and learning, and organizations and mechanisms for creating and maintaining knowledge (Lemos et al. 2007) and enabling flexible solutions (Scheffer et al. 2000; Berkes et al. 2002). Building and maintaining adaptive capacity requires a diversity of social groups with interacting networks (Pelling and High 2005), which can access a diversity of knowledge types (Folke et al. 2005) as well as shared knowledge (Redman and Kinzig 2003). Adaptive capacity also depends on a balance of power (Scheffer et al. 2003) that gives all actors a voice in decision-making, and credible information-generation processes (Yohe and Tol 2002).

2.2.3 *From Theoretical Definitions to Operational Measures*

Increasingly, attempts are being made to move the discussion of adaptive capacity beyond theoretical definitions to operational measures that can support environmental governance on the ground. So far, operational definitions of adaptive capacity have been largely in the form of conceptual frameworks and indicators, often used in comparative assessments of adaptive capacity (e.g., Turton 1999; Adger et al. 2004; Nelson et al. 2007; McClanahan et al. 2008). An alternative approach is to elicit perceptions of individuals, communities or societies of their own capacity to adapt (e.g., Bryant et al. 2000; Hertin et al. 2003; Grothmann and Patt 2005; Ford et al. 2009). Such perceptual measures of adaptive capacity can feed into quantitative approaches; for example, Adger et al. (2004)'s indicators were informed by expert judgment and validation as well as literature review.

We believe the concepts that formed our conceptual lens above – context-dependence, scale-dependence, and information and knowledge creation and transmission – all argue for a perceptual approach to defining adaptive capacity. For adaptive capacity to have resonance and meaning in "reality," it is pertinent if not imperative to consider how and by whom adaptive capacity is defined.

We note that perceptions of adaptive capacity can both enhance adaptive capacity and constrain it. Adaptive capacity is enhanced by appropriate understanding of a problem and possible responses (Bohensky and Lynam 2005). By contrast, it is constrained when, for example, deep-rooted attitudes and behaviors of an individual or society undermine the ability to adapt to new situations (Scheffer and Westley 2007). For this reason, perceptual analyses of adaptive capacity benefit from the use of both deductive and inductive measures. The former are drawn from theory or conceptual frameworks of researchers, while the latter are defined by the actors themselves. There are trade-offs associated with each. Deductive approaches may fail to capture factors that are relevant on the ground. Inductive approaches run the risk of missing factors that are critically important to understanding adaptive capacity but not readily recognized by actors in the system. However, using both together enables a more complete picture of the factors, including perceptions that are likely to enhance or erode adaptive capacity.

2.3 Adaptive Capacity in "Reality": Examples from the GBR

The adaptive capacity of the GBR is an issue that has attracted significant local and global interest (Olsson et al. 2008). However, to our knowledge, there have been no efforts to synthesize studies of adaptive capacity as it applies to the GBR across a range of contexts, scales, and methodologies.

In this chapter, we recognize the GBR as a regional SES, as defined above. However, we tend to agree with Walker et al. (2004) that "because human actions dominate in SESs, adaptability of the system is mainly a function of the social component – the individuals and groups acting to manage the system." We recognize that the GBR SES comprises heterogeneous individuals, communities, and industries, and its adaptive capacity therefore needs to be addressed across different categories of social actors. We acknowledge the inherent complexities in trying to satisfy all actors in the GBR while meeting the objectives set for the GBR as a whole, and do not assume that enhancing the adaptive capacity of one type of actor results in enhanced adaptive capacity of another, nor of the whole. We also recognize that several change processes are affecting the GBR, such as overharvesting, water quality decline and climate change, and that all of these processes are linked in various ways to economic and policy change. In keeping with our point above about context, we do not assume that the adaptive capacity of the region to deal with each of these types of change is the same, but we acknowledge that these changes need to be addressed in an integrated way.

Below we discuss four separate research efforts and draw some lessons from the collective efforts on how adaptive capacity is perceived in the GBR region by different actors: resource users, natural resource managers, community residents, government bodies, and leaders from academia, government, and business working at the scale of the whole region. Each study was undertaken to address a specific research question related to one or more of the main threats to the GBR noted above (e.g., Hughes et al 2007). As such, aspects of adaptive capacity studied and the methodologies used differed. We have relied on both inductive and deductive approaches to define and understand adaptive capacity. In some cases, the research focused on the related concept of resilience.

2.3.1 Coping with Policy Change in the Fishing Industry

The commercial fishing industry in the GBR region and the sustainability of its activities has been at the center of public debate for many years and continues to experience changes in regulatory policies to minimize its impacts on the environment. Marshall and Marshall (2007) and Marshall et al. (2007) looked at how commercial fishers in the GBR respond and adapt to changes in these policies. Marshall and Marshall (2007) examined the capacity of commercial fishers to cope and adapt to changes in resource policy in a survey of more than 60 questions to elicit the likely response of commercial fishers and their families to changes in resource policy. The survey questions were based on the results of a literature review and scoping study and reflected concepts such as flexibility, strategic skills, coping mechanisms, capacity to reorganize and willingness to experiment and learn, as well as demographic characteristics. The survey was administered to 100 fishers and their families in their homes in five towns in northern Queensland, representing between 46 and 66% of the fishing population in each town.

Four main dimensions of the capacity to cope and adapt to policy change were identified by a Principal Components Analysis. These were, in order of importance:

- How fishers interpret risk associated with change?
- Fishers' capacity to plan, learn, and reorganize, which describes their level of financial preparedness and willingness to experiment.
- The capacity to cope, which reflects each fisher's perception of their ability to cope with changes required by a new policy, their level of financial stress, marital stress, and their ability to cope relative to other fishing families.
- Interest in adapting, which reflects the level of interest in learning new skills.

The combined effect of age, education, and attachment to occupation were important for fishers' understanding of risk and coping (Marshall et al 2007). For instance, younger, better educated fishers who were not as attached to their occupation were more optimistic about the risks associated with policy and their capacity to cope. Also, fishers with larger businesses and those that had business plans, were knowledgeable about their financial positions and had business skills, were more likely to perceive themselves as being able to cope and adapt. In addition, individuals' perceptions of risk and their capacity to cope were related to the combination of the level of involvement in the decision-making process, the perceived rate of implementation, and the interpretation of regulatory change.

Some fishers perceived themselves as lacking adaptive capacity because of a perceived lack of strategic skills, whereas others perceived a lack of options. This suggests that individual fishers may respond to change events quite differently. Policy implementers need to be cognizant of this heterogeneity and of the potential inequities that may result. Better understanding of differences between individuals can help with the design of solutions to assist resource users in adapting to change.

Heterogeneity in individual behavior can be an important source of adaptive capacity and resilience at other scales. Research on the effects of policy change on individual enterprises may progress our understanding of how resource-dependent industries can increase their capacity to cope with future policy change (Smith 1995; Salz 1998; Smith et al. 2003; Bradley and Grainger 2004).

2.3.2 *Natural Resource Managers' Perceptions of Social Resilience to Water Quality Change*

Water quality change in the GBR region and the corrective actions of the Reef Plan both have potential to impact the region's social systems. In this context, research is being undertaken to understand the region's social resilience: will individuals, communities, industries, and organizations be able to absorb and withstand various environmental and policy changes without undergoing fundamental change themselves? This section briefly describes an approach to develop indicators of social resilience to water quality change and management interventions to achieve water quality targets at a regional scale (Lynam et al. 2007).

The development of the indicators was informed by subjective understandings of social resilience and its determinants based on interviews with natural resource managers in key government and nongovernmental organizations (NGOs) operating in the GBR region. The main purpose of the interviews was to ensure that the indicators developed were appropriate, meaningful, and useful for NRM agencies. As such, the interviews focused specifically on capturing and incorporating managers' understanding, experiential knowledge and perceptions of (1) what social resilience means, specifically as it relates to water quality change at the catchment-to-reef scale, (2) what enhances or erodes social resilience to water quality change, and (3) the usefulness of social resilience indicators to inform and adapt water quality policy and planning strategies.

Semistructured interviews were carried out in May–July 2007 with 20 representatives of different stakeholder groups in the GBR region. Stakeholder groups were selected according to the following criteria: (1) had a prominent role in water quality policy and/or management, and/or have an economic stake in water quality issues in the GBR; (2) operated at either the catchment-to-reef scale or across multiple catchments within the GBR catchment area, and (3) had formal, established linkages (e.g., via participation in committees) to other key GBR stakeholders (e.g., community and industry). Within each stakeholder group selected, particular attention was paid to interviewing individuals who worked on water quality issues.

Participants were asked to explain their understanding of the concept of "social resilience." The majority of interviewees viewed adaptive capacity as a core element of social resilience. In general, they perceived social resilience to consist of the capacity of the social system or different segments of the social system (e.g., society, communities) to respond, react, or adapt to changes and perturbations for the purpose of either creating and maintaining stability and the status quo, or enhancing quality of life or the economy. Some interviewees perceived the changes or perturbations to be of an unspecified or general nature, while others referred specifically to negative impacts and changes in water quality. They also differentiated between reactive and proactive adaptations.

Participants were also asked to reflect on "what would enhance social resilience to water quality change in the GBR?" They identified four broad categories of socio-cultural, economic, and ecological factors that were perceived as essential for developing, maintaining, or enhancing the GBR's social stability, status quo, quality of life or economy. These were social–cultural attributes; capacity building structures, processes, and tools; the economy, and the ecosystem. The greatest emphasis was given to socio-cultural attributes and capacity building structures, processes, and tools.

With regards to socio-cultural attributes, participants stated there was a need for:

- Values and attitudes emphasizing, for example, water protection and improvement, and desire to solve water quality problems
- Consensus and cohesion on how to approach and solve the problem

- Diversity of stakeholders, knowledge and expertise, and institutional responses to solve water quality problems, as well as the need for economic diversity
- Positive stories of successful examples and efforts underway disseminated by the media and NRM agencies and organizations

With regards to capacity building structures, processes, and tools, participants emphasized accessible, sound, "good and honest," and diverse information and knowledge that is easily and appropriately communicated, exchanged, shared, and debated, and that contributes to reducing uncertainties and surprises. Other factors that were highlighted were the need for both informal and formal social networks and partnerships that are horizontally and vertically linked, bring together a mix of stakeholders and are characterized by flexibility and innovation.

Having a stable source of income, or economic stability, was mentioned repeatedly as key to enhancing social resilience to water quality change. Rural and urban communities that suffer from production booms and busts (such as graziers), or high unemployment rates, were seen as incapable of dealing with other issues, particularly issues such as water quality changes that are generally perceived as being secondary to meeting basic needs. While economic stability was mentioned mostly in terms of individual and family economic wellbeing and economic prosperity in towns, a stable global economy was also viewed as being key to enhancing social resilience of people in the GBR region. In addition, market-based instruments to promote the adoption of best management practices, and a diversity of economic activities at the scale of both individual landholdings and the catchment-to-reef system, were also highlighted.

Participants emphasized the need to have a healthy ecological system and the implementation of land use practices that minimized impact on the environment. Also mentioned was the need to have greater control over the environment, through technologies and predictive tools (e.g., models to predict climate change and its impacts).

An understanding of how natural resource managers conceive social resilience and how they would use the concept is critical for informing management of resilience and capacity to adapt at a regional scale. However, inductively-derived, subjective measurements of social resilience such as those elicited in this study are only as comprehensive as the depth and breadth of experiences and knowledge of those interviewed. As such, it is equally vital that subjective understandings of resilience and its determinants are balanced with empirical data and theory to ensure the reliability and validity of indicators.

2.3.3 Public Perceptions of Institutional Roles in Australian Water Management

Public engagement in water planning activities is a legislated requirement at all levels of policy making in Australia (McKay 2005). However, water management involves a complex set of institutional arrangements, and catchments in the GBR region are currently regulated by a dozen statutory plans, based on various acts, as well as an equal number of relevant nonstatutory plans (Queensland Environmental Protection Agency 2006).

As Ostrom (2007) notes, multiple institutions and actors are required to create and enhance adaptive capacity, because "a mess of interactions forms the social raw material that shapes capacity to identify new information, learn and cope with change" (Pelling and High 2005). However, complex arrangements can create a confusing and thus disabling environment. Actors find it difficult to untangle complex webs of information and identify parties responsible for helping residents respond to impact. Ultimately, it is difficult for the general public to be meaningfully engaged in planning processes if the system is misunderstood. As Marshall (2008) observes, current regional-scale strategic processes and delivery models in Australia have considerably increased the complexity and the difficulty of the issues with which communities and individuals currently grapple.

A study was undertaken in the Whitsundays Shire, a local government area in the GBR region and a significant tourism destination, to explore the understanding of institutional arrangements of local residents who are expected to engage in water planning processes (Larson and Stone-Jovicich 2008). The principal goals were to investigate local residents' perceptions of a range of water quality issues and institutional responses and responsibilities for these issues, and compare these to actual institutional responses and responsibilities. Interviews were conducted with community residents and secondary data collected on water management institutions at all relevant scales. The Driving forces-Pressure-State-Impact-Response (DPSIR) framework (OECD InterFutures Study Team 1979) was used as a guide for data collection and structuring of emerging themes and perceived linkages between sources of water quality deterioration and impacts on human wellbeing as identified by interviewees (Larson and Stone-Jovicich 2008). These themes and linkages were then compared with current institutional arrangements relevant to water management.

Significant gaps were found with respect to institutional responsibilities for water quality. Residents perceived their local government body (Shire Council) as accountable for responding to water-related pressures and impacts in their Shire, whereas the responsibility lies primarily with a range of government agencies and organizations at federal, regional, and state levels. In particular, in this case study, the local council is held responsible for several water quality problems for which it has either limited or no responsibility or ability to take action. Local governments in Australia are expected to deal with an increasing number of social, ecological, and economic issues and, given budgetary constraints, are continuously facing tradeoffs between priorities for improvement (Larson 2009, Brown 2007). It is estimated that local governments receive only about 5% of total government expenditure, yet contribute some 53% of total government environmental spending (Dovers and Wild River 2008).

The findings of this study suggest the need for better communication between the various parties in water planning processes. In addition, there is a need for more research on the roles of the following in limiting or supporting adaptive capacity:

- Capacity of local government to meet expectations of its expanding role (Dovers and Wild River 2008; Larson and Stone-Jovicich 2008)
- Knowledge and understanding of the institutional system by stakeholders (Measham et al. 2009; Larson and Stone-Jovicich 2008)

Other relevant research needs have been identified in the areas of:

- Devolution of centralized power and devolution of resources (Lane 2003, 2005; Lane and McDonald 2005; Lane et al. 2004)
- Ability to link and manage multiple sources of knowledge (Measham et al. 2009; Stafford Smith 2008)
- Levels of trust in institutions by stakeholders (Marshall 2008; Larson 2006)

The multi-layered institutional system emerging in Australia is revealing some opportunities for a more adaptive, participative, and deliberative regional style of governance. In particular, progress has been noted in areas such as broadening the scope and scale of institutional collaborations, emergence of new network configurations or arrangements, fostering of new forms of participation among regional communities and increased capacity of social actors to coordinate amongst themselves (Bellamy 2007).

Effective engagement and social learning are crucial for long-term improvement of adaptive capacity (Bellamy 2007; Larson and Williams 2009). In turn, capacity building and social learning that develop during engagement processes play a role in legitimizing new organizations or rules (Lemos and Oliveira 2004; Larson 2006; Ostrom 2007).

2.3.4 The Future Great Barrier Reef: Adaptive Capacity in the Eyes of the Region's Leaders

Adaptive capacity may be built through processes such as scenario planning that stimulate thinking about the future, how different institutions shape it, and how surprises, unexpected consequences, and possible responses may unfold (Peterson et al. 2003; Bohensky et al. 2006). As part of a scenario planning exercise for the GBR region conducted by Bohnet et al. (2008), 47 leaders representing Australian and Queensland government agencies, local government, regional NRM bodies, NGOs, industry and research organizations were interviewed about their perceptions and aspirations for the future of the region. Leaders were selected as individuals in influential positions, as they are often instrumental in making change processes happen (Olsson et al. 2004). Participants were selected on the basis of their past or present involvement in the whole GBR catchment-to-reef system rather than specific subregions, urban centers, or communities (see also Chap. 13).

In a semi-structured interview process, interviewees were asked to describe in their own words the adaptive capacity of the GBR's different subregions, communities, industries, and government in 2050, and the extent to which they would be prepared for change as opposed to being reactive. Interviewees were asked the following questions related to adaptive capacity of the GBR catchment in 2050: (1) How do you think communities, industries, and government will respond to environmental problems? Will they be prepared to respond to environmental problems or will they only react once they happen? How will they prepare or react to change? (2) Will there be differences between subregions within the GBR catchment in terms of their responses to environmental problems? Will some subregions be better prepared than others? (3) What capacity to adapt to change will exist in different regions, communities, industries, and government in the GBR by 2050?

The most frequently mentioned issues by interviewees relate to:

- Timing of change processes
- Comparative adaptive capacity in subregions, industries, communities, and government
- Scales of adaptive action and governance
- Determinants of and constraints on adaptive capacity

Several respondents noted the difficulty of evaluating adaptive capacity in the absence of actual adaptation. One interviewee stated that adaptive capacity needs to exist in sufficient amounts to be appropriately matched to the problem, but noted that it is difficult to define this in practice.

Most interviewees agreed that all sectors of society, apart from a few forward-thinking individuals, tend to be reactive rather than proactive in responding to change, and that crisis or catastrophic change is usually required to shift society to new ways of thinking and modes of operation (however, some noted that changes tend to be made incrementally). Most mentioned the importance of leaders or champions, the role of education and information, and economic, demographic, biophysical, and geographic factors as determinants of adaptive capacity and ability to effect change. Uncertainty of scientific information required to understand and guide action was noted as a constraint to adaptive capacity. Other constraints included the lack of extension officers and coordinated research.

Interviewees also agreed that adaptive capacity differs among industries, with most suggesting the sugar industry is least adaptable because it exports into a global market that it cannot control. Some argued that tourism is most adaptable and regularly demonstrates its ability to reinvent itself, while others observed that tourism is also vulnerable to global changes such as the economy and climate. It was also suggested that adaptive capacity differs between different types of tourism enterprises, depending on their mobility, resources, and other factors.

The scale or organizational level of adaptive capacity was a common theme throughout the interviews. Some suggested that adaptation occurs in parallel at each scale or level (individual, community, industry, government, and region), while others suggested that adaptation begins at the community level and triggers change by the government, and still others suggested that government regulation is necessary to achieve change in communities.

Four future scenarios for the GBR region were developed from the interviews and refined in a workshop. The scenarios reflected two key uncertainties: (1) the nature and timing of climate change impacts, and responses to the impacts and (2) the type of governance in the region and the extent to which it is influenced by regional leadership or global economic forces. In effect, scenario building and exploring responses to change is a process of operationalizing adaptive capacity, intended to assist participants to deal proactively with the complexity and uncertainty of the region's future.

2.4 Adaptive Capacity in Theory and Reality: Matches, Mismatches, and Future Governance of the GBR

How well do adaptive capacity in theory and in "reality" match? In this section, we revisit our conceptual lens (the context of adaptive capacity, scale and information and knowledge), and use it to compare and contrast the definitions and measures of adaptive capacity that we have found in the literature and our empirical research. We conclude by remarking on implications for addressing the complex governance challenges in the GBR region.

The literature suggests that adaptive capacity is context-dependent. Our four studies analyzed adaptive capacity in the GBR in different contexts. Does adaptive capacity differ between different actors, or between actors confronted by different change processes? Although we did not design our research to address this question explicitly, our findings suggest that it often does. Regional leaders noted that adaptive capacity differs between industries and in response to different types of change, although their views diverged on determinants of adaptive capacity in industries such as tourism. In interpreting statements of interviewees, we note that perceptions of adaptive capacity can be influenced by the stage of the change process. Where policies are already in place, it is possible to observe adaptation that is already occurring, such as that of fishers to policy change. Perceptions of adaptive capacity to climate change are based more on inference, as there remains much uncertainty about specific adaptive behaviors and actions that may be undertaken.

The literature suggests adaptive capacity is scale-dependent. We were interested in identifying differences in perceptions between individuals and stakeholder groups operating at various scales. The analysis of resource users and organizations and their perceptions of rules identified gaps that may inhibit effective local participation in planning processes. However, there were also common perceptions across scales, in particular amongst natural resource managers working at different scales and leaders working at the broad regional scale. These commonalities may reflect the relatively high levels of education, knowledge, communication, and networks in this region compared to other regions in which there are greater inequalities in access to information. Better understanding of similarities and differences in perceptions across scales would be valuable for future management and capacity building of managers. Ultimately, the effects of scale-specific perceptions in creating synergies or trade-offs in adaptive capacity across scales need further research: does individual adaptive capacity come at a cost to the adaptive capacity of the GBR region, and vice versa?

Information and knowledge is another theme common to the theoretical and empirical descriptions. While different types of knowledge are considered important for adaptive capacity in the literature (e.g., Folke et al. 2005), natural resource managers expressed a need for "honest" information about water quality change that is easily communicated. The significant barriers that exist between knowledge creators, holders, and users are noted in the institutional analysis, echoing challenges found elsewhere (Roux et al. 2006). We also note that differences in either the fundamental concepts or the language of theory and "reality" may prohibit effective communication between researchers, government agencies, and resource users. Clearly the language used in the theoretical and empirical definitions differs, and it is not always possible to discern the extent to which this reflects a mismatch in the underlying understanding.

The theory of adaptive capacity embodies ideas about complex SES, which requires great care to be conveyed successfully to nontechnical audiences. These concepts seem to be only partially comprehended in the GBR region. For example, some natural resource managers noted the need to have greater control over the environment, through technologies and predictive tools. This contradicts the theoretical views that acknowledge the need for governance to maintain the self-organizing capacities of ecological systems, and allow variation and adaptation, while seeking to keep the system within critical thresholds (Gunderson and Holling 2002). In addition, the managers' view that a stable global economy is an important factor in enhancing adaptive capacity is in contrast to the view in the literature that adaptive capacity is itself part of an adapting and largely uncontrollable system. While the theory notes that taking advantage of opportunities and novelty are part of adaptive capacity, the perceptual definitions focus more on coping with consequences. In fact, most of the latter reflect a passive, reactive approach to adaptation, as was noted in the interviews with GBR leaders, and has also been found in other research (Olsson et al. 2004). Furthermore, most of the constraints to adaptive capacity mentioned in these interviews were related to information and understanding, yet interviewees also indicated a lack of will or responsibility to take adaptive action.

It is this point that we feel is noteworthy as a key message from this analysis. Overall, our findings indicate that the GBR region has high adaptive capacity, or high potential to develop it: despite the prospect of substantial change in the region, much of it uncertain, many of the individuals and organizations we interviewed remarked positively about forthcoming change, and some even welcome it. Yet they also indicated that this adaptive capacity may not be turned into active adaptation until a crisis occurs, and acknowledged the significant danger that this may come too late. Indeed, the theoretical definitions suggest that adaptive capacity is latent and harnessed in response to stimuli. We wonder at which point the stimuli will be sufficient to provoke a response, and what this response might be. The prevailing belief among the GBR leaders who were interviewed that a catastrophe is the most likely pathway to change is unsettling. Reversing this *modus operandi* is probably the region's greatest, most complex governance challenge of all.

As a future applied research direction, we recommend that more emphasis be given to understanding motivations that underlie adaptive capacity and indeed adaptation, as argued by Haddad (2005). Empirical research requires analysts to observe adaptive capacity through adaptation processes, which is difficult to do in the absence of adaptation. The alternative approach of eliciting perceptions of adaptive capacity is also problematic as perceptions are couched in the unique views of each adapting actor, and importantly, he or she may not always act in accordance with stated perceptions. Combining inductive and deductive approaches to define and understand adaptive capacity as we have discussed in this chapter is therefore key to understand the motivational context of adaptive capacity.

Solving the complex governance challenges of the GBR requires nothing less than a multi-faceted, multi-perspective approach. It requires nothing less than a combination of theoretical and empirical analyses, at multiple scales, with heterogeneous social groups and through multiple methods, in order to confront disparities and improve both the theory and the development of practical and relevant operational measures of adaptive capacity. Governance that is based on theory or empirical analysis alone is not enough.

References

Economics A (2008) Economic value of the Great Barrier Reef Marine Park 2006–07. Access Economics Pty Limited, Canberra

Adger WN (2003) Social aspects of adaptive capacity. In: Smith J, Klein R, Huq S (eds) Climate change, adaptive capacity and development. Imperial College Press, London

Adger WN, Brooks N, Bentham G, Agnew M, Eriksen S (2004) New indicators of vulnerability and adaptive capacity. Tyndall Project IT1.11: July 2001 to June 2003, Final Project Report. Tyndall Centre for Climate Change Research, Norwich

Adger WN, Vincent K (2005) Uncertainty in adaptive capacity. Cr Geosci 337:399–410

Adger WN, Arnell NW, Tompkins E (2005) Successful adaptation to climate change across scales. Glob Environ Change 15:77–86

Allison HE, Hobbs RJ (2004) Resilience, adaptive capacity, and the "Lock-in Trap" of the Western Australian agricultural region. Ecol Soc 9(1):3

Bellamy J (2007) Adaptive governance: the challenge for regional natural resource management. In: Brown AJ, Bellamy JA (eds) Federalism and regionalism in Australia: new approaches, new institutions? Australian National University E Press, Canberra

Bengtsson J, Angelstam P, Elmqvist T, Emanuelsson U, Folke C, Ihse M, Moberg F, Nyström M (2003) Reserves, resilience and dynamic landscapes. Ambio 32:389–396

Berkes F, Colding J, Folke C (2002) Navigating social–ecological systems: building resilience for complexity and change. Cambridge University Press, Cambridge

Bohensky E, Lynam T (2005) Evaluating responses in complex adaptive systems: insights on water management from the Southern African Millennium Ecosystem Assessment (SAfMA). Ecol Soc 10(1):11

Bohensky EL, Reyers B, van Jaarsveld AS (2006) Future ecosystem services in a Southern African river basin: a scenario planning approach to uncertainty. Conserv Biol 20:1051–1061

Bohnet I, Bohensky E, Gambley C, Waterhouse J (2008) Future scenarios for the Great Barrier Reef Catchment. CSIRO, Water for a Healthy Country National Research Flagship, Canberra

Bradley D, Grainger A (2004) Social resilience as a controlling influence on desertification in Senegal. Land Degrad Dev 15:451–470

Brown AJ (2007) Federalism, regionalism and the reshaping of Australian governance. In: Brown AJ, Bellamy JA (eds) Federalism and regionalism in Australia: new approaches, new institutions? Australian National University E Press, Canberra

Bryant CR, Smit B, Brklacich M, Johnston TR, Smithers J, Chiotti Q, Singh B (2000) Adaptation in Canadian agriculture to climatic variability and change. Clim Change 45:181–201

Carpenter S, Walker B, Anderies J, Abel N (2001) From metaphor to measurement: resilience of what to what? Ecosystems 4:765–781

Cinner J, Fuentes MMPB, Randriamahazo H (2009) Exploring social resilience in Madagascar's marine protected areas. Ecol Soc 14(1):41

Day S, Dwyer A (2003) How vulnerable is your community to a natural hazard? Using synthetic estimation to produce spatial estimates of vulnerability. Aust J Reg Stud 9:299–316

Dovers S, Wild River S (2008) Institutions for sustainable development. Centre for resource and environmental studies. The Australian National University, Canberra

EPA – Environmental Protection Agency (2006) Draft Mackay-Whitsunday regional coastal management plan. Environmental Protection Agency, Rockhampton

Folke C, Carpenter S, Elmqvist T, Gunderson L, Holling CS, Walker B (2002) Resilience and sustainable development: building adaptive capacity in a world of transformations. Ambio 31:437–440

Folke C, Hahn T, Olsson P, Norberg J (2005) Adaptive governance of social–ecological systems. Annu Rev Environ Resour 30:441–473

Ford J, Gough B, Laidler G, MacDonald J, Qrunnut K, Irngaut C (2009) Sea ice, climate change, and community vulnerability in northern Foxe Basin, Canada. Clim Res 38:138–154

Gallopín GC (2006) Linkages between vulnerability, resilience, and adaptive capacity. Glob Environ Change 16:293–303

Gunderson L, Holling CS (2002) Panarchy: understanding transformations in human and natural systems. Island Press, Washington, DC

Grothmann T, Patt A (2005) Adaptive capacity and human cognition: the process of individual adaptation to climate change. Glob Environ Change 15:199–213

Haddad BM (2005) Ranking the adaptive capacity of nations to climate change when sociopolitical goals are explicit. Glob Environ Change 15:165–176

Hertin J, Berkhout F, Gann D, Barlow J (2003) Climate change and the UK house building sector: perceptions, impacts and adaptive capacity. Build Res Inf 31:278–290

Hughes TP, Baird AH, Bellwood DR, Card M, Connolly SR, Folke C, Grosberg R, Hoegh-Guldberg O, Jackson JBC, Kleypas J, Lough JM, Marshall P, Nyström M, Palumbi SR, Pandolfi JM, Rosen B, Roughgarden J (2003) Climate change, human impacts, and the resilience of coral reefs. Science 301:929–933

Hughes T, Rodrigues MJ, Bellwood DR, Ceccarelli D, Hoegh-Guldberg O, McCook L, Moltschaniwskyj N, Pratchett MS, Steneck RS, Willis B (2007) Phase shifts, herbivory, and the resilience of coral reefs to climate change. Curr Biol 17:360–365

IPCC – Intergovernmental Panel on Climate Change (2001) Climate change 2001: impacts, adaptation and vulnerability. Summary for policy makers. World Meteorological Organisation, Geneva

Janssen MA, Schoon ML, Kee W, Börner K (2006) Scholarly networks on resilience, vulnerability and adaptation within the human dimensions of global environmental change. Glob Environ Change 16:240–252

Lane MB (2005) Public participation in planning: an intellectual history. Aust Geogr 36:283–299

Lane MB (2003) Decentralization or privatization of environmental governance?: case analysis of bioregional assessment in Australia. J Rural Stud 19:283–294

Lane MB, McDonald G (2005) Community-based environmental planning: operational dilemmas, planning principles and possible remedies. J Environ Plann Manag 48:709–731

Lane MB, McDonald GT, Morrison T (2004) Decentralisation and environmental management in Australia: a comment on the prescriptions of the Wentworth Group. Aust Geogr Stud 42: 102–114

Larson S (2009) Communicating stakeholder priorities in the Great Barrier Reef region. Soc Nat Resour 22(7):650–664

Larson S (2006) Analysis of the water planning process in the Georgina and Diamantina catchments: an application of the institutional analysis and development (IAD) framework. Desert Knowledge CRC, Alice Springs

Larson S, Williams LJ (2009) Monitoring the success of stakeholder engagement: literature review. In: Measham TG, Brake L (eds) People, communities and economies of the Lake Eyre Basin, DKCRC Research Report 45. Desert Knowledge Cooperative Research Centre, Alice Springs, Australia

Larson S, Stone-Jovicich S (2008) Community perceptions of water quality and current institutional arrangements in the great barrier reef region of Australia. Paper presented at International water resources association world water congress, Montpelier, France, 1–4 Sept 2008

Lebel L, Anderies JM, Campbell B, Folke C, Hatfield-Dodds S, Hughes TP, Wilson J (2006) Governance and the capacity to manage resilience in regional social–ecological systems. Ecol Soc 11(1):19

Lemos MC, Oliveira JL (2004) Can water reform survive politics? Institutional change and river basin management in Ceará, northeast Brazil. World Dev 32:2121

Lemos MC, Boyd E, Tompkins EL, Osbahr H, Liverman D (2007) Developing adaptation and adapting development. Ecol Soc 12(2):26

Lynam T, Gooch M, Ross H (2007) Marine Tropical Sciences Research Facility (MTSRF) Milestone Report June 10. Project 4.9.7 Understanding and enhancing social resilience: science and management integration project. CSIRO, James Cook University, University of Queensland, Townsville

Marshall GR (2008) Community-based regional delivery of natural resource management – Building system-wide capacities to motivate voluntary farmer adoption of conservation practices. Rural Industries Research and Development Corporation Publication No 08/175, Canberra

Marshall NA, Marshall PA (2007) Conceptualizing and operationalizing social resilience within commercial fisheries in northern Australia. Ecol Soc 12(1):1

Marshall NA, Fenton DM, Marshall PA, Sutton S (2007) How resource-dependency can influence social resilience within a primary resource industry. Rural Sociol 72:359–390

McClanahan TR, Cinner JE, Maina J, Graham NAJ, Daw TM, Stead SM, Wamukota A, Brown K, Ateweberhan M, Venus V, Polunin NVC (2008) Conservation action in a changing climate. Conserv Lett 1:53–59

McKay J (2005) Water institutional reforms in Australia. Water Policy 7:35–52

Measham TG, Robinson C, Richards C, Larson S, Stafford Smith M, Smith T (2009) Tools for successful NRM in the Lake Eyre Basin: achieving effective engagement. In: Measham TG, Brake L (eds) People, communities and economies of the Lake Eyre Basin, DKCRC Research Report 45. Desert Knowledge Cooperative Research Centre, Alice Springs

MA – Millennium Ecosystem Assessment (2005) Ecosystems and human well-being: synthesis. Island Press, Washington, DC

Nelson R, Brown PR, Darbas T, Kokic P, Cody K (2007) The potential to map the adaptive capacity of Australian land managers for NRM policy using ABS data. CSIRO, Australian Bureau of Agricultural and Resource Economics, prepared for the National Land and Water Resources Audit, Canberra

Norris FH, Stevens SP, Pfefferbaum B, Wyche KF, Pfefferbaum RL (2008) Community resilience as a metaphor, theory, set of capacities, and strategy for disaster readiness. Am J Community Psychol 41:127–150

OECD InterFutures Study Team (1979) Mastering the probable and managing the unpredictable. Organisation for Economic Co-operation and Development and International Energy Agency, Paris

Olsson P, Folke C, Hahn T (2004) Social–ecological transformation for ecosystem management: the development of adaptive co-management of a wetland landscape in southern Sweden. Ecol Soc 9(4):2

Olsson P, Folke C, Hughes TP (2008) Navigating the transition to ecosystem-based management of the Great Barrier Reef, Australia. Proc Natl Acad Sci USA 105:9489–9494

Ostrom E (2007) Multiple institutions for multiple outcomes. In: Smajgl A, Larson S (eds) Sustainable resource use: institutional dynamics and economics. Earthscan, London

Pelling M, High C (2005) Understanding adaptation: what can social capital offer assessments of adaptive capacity? Glob Environ Change 15:308–319

Peterson GD, Allen CR, Holling CS (1998) Ecological resilience, biodiversity and scale. Ecosystems 1:6–18

Peterson GD, Cumming GS, Carpenter SR (2003) Scenario planning: a tool for conservation in an uncertain world. Conserv Biol 17:358–366

Redman CL, Kinzig AP (2003) Resilience of past landscapes: resilience theory, society, and the *longue durée*. Conserv Ecol 7(1):14

Roux DJ, Rogers KH, Biggs HC, Ashton PJ, Sergeant A (2006) Bridging the science – management divide: moving from unidirectional knowledge transfer to knowledge interfacing and sharing. Ecol Soc 11(1):4

Salz RJ (1998) Social justice and the Florida net ban controversy. Human dimensions research unit. Department of Natural Resources Conservation, Amherst

Scheffer M, Brock W, Westley F (2000) Mechanisms preventing optimum use of ecosystem services: an interdisciplinary theoretical analysis. Ecosystems 3:451–471

Scheffer M, Westley F, Brock W (2003) Slow response of societies to new problems: causes and costs. Ecosystems 6:493–502

Scheffer M, Westley FR (2007) The evolutionary basis of rigidity: locks in cells, minds, and society. Ecol Soc 12(2):36

Smit B, Wandel J (2006) Adaptation, adaptive capacity and vulnerability. Glob Environ Change 16:282–292

Smit B, Burton I, Klein R, Wandel J (2000) An anatomy of adaptation to climate change and variability. Clim Change 45:223–251

Smith S (1995) Social implications of changes in fisheries regulations for commercial fishing families. Fisheries 20:24–26

Smith S, Jacob SJ, Jepson M, Israel G (2003) After the Florida net ban: the impacts on commercial fishing families. Soc Natur Resour 16:39–59

Stafford Smith M (2008) The 'desert syndrome': causally linked factors that characterise outback Australia. Rangeland J 30:3–14

Turton AR (1999) Water scarcity and social adaptive capacity: towards an understanding of the social dynamics of managing water scarcity in developing countries. MEWREW Occasional Paper No. 9 Water Issues Study Group. School of Oriental and African Studies, London

Walker B, Carpenter S, Anderies J, Abel N, Cumming GS, Janssen M, Lebel L, Norberg J, Peterson GD, Pritchard R (2002) Resilience management in social–ecological systems: a working hypothesis for a participatory approach. Conserv Ecol 6(1):14

Walker B, Holling CS, Carpenter SR, Kinzig A (2004) Resilience, adaptability and transformability in social–ecological systems. Ecol Soc 9(2):5

Walker BH, Abel N, Anderies JM, Ryan P (2009) Resilience, adaptability, and transformability in the Goulburn-Broken Catchment, Australia. Ecol Soc 14(1):12

Westley F, Carpenter SR, Brock WA, Holling CS, Gunderson LH (2002) Why systems of people and nature are not just social and ecological systems. In: Gunderson L, Holling CS (eds) Panarchy: understanding transformations in human and natural systems. Island Press, Washington, DC

Yohe G, Tol RSJ (2002) Indicators for social and economic coping capacity: moving toward a working definition of adaptive capacity. Glob Environ Change 12:25–40

Chapter 3
Building Adaptive Capacity in Systems Beyond the Threshold: The Story of Macubeni, South Africa

Christo Fabricius and Georgina Cundill

3.1 Introduction

Adaptive co-management theory is based predominantly on case study examples from developed countries where resources are abundant and communities and other stakeholders are, in comparison with their developing country counterparts, well organised and highly educated. Resource-poor developing areas present different challenges to communities, facilitators and other professionals. Because infrastructure is weakly developed, literacy levels are low, and people are reliant on their own knowledge and local resources (CEPSA Consortium for Ecosystems and Poverty in sub-Saharan Africa 2008), developing rural areas are notoriously neglected and have received scant attention from officials and facilitators. People are typically uninformed about their rights and opportunities, and are politically marginalised with decisions taken on their behalf in capitals and head offices. However, two-thirds of sub-Saharan Africa's 770 million people live in rural areas (Anderson et al. 2004), where common property regimes abound and efforts at community-based interventions have largely failed (Campbell et al. 2001; Blaikie 2006). Adaptive co-management is an attractive metaphor for common property management in such areas. The 'collaborative' part of adaptive co-management might overcome some of the capacity challenges by supplementing the pool of available expertise, linking local people to higher level institutions and bringing in financial and human capital. The 'adaptive' part is ubiquitous in rural areas and requires little intervention. Rural people are of necessity adaptive and opportunistic to secure their livelihoods under challenging conditions (Berkes et al. 2000; Ellis 2000; Gadgil et al. 2003).

C. Fabricius
Nelson Mandela Metropolitan University, X 6531, George 6530, South Africa
e-mail: Christo.Fabricius@nmmu.ac.za

G. Cundill
Rhodes University, 94 Grahamstown 6140, South Africa
e-mail: Georgina.Cundill@gmail.com

South Africa's former separate development 'homelands' are examples of rural areas with weak infrastructure and low capacity, with very low endowments of human, physical and financial capital (sensu Carney 1998). During an exploratory field trip in August 2001 (Fabricius et al. 2002), local community members, government officials and political representatives asked academics at a regional university to initiate a project aimed at enhancing rural livelihoods, repairing ecosystem services and strengthening local capacity to manage natural resources and generate income at Macubeni in South Africa's former Transkei 'homeland'. This area presented a particular challenge to the proponents of adaptive co-management because of heavy levels of resource degradation, very weak human capacity, extreme poverty and poor infrastructure. The community relies heavily on natural resources, remittances and social capital (reciprocity and kinship networks) to make a living (Fabricius and Collins 2007). A window of opportunity (sensu Olsson et al. 2006) opened when a potential funder, the German Agency for Technical Cooperation (GTZ) provided funding, the national government wanted to test its community-based natural resource management or CBNRM guidelines and policies, and the national Department of Environmental Affairs and Tourism (DEAT) launched a number of social responsibility projects with a CBNRM emphasis. There also appeared to be champions for the project, both at local level in the form of an elected municipal Councillor for the Macubeni Ward, and a Local Economic Development official in the Emalahleni Municipality. At Provincial levels, the Department of Economic Affairs Environment & Tourism as well as the Department of Agriculture indicated their support for an intervention that would restore social and natural capital.

At about the same time, we became exposed to resilience and adaptive co-management theory through our links with the Resilience Alliance (www.resalliance.org) and our participation in the Millennium Ecosystem Assessment. This provided a solid theoretical foundation, a scholarship and funding base via the Millennium Ecosystem Assessment and GTZ, legitimacy in the eyes of government and local communities, and a reasonable time horizon of 5 years. Our goals were to (a) increase the available natural, human and financial capital in Macubeni; (b) repair the adaptive capacity of the social–ecological system, to enable it to cope with change; (c) develop the governance capacity for communal property management.

3.2 Study Area

Macubeni is regarded as one of the most degraded areas in South Africa, with extensive sheet and gully erosion on both the hill slopes and valley bottoms (ATS/iKhwezi 2004; Shackleton and Gambiza 2008). The locality 'Macubeni' refers to a cluster of 14 villages that collectively consider themselves part of one community and form part of a single electoral Ward. The population of roughly 7,344 people resides on 16,150 ha of land, held primarily under communal tenure (Fig. 3.1).

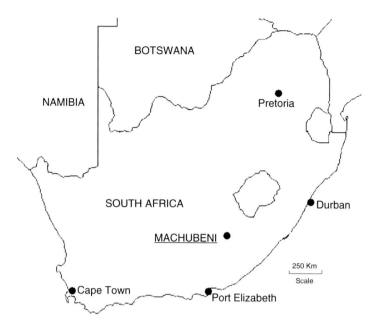

Fig. 3.1 Macubeni

The key issues affecting social–ecological resilience at Macubeni are: (a) historical and contemporary erosion of ecosystem services through land degradation, inappropriate cultivation practices, inappropriate fire management and biodiversity loss; (b) loss of ecosystem productivity through invasion of unpalatable plants; (c) loss of governance capacity through past political oppression and contemporary emigration of resourceful individuals; (d) lack of access to markets for agricultural products and livestock; (e) lack of finances to invest in agriculture, infrastructure and education; and (f) poor infrastructure, particularly related to piped water.

3.2.1 Ecosystem Services

Macubeni is characterised by hilly and mountainous terrain. The soils are generally shallow and stony, except in the valley bottoms. The area falls within the Grassland Biome, and is characterised by both Tsomo Grassland and Tarkastad Montane Shrubland (Mucina and Rutherford 2006). The average annual rainfall is 590 mm, 80% of which falls in the summer months, between October and March. This rainfall is erratic however, and dominated by convective storms (iKhwezi and Setplan 2004). Although temperatures can exceed 40°C, average day time summer temperatures are 20°C, and 12°C in winter. Temperatures of below zero do occur and winter snowfalls are not uncommon.

Ecosystem productivity is severely undermined by extensive and increasing gully and sheet erosion (iKhwezi et al. 2004). Sheet erosion is found close to roads and on the open veld, caused by trampling and over grazing. Gully erosion occurs along the water courses and in the relatively deeper soils of the valley bottoms. Because of the deeper soils and flatness of the land, these areas have historically been cultivated, and the villages are generally located in these areas too. The majority of the erosion occurs in these flatter areas due in large to inappropriate ploughing practices.

Although crop production and arable lands are highly valued in Macubeni, crop yields are extremely low (iKhwezi et al. 2004). Nevertheless, fruit trees tend to be grown within garden plots, while maize, sorghum, wheat, barley, pumpkins and a range of legumes are grown in fields. Crops are rain fed, and hand held tools are the dominant mode of cultivation. The main restraining factors for cultivation are the lack of water, infrastructure and resources such as equipment and capital.

The main water source is the eCacadu River, which flows through Macubeni and into the Macubeni Dam. This river only flows after rains and during the wet season. There are also numerous natural springs, which are utilised for domestic and stock water.

Livestock ownership is widespread within the community, with an estimated 37% of households owning stock, and average ownership across Macubeni being ten large stock units (LSU) per stock owner. There is no fencing between villages or to divide grazing camps. There is no rotational grazing system or collective system of grazing management. Macubeni's long term ecological 'carrying capacity' is 1,129 LSU, with a maximum potential of 2,691 LSU if erosion was prevented and the veld restored to its optimum. A survey conducted in 2003 suggested that there were 7,670 LSU in Macubeni, while a Department of Agriculture estimate placed the number in excess of 11,000 LSU (iKhwezi et al. 2004). Macubeni is therefore severely overstocked and the productivity of the land is constantly being reduced.

There are three planted woodlots, and also three brick making plants within the area. Grasses are harvested for roofing and for sale, as are medicinal plants. The invasive *Euryops*, a plant species that has flourished in the degraded soil, is used by the majority of households for fuel, and for kraal (enclosures for livestock) building (Shackleton and Gambiza 2008).

3.2.2 History of Land Management and Institutional Capacity

Land allocation and management has until recently taken place largely through traditional structures, with a sub-headman responsible for each of the 14 villages that make up Macubeni, and a headman to whom these sub-headmen report. The creation and enforcement of rules governing access to resources similarly took place through these structures. However, the last six decades have witnessed growing incursions into land management by the state, and the concomitant weakening of local institutional structures for resource management.

During the 1960s and 1970s, the government's 'Betterment Planning' model was implemented in Macubeni and entailed moving scattered homesteads into consolidated villages. Those living on the mountain sides were the most severely affected as they were forced to relocate with very little consultation. As in other areas, it is likely that the traditional leaders in Macubeni co-operated in the process. Two decades later, between 1986 and 1987, the Macubeni Dam was constructed to supply water to the Lady Frere district and surrounding areas. Once again, there was no consultation or negotiation with the community. Some compensation was paid to households; however, resentment still lingers over the lack of consultation over the construction of the dam and the subsequent relocations.

The 1990s witnessed a period of institutional inertia and confusion over land use and management. During this period, traditional leaders in many cases lost their ability to control land use, largely because of being associated with the state's Bantu Authorities system, whereby traditional leaders were paid by the state, and because of their apparent collusion with the state during the forced removals in the 1960s and 1980s. The result was that many people felt that leaders were accountable to the state rather than to the people. Traditional leadership in Xhosa culture has always relied on the support of the people (Pieres 1981), and therefore the lack of faith in traditional leadership severely reduced their power base. During this period, democratic processes began to gain momentum, and local democratic structures attempted to assert their authority (Manona 1995). However, there was no clear national policy or locally accepted norms that gave these democratic structures the power to manage and allocate land use rights. Thus, land use and management all but disappeared in the leadership vacuum that resulted. One symptom of this collapse was the fact that fences that formerly separated grazing areas and fields were stolen in the 1990s to fence off homesteads, with no repercussions for the perpetrators (iKhwezi et al. 2004).

The former Transkei homeland, in which Macubeni was situated, was reincorporated into South Africa in 1994. Since 1994, land allocation and management have formally fallen into the hands of local government structures, particularly the Ward Committee and an elected Councillor, although traditional leadership still has a strong position and is consulted of necessity. Indeed, an application for land by an individual would seldom reach the elected Councillor without first having gone through the hands of the sub-headmen and the headman, and rule enforcement is not considered possible by community members unless the traditional leaders are involved, as indicated by the following quote

> The major challenge [with rule enforcement] is the support from the traditional leaders, if they won't support the rules then we can't enforce them (MPASC member, March 2007).

3.2.3 Social Vulnerability

Social vulnerability has increased progressively at Macubeni through the combination of multiple forced relocations, the resultant weakening of local institutional capacity for land management, and the progressive erosion of the land as a result of this. These factors represent successive and synergistic stresses on the social–ecological system. Social vulnerability is equally influenced by the socio-economic conditions that characterise the area.

The Eastern Cape has the highest proportion of people living in poverty in South Africa, with 71.9% of the population living on less than US$87 per month, and 12.3% of the population surviving on less than US$1 a day (Kane-Berman et al. 2007). Macubeni is remote, with little infrastructure or formal employment opportunities, and therefore households are reliant on family members sending remittances from urban centres, arable production, livestock and ecosystem goods and services as described earlier (Shackleton and Gambiza 2008). Fifty seven percent of households are female headed, and 41% of households are headed by people aged above 60. Well over 50% of the population is under 20 years of age, indicating that the permanent population is made up largely of children and the elderly. Population figures from the 1995 to 2001 census suggest that the population declined by 13%, a rate of 2.6% per annum. Population growth in the area is affected by both out migration and a high death rate rather than a declining birth rate.

Ninety five percent of the local population is unemployed with only 14.8% actively looking for work (Statistics South Africa 2001). However, between 35 and 40% of households have access to some form of wage income during the year by part time, seasonal and self employment (iKhwezi et al. 2004). In terms of household income, between 1995 and 2001, the number of households declaring a nil income doubled, indicating growing poverty in the area. Forty nine percent of households receive a government welfare grant (iKhwezi et al. 2004). Two of the 14 villages have community taps for household water, and none of the households have taps inside the home.

3.3 Methods

We took a complex adaptive systems approach (Holling 2001), and conceptualised Macubeni as a linked social–ecological system (Berkes and Folke 1998). The social system consisted of the culture and traditions of the amaXhosa, their livelihood strategies, decision making structures and processes, as well as their belief systems and world views. The ecological system consisted of the soils, rangelands, livestock and landscape level processes within the Macubeni watershed (Fabricius and McGarry 2004). Knowledge, institutions and management practices formed the 'bridge' between the social and ecological systems (Berkes et al. 2003) (Fig. 3.2). We relied on Sustainable Livelihoods theory (Ashley and Carney 1999) to assess the available social, human and natural capital in the system and, ultimately, to attempt to enhance financial capital by stimulating the establishment of small enterprises. We aimed to stimulate adaptive co-management by strengthening the adaptive capacity of institutions and their ability to respond to social–ecological feedback (Berkes and Folke 1998; Olsson et al. 2004a) in an attempt to transform

Fig. 3.2 Human capital (knowledge, governance capacity, management systems and institutions) at local, national and sub-national levels acts as the link between social and ecological systems

the community from a 'powerless spectator' state to an 'adaptive co-manager' state (Fabricius et al. 2007). Through ecological restoration, we also aimed to strengthen the ecosystem's capacity to respond to shocks and surprise (Walters and Holling 1990) in particular floods and droughts, which were a major threat at Macubeni and threatened the integrity of soils and rangelands (iKhwezi et al. 2004). A multi-disciplinary team of plant ecologists, rangeland scientists, development economists, social scientists and modellers was involved and expert facilitators formed part of the core team. The capacity of community members to take control of resource management and decision making was developed by assigning specific roles and responsibilities to individuals through a process of discussion and consensus building between the different role players. However, because of the administrative requirements of the funding agencies, there were significant limitations to implementing a comprehensive community-based approach (Mitchell et al. 2007).

Our intervention focused on the trajectory: awareness → knowledge → new awareness → motivation → action with a feedback loop called 'learning' from 'action' to 'knowledge' (Fig. 3.3) (Fabricius et al. 2007). Trust building, institutional development and strengthening of governance systems were key elements of this process (Table 3.1), and every step of the process was strongly linked to cultural practises (Xu et al. 2005; Folke et al. 2005b). This will be discussed in subsequent sections, but included, for example, the integration and traditional decision making processes and authorities within the governance structures that were developed during the initiative, and the inclusion of local knowledge about indicators of ecosystem change within community-based monitoring and management plans.

We attempted to stimulate the awareness of stakeholders by first gathering background information about the social–ecological system and about people's

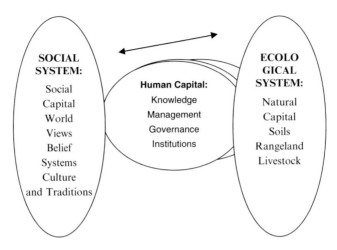

Fig. 3.3 A conceptual framework for strengthening social–ecological systems through adaptive co-management

Table 3.1 Key interventions aimed at building adaptive capacity and social–ecological resilience

Key focus within the social–ecological system	Intervention
Ecosystem services	Ecological restoration of wetlands, rangelands, natural springs and cultivated land
Governance and institutions	Development of mini-management plans for key resource areas in each village
	Formation of village land committees to implement and enforce mini-management plans
	Formation of the MPASC, the smaller technical sub-committee of the MPASC, and eventually the Section 21 Company
System dynamics – adaptive management	Community based monitoring and evaluation of both social and ecological components of the system
Financial capital	Temporary employment through government social responsibility funds
Human capital	Skills development – fencing, erosion control, natural spring development and alien plant removal, formal training in CBNRM and ecological and social monitoring, the formation of 'not for profit' organisations for land management and business and accredited training in small and medium enterprise development

understanding and perceptions of the reasons behind land degradation. In this we relied heavily on local and traditional knowledge and made extensive use of participatory methods (Chambers 1994). Initiating collaborative monitoring was a core component of efforts to raise awareness and build knowledge in an on-going way during the initiative. Through this approach, new knowledge was developed and shared among stakeholders, and solutions were sought to identified challenges and trends. Knowledge was shared and awareness further raised when community members who participated in information gathering regularly reported to villagers at general meetings. These report-backs became increasingly more sophisticated when certain members were assisted in developing computer-aided presentations.

3.3.1 Scale

We adopted a spatially and temporally multi-scale approach (Capistrano et al. 2005) by, from a temporal perspective, assessing landscape change over time and political change since 1930, and, from a spatial perspective, assessing the key institutions, and landscapes and land elements, at multiple spatial scales from the local to District level (Adger et al. 2005; Hein et al. 2006). At the local level, we matched the scale of governance and institutions to the scale of ecosystems being managed (Murphree 1993; Murphree 2000; Bohensky and Lynam 2005) by focusing on fine-grained key resources such as springs, prime grazing patches and cultivated lands on deep and nutrient-rich alluvial soils in the landscape. Together with village level leaders and community members, we developed mini management plans, to be governed at the village level and this is discussed further in the sections that follow.

3.3.2 Drivers

Through the multi-scale approach, we focused on and aimed to understand the key, or driving, social and ecological processes operating at local, sub-national and national scales that influence current social–ecological trends (MA 2005) at Macubeni. These key drivers were political, economic and biophysical. In conjunction with this analysis, we combined an assessment of the roles of critical social actors, such as leadership figures and visionaries and their ability to steer the system toward a different trajectory (Schultz et al. 2007) with an analysis of the key ecological elements that maintained the integrity of the social–ecological system (Machado 2004).

3.3.3 Governance and Co-Management

We viewed knowledge, institutions and management practices as the 'bridges' between the social and ecological components of the system (Berkes et al. 2003) and aimed to strengthen them by augmenting the 'soft' and flexible traditional institutions (Becker and Ghimire 2003) with 'harder' and more formal institutions that would be compatible with formal government policies and local governmental processes. We drew substantially on common property theory (Ostrom et al. 1999;

Dietz et al. 2003) by putting much effort into strengthening the institutions that govern human behaviour.

We aimed to strengthen local governance systems and institutions and worked closely with a new structure, the Macubeni Project Advisory and Steering Committee (MPASC), an offshoot of an existing organisation, the Macubeni Ward Committee. The MPASC was established by the local municipality and national government as a link between the community and local political representatives in the implementation of the project. The MPASC formed the primary stimulus for co-management. The MPASC was established as a multi-stakeholder steering body aimed at improving communication between role players operating at different spatial scales (Table 3.2). Two elected representatives from each of the 14 villages in Macubeni sat on the MPASC, along with representatives from community institutions, government departments and implementing agents.

The MPASC met monthly and was intended to support the project management team in various ways. First, the MPASC played a guiding, advisory and sometimes decision making role. Second, the MPASC aimed at improving relationships between the various project partners by creating a forum for conflict management and discussion. Third, the MPASC played a critical communication role between all partners, but especially between the steering committee and the larger community. Community representatives on the MPASC were expected to regularly inform their constituencies about the progress of the initiatives. A community working group, the Macubeni Technical Committee was also established, consisting of eight members elected on the basis of their exceptional local knowledge and dedication (Chalmers and Fabricius 2007), and closely involved the traditional leadership structures at Macubeni.

Noticing an institutional vacuum, we set out to establish management plans as institutions for natural resource management. Conscious of the need to create congruence between institutional and ecological scales (Carpenter et al. 2001; Bohensky et al. 2004), we focused on fine-grained 'key resource areas', which were identified by local villagers and assisted the community in creating mini management plans for each key resource area. Each of the 14 villages identified three key resource areas: a grazing area, a spring and a cultivation area.

Table 3.2 Role players in the Macubeni MPASC

Community institutions	Ward committee, headmen, sub-headmen, youth forum, farmers association
Local and district government departments	Emalahleni local municipality, Chris Hani District Municipality, Department of Agriculture, Department of Water Affairs and Forestry
Implementing agents and supporting organisations	GTZ transform, Ruliv (Eastern Cape Rural Livelihoods Support programme), iKhwezi (consultancy sub-contracted for project implementation)
Provincial and national government departments	Department of Agriculture, Department of Environmental Affairs and Tourism

Seven steps were followed in the construction of these plans, which consisted of the following seven sections:

1. Where is it? This section contains a general description of the key resource area, its locality and importance for the village including a locality map.
2. What does it produce? The goods and services it produces for the community, including tangible and intangible services.
3. Who has access to the key resource area?
4. Who is responsible for its management?
5. What are the rules and regulations, with sub-sections containing general and specific rules, and sanctions?
6. Management recommendations.
7. Revision intervals.

The document used by the village was in isiXhosa, with a translation into English for the purposes of reporting to donors. Each village received a file containing its management plans, monitoring data sheets and maps.

Towards the end of the project, Village Land Committees were established, tasked with implementing these fine-scale management plans for key resource areas. The traditional leaders (sub-headmen) operating at village level headed these committees.

3.3.4 Capacity Development

Because of the acute multi-dimensional poverty at Macubeni, and the shortages of financial, human, natural and physical capital, we aimed to strengthen the quantities of these four capitals (Fabricius and Collins 2007). To some extent, the physical capital was strengthened by the introduction of critical infrastructure such as water storage tanks, nurseries and fences to regulate livestock and fence off key resource areas. Material incentives were improvements in people's livelihoods through restoring the ecosystem's capacity to produce goods and services; training to improve people's management and technical skills; and financial incentives in the form of short term income from government's social responsibility funds, as well as longer term income from the establishment of small and medium enterprises (Mitchell et al. 2007) by individuals and consortiums. Natural capital restoration included the repair of ecosystem services through erosion control works, the removal of alien species, re-vegetation and management of key resource areas, establishment of nurseries and restoration of some of the key natural springs.

Participants received training in a wide variety of areas. At a basic level, those employed in the work teams aimed at building physical infrastructure as part of the Social Responsibility programme received training in fencing, erosion control, natural spring development and alien plant removal. Within the MPASC, participants received formal training in CBNRM, ecological and social monitoring, giving presentations, and the formation of organisations for land management and

Table 3.3 Types of formal training received and percents o respondents who received training ($n = 110$)

Formal training received	Percentage of respondents
Land Management	49
Technical (Basic, e.g. fencing, hole digging)	43
Construction	15
Leadership	13
Health and Safety	12
First Aid	4
Financial Management	3
Tourism	0.5

business. Accredited training in small and medium enterprise development was offered to a core group of individuals following an application and selection process. This training included aspects of financial management and leadership. Table 3.3 summarises the types of training provided and the percentage or respondents who benefited.

3.3.5 Motivation

Awareness raising does not automatically lead to action and adaptation. People need to be motivated to transform their social–ecological system. Motivation consists of two components: vision and leadership; and incentives for transformation (Lambin 2005). Leadership existed in the form of the locally elected Ward Councillor, the traditional leaders (Headman and sub-Headmen), and elected members of the Macubeni Technical Committee. The Macubeni vision was established very early on and reiterated at every important community meeting.

3.3.6 Adaptive Management and Monitoring

Adaptive management implies an ability to learn from mistakes (Colding et al. 2003) and to respond to feedbacks and innovation (Armitage 2005). This requires the establishment of participatory community-based monitoring systems (Fleming and Henkel 2001). Although acknowledging that communities have their own informal monitoring systems, and regularly track changes in ecosystem services (Moller et al. 2004), community-based monitoring is however less prevalent in tracking change in the social domain of social–ecological systems. Therefore, we initiated a community-based monitoring system to track change in both the social and the ecological domains of the system by focussing simultaneously on the process of implementation, governance, livelihoods and ecosystem services. The monitoring indicators were selected collaboratively with community representatives and contained as many as possible traditional indicators already in use or widely recognised within the community (Berkes and Jolly 2001).

The frequency of monitoring was matched to the rate of change of different aspects of the social–ecological system. Because MPASC meetings took place monthly, governance and process monitoring took place on a three monthly basis as a means of capturing change and allowing space for adaptive decision making whilst not overloading the already busy MPASC members. Livelihood monitoring took place during and after the employment phase of the Social Responsibility initiative, and included 'hindsight' questions about activities prior to the initiative. These 'hindsight' questions provided the baseline from which changes could be measured. Natural resource monitoring took place primarily as training exercises aimed at developing tools that each Village Land Committee could use to adaptively manage key resource areas using the mini management plans, and was intended to be conducted seasonally. A monitoring toolkit was developed, containing descriptions of monitoring methods and data sheets. Each village was provided with a file containing the toolkit, monitoring data sheets and its management plan. The toolkit can be accessed at (http://oldwww2.ru.ac.za/academic/departments/environsci/CBNRM/index.html). Results and insights from this ongoing collaborative monitoring toolkit are presented throughout the following section.

3.4 Outcomes

3.4.1 Identifying Drivers

Our initial assessment indicated that the ultimate drivers of the system were as follows: (a) emigration of human capital, mainly as migrant workers, thereby depleting the capacity for governance; (b) weakly developed infrastructure such as road networks, piped water and electricity supply; (c) human population density that was too high for the capacity of the resource base to provide goods and services; and (d) the historical impact of separate development policies which affected traditional institutions, infrastructure, human population density and local and municipal governance systems.

Proximate drivers were the following: (a) depletion of ecosystem services, mainly through overgrazing and inappropriate cultivation, which increased soil erosion and decreased soil fertility; (b) the collapse of local institutions, leaving an institutional vacuum and leading to open access resource management; (c) extreme poverty, which resulted in short term planning and 'struggle and survive' attitudes due to perceptions of an uncertain future; and (d) weak capacity of municipal, national and provincial government structures to promote rural development.

3.4.2 Strengthening of Governance

Governance gradually improved between 2001 and 2007 when the intervention ended. The Village Land Committees became key organisational structures, with their activities centering on monitoring, and the development and implementation of mini management plans.

Mini management plans were developed for 13 of the 14 villages in the Macubeni Ward, and people were trained in natural resource management, monitoring and rangeland management. Village Land Committees were formed, and they took responsibility for implementation of the mini management plans. To begin with, these Village Land Committees worked through the MPASC and were specifically supported by the Technical Committee. During the course of the project, people were introduced to different types of legal entities for contractual and business purposes. In 2007, a legal entity in the form of a Section 21 company was formed with representatives of national, provincial and local government and with two representatives from each of the 14 Macubeni villages. People also received training in entrepreneurship and small enterprise development, and small enterprises were established for meat production, to run nurseries and to produce and sell firewood from local woodlots.

However, a number of governance surprises were experienced. At the beginning of the initiative, the locally elected Ward Councillor played a critical role in garnering community and local government support for the initiative. As a result, the members of the MPASC, although elected by each village, were drawn largely from this individual's political party. However, in early 2007, this councillor's term came to an end, and the democratic elections were won by a political rival from a different party. This candidate won the support of the traditional leadership. Many community members were suspicious about the benefits that the MPASC members had accrued over the years, for example through training and access to information about employment possibilities. As a result, when the Section 21 Company was finally formed later in 2007, the elections for village representatives resulted in all but one of the original members being replaced. None of these new members had received land management or leadership training over the course of the initiative. The build up of human and social capital envisioned at the outset of the initiative, and pursued over the course of a number of years, was effectively lost in just a few days, although it could be argued that the skills remained within the community.

Nonetheless, analysis of the outcomes from on-going collaborative monitoring, based on self-administered rating systems, suggested that governance trends were positive with a 50% improvement in social capital (with indicators such as trust building, common rules and norms and incentives for collective action c.f. Pretty 2003) and a 100% improvement was reported for adaptive governance (with indicators such as access to information, conflict resolution mechanisms, compliance with rules and being prepared for change c.f. Dietz et al. 2003). No change was identified in the capacity for self-organisation (with indicators such as enabling legislation, access to long term funding, effective leadership and social networks c.f. Olsson et al. 2004a) or adaptive capacity (with indicators such as maintaining options and the willingness to learn from mistakes and to engage in collaborative decision making c.f. Armitage 2005).

3.4.3 Capacity Development

Although many people received formal training, most of the training was in land management (49%), and in technical skills such as fencing and erosion control (43%). Seventeen percent received training in leadership, and health and safety. Very few individuals were trained in financial management and leadership (16%). As part of collaborative monitoring activities, random surveys of those involved in the training activities suggested that although the majority of participants (64%) were happy with the training they received, and 60% believed that the training will improve their livelihoods, only 4% had used the training to find employment.

The capacity of the Macubeni Technical Committee was however strongly developed through their on-going involvement in management planning, monitoring and reporting back to the community. Most of them became involved in small enterprises, although this also contributed to their institutional downfall when it came to election of the Section 21 Company, as described previously.

3.4.4 Motivation

Local actors were highly motivated to develop their capacity and improve the ecosystem's capacity to deliver goods and services. They developed a strong vision, expressed as 'our dream is to have the maximum benefit for the Macubeni community from our natural resources, and to secure its sustainability by forming partnerships with all relevant structures and create a production market from these resources and encourage youth to participate and to motivate the home coming of those who had left to study'. This vision was confirmed at every stakeholder meeting and placed on a flip-chart in a prominent place.

Material incentives in the form of employment in the social responsibility project played an important role in motivating people, but only temporarily. Although people were aware that these stipends would be terminated within 24 months, they continued to hope that the SRP would be renewed. A number of them hoped to find permanent jobs, while a small minority, who were already entrepreneurs, used opportunities to establish small enterprises.

One of the main motivating forces was ecosystem repair and restoration of ecosystem services, particularly soil and rangeland conservation measures, and measures to protect springs for water security. There was visible evidence of improvements in rangelands, erosion and water provision, but this was limited to small areas with the largest part of the catchment still remaining in a degraded state. This was because of financial and time constraints.

3.4.5 Adaptive Management and Monitoring

Between mid 2006 and the end of 2007, members of the MPASC and the Technical Committee regularly monitored the process of implementation, governance, livelihoods and (to a lesser extent) the natural resource base and kept records of trends. This process enabled community members to hold implementing agents responsible for their actions, and lead to the active questioning of the status quo.

Active adaptive management took place during monitoring workshops, where participants identified 'actions' that should be taken to rectify problems and challenges identified. These actions became 'the way forward' at the end of the workshops, and individuals were assigned by their fellow community members to follow up and report back. At the beginning of the following workshop, the way forward was returned to and the successes or failures in solving the problems were reflected upon. Generally, the recommendations that emanated from monitoring were presented and discussed at the monthly MPASC meetings with all stakeholders. Process monitoring revealed clear of evidence of learning having taken place, and of active questioning of the values that underpinned existing institutions. This was revealed particularly through the creation of the Section 21 Company, which was based equally on traditional norms of governance, as evidenced by the inclusion of traditional leaders at all levels of the organisation, and more conventional norms in keeping with legal requirements. Community-based monitoring was therefore critical in raising awareness of changes that were taking place, generating new knowledge and understanding, and leading to appropriate action (Fig. 3.3), as the following quote from community members attests:

> Before we started monitoring, things just happened here, now we are able to plan. Monitoring opens our eyes to see forthcoming crises, so that when those crises arrive we are not surprised (MPASC member, September 2007)

3.4.6 Co-Management

As a forum for conflict resolution and information sharing, the MPASC model was very effective. For example, it soon became apparent that the roles of the various government departments overlapped considerably, which caused confusion and highlighted the importance of the MPASC. While irrigation water was provided by the Department of Water Affairs and Forestry, once the water reached the fence of a farmer responsibility for irrigation infrastructure fell to the Department of Agriculture. In Macubeni, it was therefore critical that these two departments be engaged concurrently in long term land use planning. However, critical shortages in staff, capacity and resources, such as a vehicle to reach Macubeni, meant that all relevant departments were not able attend every meeting. Indeed, the local agricultural extension officer soon proved to be a vital link between the community and other government departments.

Some of the challenges with co-management, which we observed, were the dominance of government and municipal officials' views (when they were present), and the frequent absences of officials from meetings due to their limited budgets and staff shortages, but also because they seemed to be very involved in internal meetings and processes, the importance of which seemed to override community development considerations.

3.5 Discussion

3.5.1 Linking Resilience and Adaptive Co-Management Theory to Practise

The approach described in Fig. 3.3 is a useful structure for the facilitation of adaptive co-management initiatives. It provides a logical, sequential approach to focus the attention of facilitators and communities. The initial knowledge generation step was essential, as it exposed the key drivers of the system, and provided a focus and catalyst for the formation of knowledge networks, through for example the formation of the MPASC, which became one of the mainstays of the project. The challenge in this phase was to balance the powers between informal local knowledge and formal scientific knowledge (Healy 2003; Libel et al. 2006). Our regular local workshops, where information was consistently shared and all stakeholders were given the opportunity to present their views, guarded against imbalances. We also made specific efforts to strengthen the confidence of local people in their own knowledge by assisting them with developing their own presentations, and providing numerous opportunities for community members to present their own ideas and solutions in community and broader public forums. Community knowledge about the local uses and distribution of natural resources was readily accepted, but local knowledge about management, and processes, was less readily trusted by outsiders.

The community's awareness about their responsibility for ecosystem management increased dramatically between 2006 and 2007, with the percentage of people believing it was the community's responsibility, in collaboration with government, almost doubling from the 2006 to the 2007 survey. The motivation of participants was stimulated through the training they received, personal recognition received from the community and the prospects of improvements in ecosystem services, improved fecundity and condition of livestock, and, most importantly, greater water security because of the repair of deteriorating springs.

As an attempt to match the scale of institutions and decision making to the ecosystems being managed (Bohensky and Lynam 2005), the development of fine-grained or mini management plans at the village level for key resources was effective. The seven elements made them easy to conceptualise and implement: a typical mini management plan took one working day for preparation and training, and one day to complete. These management plans focused people's attention on land elements that were directly relevant to their respective villages, and were small enough to make people feel capable of managing them. Most of the key resource areas were within visible distance of the villages, and were therefore constantly in people's minds. The involvement of sub-headmen and direct users of ecosystem services proved to be extremely effective: people were motivated, and felt they had the ability, to manage their own resources. The facilitators assisted by documenting the management plans and providing the necessary stationary and equipment such as files, data sheets, and, most importantly, training. These training sessions were attended by a diversity of village members such as the youth and people of different ages, and many more people than those who served on the Village Land Committees attended the training sessions, which greatly benefited broad-based awareness raising.

Power-knowledge dynamics are an inherent feature of adaptive co-management in resource poor communities. This dynamism was evidenced several times during the project. For example, the power dynamics that eventually resulted in the 'overthrow' of the MPASC, during the critical change-over phase of becoming a Section 21 Company, were centered largely around access to information. MPASC members were accused of holding on to knowledge for personal gain. This conflict over access to information was overlaid by political power struggles, which, together, produced a wholly unexpected outcome in the election of the members of the Section 21 Company. Another example was the fact that the Technical Committee formed out of the MPASC to receive intensive training in monitoring and management became increasingly alienated by other MPASC members, who felt that they were receiving undue attention from facilitators. Since these technical committee members were originally voted for by the MPASC because they were regarded locally as experts in several fields, their loss during the elections for the Section 21 Company represented the loss of a substantial amount of skills and knowledge from the system. The interaction between power and knowledge is one of the sources of novelty and surprise in complex systems, and this is explored further in the section that follows.

3.5.2 *The Impact of Surprise in Systems Beyond the Threshold: Politics, Conflict, Government Decisions*

Despite the existence of a window of opportunity, and the benefits of adaptive co-management coupled with several decades' experience working in rural African communities in community-based natural resource management settings, things still went wrong. It seems that, in the early stages of capacity development when capacity is low, social–ecological systems are extremely vulnerable to surprise. A relatively 'normal' occurrence of local political change, when the local councillor and champion of the project was ousted, catalysed an entire cascade of events culminating in the rejection of almost every individual trained during the project when the time arrived to re-elect community members. However, this turn of events was paralleled by political change at national level, with the split within the ruling party and the development of two political 'camps' at national, provincial and, ultimately, local level (Majova 2007). The consequence of these events was that the higher level co-management institution faltered due largely to a lack of understanding of the long term vision and intended role of the institution by its new members, and a complete lack of training. Since there was no structure to report to, the Village Land Committees, while continuing to exist, have not progressed further than developing the mini management plans, with negative implications for monitoring, institutions and, ultimately, ecosystem resilience. These political upheavals were beyond the control of anyone inside the project but, we contend, would have been overcome had the capacity to cope with change been more fully developed. On the positive side, no major or life-threatening conflicts occurred during this period and the events demonstrated two key elements of a community's capacity for governance: democracy and decisiveness.

3.5.3 *Implications for Adaptive Co-Management Theory*

Conceptualising Macubeni as complex, adaptive social–ecological system enabled the team to, from the outset, interrogate the interactions between the different components of the system, their change over time and the complex drivers of change. This informed the nature of interventions proposed to government and international development agencies, and also guided awareness raising and training programmes. Using resilience as a lens enabled us to focus on repairing the adaptive capacity of the social–ecological system, rather than merely adding or restoring capital as advocated by many development practitioners. The adaptive co-management approach meant that 'learning by doing' and constant reflection was deliberate. This was communicated up-front to project funders to obtain their buy-in into a less rigid but more informative approach to development. The local stakeholders readily and intuitively accepted this approach, as this reflected their traditional and contemporary livelihood strategies and natural resource management and governance systems. The learning and reflecting aspects increased awareness amongst local people and government officers of flaws in their current resource management practises, and the reasons for degradation. This catalysed innovation and adaptive governance.

Adaptive co-management under resource-poor conditions has six challenges: (1) Maintaining key individuals and balancing power relations; (2) Motivating all actors to collaborate; (3) Making the most of available capacity and resources; (4) Overcoming and coping with disturbances during the early stages of capacity development; (5) Focusing on the finest resolution that time and budget allows; (6) Persisting long enough to facilitate long-term change.

3.5.3.1 Maintaining Key Individuals and Balancing Power Relations

A key challenge in initiating adaptive co-management under resource poor conditions is finding a balance between nurturing key individuals and managing the power dynamics that emerge from this process. Jealousies flared at all levels in Macubeni when certain individuals took the lead in the initiative: for example, the MPASC felt that the Technical Committee was not sharing information and was receiving undue attention and possibly benefits, whilst the broader community felt that the MPASC was not sharing information effectively and was holding on to information about employment possibilities. Political leaders capitalised on this distrust and ushered in an entirely new set of committee members from their own political party.

Attempts to manage these power relationships should not however undermine community self-determination. It would have been tempting, for example, to insist that at least some of the members of the MPASC should sit on the Section 21 Company. It is even more tempting to state that the initiative has failed. However, it is in the reorganisation phase of the adaptive cycle, immediately following collapse, that learning and novelty are thought to emerge (Folke et al. 2003). Indeed, community members are already suggesting that in the future only 50% of the committee can be rotated during any one election cycle. Balancing power relations relies on patience, and an understanding that initiatives will experience periods of rapid development and periods of comparatively little change.

Our observations that officials' views tended to dominate meetings are not new. This resonates with the critiques of co-management raised by other authors, for example, Nadasdy (2003) and Reid and Turner (2004). Other challenges with co-management are that officials who are crucial to the notion of adaptive co-management take their natural resource management obligations less seriously than communities do, lack of resources for officials to actively participate, and some powerful role players within the community and outside it abusing the co-management process to promote their own materialistic objectives.

3.5.3.2 Motivating all Actors to Collaborate

A key challenge to adaptive co-management is to motivate those collaboration partners who are not directly affected by or benefiting from ecosystem services, to commit themselves to co-management. It is relatively easy to mobilise and motivate local people to restore ecosystem services, but their capacity is seldom adequate to go it alone. Therefore, other partners have to be motivated to actively participate and contribute. However, the capacity of local municipalities and government officials to perform this function is relatively low. Therefore, academic institutions have crucial roles to play as sources of knowledge and intellectual capital in adaptive co-management. Action research (Blythe et al. 2008) therefore has a crucial function in resource-poor countries, and academic entities should explicitly pursue this in their strategic objectives (Tyler 2006).

3.5.3.3 Making the Most of Available Capacity and Resources

Under resource-poor conditions all forms of capacity matter. This relates to knowledge of both social and ecological processes; infrastructure; finances; and social relationships in particular. These assets are the 'seeds of renewal' and, if nurtured, can grow to increase the resilience of the social–ecological system (Gunderson et al. 1995). We made abundant use of local and traditional knowledge, and made the most of government, development agency and academic funds as well as municipal resources. We collaborated with government agricultural experts, international development advisors, other academics from different disciplines, NGOs and other universities. We used community halls, schools, clinics and people's homes to conduct workshops and meetings, and stayed in with local families during visits to the area. The key resource areas repaired and managed by the community were not extremely rich in biodiversity or nutrients, and were relatively small but important patches of 2–10 ha in size, and the springs did not deliver more than a few hundred litres per hour, but they were critical to the livelihoods of the community. Existing institutions such as the traditional leadership were in disarray and had all but collapsed at the start of the project. They were, however, critical to the establishment of the Village Land Committees committees, and we can assume that the adaptive co-management process strengthened their legitimacy and played a catalytic role in reviving them.

3.5.3.4 Overcoming Disturbances During the Early Stages of the Project

Because of the shortages of capital during the early stages of adaptive co-management, small disturbances can have a large impact on the trajectory of transformation (Folke et al. 2002; Olsson et al. 2004b). Because of the complexity of system interactions and the plurality of actors involved (Fabricius et al. 2007), adaptive co-management is prone to disturbance and surprise throughout. When surprises such as political upheaval, institutional challenges such as transgressions of rules, conflicts, ecological disturbances such as flash floods or fires occur in the early stages of adaptive co-management they can de-rail the system, especially in the absence of facilitators (Jones 1999). In well-resourced systems such as those described by Olsson et al.

(2004a, b), the system has enough 'buffering capacity' (Folke et al. 2005a) to cope with disturbance.

3.5.3.5 Focusing on the Finest Resolution Within Time and Budget Constraints

The fine resolution approach was advocated by Murphree (2000) in his analysis of success factors in Zimbabwe's CAMPFIRE (Communal Areas Management Programme for Indigenous Resources). One of the principles advocated by Murphree is that the unit of management should be at the same level as the unit of resource utilisation (Fabricius 2004). In the case of Macubeni, the shift in focus to fine-grained key resource areas, 'owned' and managed by individual villages, created new impetus and maintained the continued provision of ecosystem services. These resource patches were highly visible, and in daily use by the people directly responsible for their management. Matching the unit of management to the resolution of the ecosystem by facilitating the formation of Village Land Committees, further promoted the functioning of lasting institutions. When political turmoil occurred, these Village Land Committees survived, because of the close relationships between their respective members, their focus on ecosystem services that mattered to the community, and their ability to regularly meet without much cost to their members.

3.5.3.6 Persistence

Our intervention started in mid 2001 and ended in early 2007, when key individuals left, government's social responsibility funding cycle ended, and the international development agency withdrew. We contend that 6 years is not enough to facilitate transformation and the restoration of human, physical, natural and financial capital in systems beyond the threshold. This is because of the hysteresis effect (www.resalliance.org), where the 'memory' of past impacts (Folke et al. 2003) linger on in the system. In the Macubeni case, the impacts of historical social engineering and ecosystem degradation are still highly visible in the form of extreme poverty, low formal qualifications, demographic distortions with weak representation by middle-aged males, and large areas that remain eroded despite the ecosystem restoration efforts. Close to two million US$ were spent over the 6 years (Mitchell et al. 2007), with many organisations contributing to the funding base. We maintain that this represents only a fraction of the investment and effort required to truly transform these resource-poor, degraded systems. This has major implications for rural development in South Africa's former 'homeland' areas, as Macubeni represents just one of the hundreds of Wards in a similar state. The shortage of skilled facilitators, funds and weak project management capacity in South Africa make this a formidable challenge, requiring the formation of a dedicated Ministry for Rural Development with a significant budget.

3.6 Conclusions

Despite the substantial investment of expertise, time and money at Macubeni over a 6 year period, the social–ecological resilience of the system could not be adequately strengthened to survive a relatively small disturbance in the form of orderly political change. We believe, however, that the use of the theoretical principles of adaptive co-management maximised the impact of the intervention, and that the Macubeni community is more knowledgeable, skilled and aware of the challenges facing them, and much more resilient than before. To overcome the legacy of historical policies will, however, require prolonged interventions and investments by local authorities, national and provincial government, academic institutions and international development agencies and the local community, who understands that ecosystem management is primarily their responsibility.

References

Adger WN, Brown K, Tompkins EL (2005) The political economy of cross-scale networks in resource co-management. Ecol Soc 10(2):9

Anderson J, Bryceson D, Campbell B, Chitundu G, Clarke J, Drinkwater M, Fakir S, Frost P, Gambiza J, Grundy I, Hagmann J, Jones B, Jones GW, Kowero G, Luckert M, Mortimore M, Phiri ADK, Potgieter P, Shackleton S, Williams T (2004) Chance, change and choice in Africa's drylands. A new perspective on policy priorities? Centre for International Forestry Research (CIFOR), Bogor

Armitage D (2005) Adaptive capacity and community based natural resource management. Environ Manage 35:703–715

Ashley C, Carney D (1999) Sustainable livelihoods: lessons from early experience. Department for International Development, London

Becker CD, Ghimire K (2003) Synergy between traditional ecological knowledge and conservation science supports forest preservation in Ecuador. Ecol Soc 8:1

Berkes F, Folke C (1998) Linking social and ecological systems: management practices and social mechanisms for building resilience. Cambridge University Press, Cambridge

Berkes F, Jolly D (2001) Adapting to climate change: social-ecological resilience in a Canadian western arctic community. Conserv Ecol 5:18

Berkes F, Colding J, Folke C (2000) Rediscovery of traditional ecological knowledge as adaptive management. Ecol Appl 10:1251–1262

Berkes F, Colding J, Folke C (2003) Navigating social–ecological systems: building resilience for complexity and change. Cambridge University Press, Cambridge

Blaikie P (2006) Is small really beautiful? Community based natural resource management in Malawi and Botswana. World Dev 34:1942–1957

Blythe S, Grabill J, Riley K (2008) Action research and wicked environmental problems: exploring appropriate roles for researchers in professional communication. J Bus Tech Commun 22:272–298

Bohensky E, Lynam T (2005) Evaluating responses in complex adaptive systems: insights on water management from the Southern African Millennium Ecosystem Assessment (SAfMA). Ecol Soc 10:11

Bohensky E, Reyers B, van Jaarsveld AS, Fabricius C (2004) Ecosystem services in the Gariep Basin: a basin-scale component of the Southern African millennium assessment. Sun, Stellenbosch

Campbell B, Mandondo A, Sithole B, De Jong W, Luckert M, Matose F (2001) Challenges to the proponents of common property resource systems: despairing voices from the social forests of Zimbabwe. World Dev 29:589–600

Capistrano D, Samper C, Lee M, Raudsepp-Hearne C (2005) Ecosystems and human well-being: multiscale assessments, vol 4. Island, Washington

Carney D (1998) Implementing the sustainable livelihoods approach. In: Carney D (ed) Sustainable rural livelihoods. What contribution can we make? Department for International Development, London, pp 3–26

Carpenter S, Walker B, Anderies JM, Abel N (2001) From metaphor to measurement: resilience of what to what? Ecosystems 4:765–781

Chalmers N, Fabricius C (2007) Expert and generalist local knowledge about land-cover change on South Africa's Wild Coast: can local ecological knowledge add value to science? Ecol Soc 12:10

Chambers R (1994) Participatory rural appraisal (PRA): analysis and experience. World Dev 22:1253–1268

Colding J, Elmqvist T, Olsson P (2003) Living with disturbance: building resilience in social–ecological systems. In: Berkes F, Colding J, Folke C (eds) Navigating social–ecological systems: building resilience for complexity and change. Cambridge University Press, Cambridge, pp 163–186

Consortium for Ecosystems and Poverty in sub-Saharan Africa (CEPSA) (2008) Situation analysis of ecosystem services and poverty alleviation in arid and semi-arid Africa. Khanya-aicdd, Johannesburg, pp 1–141

Dietz T, Ostrom E, Stern P (2003) The struggle to govern the commons. Science 302:1907–1912

Ellis F (2000) Rural livelihoods and diversity in developing countries. Oxford University Press, Oxford

Fabricius C (2004) The fundamentals of community-based natural resource management. In: Fabricius C, Koch E, Magome H, Turner S (eds) Rights, resources and rural development. Earthscan, London, pp 3–43

Fabricius C, Collins S (2007) Community-based natural resource management: governing the commons. Water Policy 9:83–97

Fabricius C, McGarry D (2004) Frequently asked questions at Macubeni. GTZ Transform, Pretoria, pp 1–12

Fabricius C, Matsiliza B, Buckle J (2002) Community-Based Natural Resource Management (CBNRM) in RURAL Livelihoods (RULIV) – Eastern Cape planning process. GTZ Transform, Pretoria, pp 1–44

Fabricius C, Folke C, Cundill G, Schultz L (2007) Powerless spectators, coping actors, and adaptive co-managers: a synthesis of the role of communities in ecosystem management. Ecol Soc 12:29

Fleming B, Henkel D (2001) Community-based ecological monitoring: a rapid appraisal approach. J Am Plann Assoc 67:456

Folke C, Carpenter S, Elmqvist T, Gunderson L, Holling CS, Walker B (2002) Resilience and sustainable development: building adaptive capacity in a world of transformations. Ambio 31:437–440

Folke C, Colding J, Berkes F (2003) Synthesis: building resilience and adaptive capacity in social–ecological systems. In: Berkes F, Colding J, Folke C (eds) Navigating social–ecological systems: building resilience for complexity and change. Cambridge University Press, Cambridge, pp 352–387

Folke C, Fabricius C, Cundill G, Schulze L (2005a) Communities, ecosystems and livelihoods. In: Capistrano D, Samper C, Lee M, Raudsepp-Hearne C (eds) Ecosystems and human well-being: multiscale assessments, vol 4. Island, Washington, pp 261–277

Folke C, Hahn T, Olsson P, Norberg J (2005b) Adaptive governance of social–ecological resources. Ann Rev Environ Resour 30:441–473

Gadgil M, Olsson P, Berkes F, Folke C (2003) Exploring the role of local ecological knowledge in ecosystem management: three case studies. In: Berkes F, Colding J, Folke C (eds) Navigating

social–ecological systems: building resilience for complexity and change. Cambridge University Press, Cambridge, p 189
Gunderson LH, Holling CS, Light S (1995) Barriers and bridges to the renewal of ecosystems and institutions. Columbia University Press, New York
Healy S (2003) Epistemological pluralism and the 'politics of choice'. Futures 35:689–701
Hein L, van Koppen K, de Groot R, van Lerland E (2006) Spatial scales, stakeholders and the valuation of ecosystem services. Ecol Econ 57:209–228
Holling CS (2001) Understanding the complexity of economic, ecological and social systems. Ecosystems 4:390–405
iKhwezi ATS, Setplan (2004) Macubeni land use plan situation assessment report, GTZ Transform, Pretoria, pp 1–50
Jones BTB (1999) Community-based natural resource management in Botswana and Namibia: an inventory and preliminary analysis of progress. Evaluating Eden Discussion Paper No 6, International Institute for Environment and Development (IIED), London
Kane-Berman J, Macfarlane M, Tempest J (2007) South Africa Survey 2006/2007. South African Institute for Race Relations, Johannesburg
Lambin EF (2005) Conditions for sustainability of human-environment systems: information, motivation, and capacity. Glob Environ Change A 15:177–180
Libel L, Anderies JM, Campbell B, Folke C, Hatfield-Dodds S, Hughes TP, Wilson J (2006) Governance and the capacity to manage resilience in regional social–ecological systems. Ecol Soc 11:230–250
Machado A (2004) An index of naturalness. J Nat Conserv 12:95–110
Majova Z (2007) ANC war of the bulldogs – part one. Mail & Guardian Online, thought leader blog. http://www.thoughtleader.co.za/zukilemajova/2007/10/17/anc-war-of-the-bulldogs-part-one/. Cited 17 Oct 2007
Manona C (1995) The collapse of the 'tribal authority' system and the rise of the civic organisations. Institute for Social & Economic Research (ISER), Rhodes University South Africa, Seminar Series
Millennium Ecosystem Assessment (2005) Ecosystems and human well-being: synthesis. Island, Washington, DC
Mitchell D, Coelho L, Baumgart J, Snel H (2007) Training and support for resource management Transform programme: lessons learnt from implementing community based natural resource management projects in South Africa. GTZ Transform and Dept of Environmental Affairs and Tourism, Pretoria, pp 1–70
Moller H, Berkes F, O'Brian Lyver P, Kislalioglu M (2004) Combining science and traditional knowledge: monitoring populations for co-management. Ecol Soc 9:2
Mucina L, Rutherford MC (2006) The vegetation of South Africa, Lesotho and Swaziland. South African National Biodiversity Institute, Pretoria
Murphree MW (1993) Communities as resource management institutions, Sustainable Agriculture and Rural Livelihoods Programme. International Institute for Environment and Development, Gatekeeper series No. 36
Murphree MW (2000) Boundaries and borders: the question of scale in the theory and practice of common property management. Bloomington, Indiana
Nadasdy P (2003) Reevaluating the co-management success story. Arctic 56(4):367–380
Olsson P, Folke C, Berkes F (2004a) Adaptive co-management for building resilience in social–ecological systems. Environ Manage 34:75–90
Olsson P, Folke C, Hahn T (2004b) Social–ecological transformation for ecosystem management: the development of adaptive co-management of a wetland landscape in southern Sweden. Ecol Soc 9:2
Olsson P, Gunderson L, Carpenter S, Ryan P, Lebel L, Folke C, Holling CS (2006) Shooting the rapids: navigating transitions to adaptive governance of social–ecological systems. Ecol Soc 11:18

Ostrom E, Burger J, Field CB, Norgaard RB, Policansky D (1999) Revisiting the commons: local lessons, global challenges. Science 284:278–282

Pieres J (1981) The house of Phalo: the history of the Xhosa people during the time of their independence. Ravan, Johannesburg

Pretty J (2003) Social capital and the collective management of resources. Science 302:1912–1914

Reid H, Turner S (2004) The Richtersveld and Makuleke Contractual Parks in South Africa: win–win for communities and conservation? In: Fabricius C, Koch E, Magome H, Turner S (eds) Rights, resources and development: community-based natural resource management in Southern Africa. Earthscan, London, pp 223–234

Schultz L, Folke C, Olsson P (2007) Enhancing ecosystem management through social–ecological inventories: lessons from Kristianstads Vattenrike, Sweden. Environ Conserv 34:141–152

Shackleton C, Gambiza J (2008) Trade offs in combating land degradation resulting from invasion of a woody shrub (*Euryops floribundus*) at Macubeni, South Africa. Land Degrad Dev 19:454–464

Statistics South Africa (2001) Community profiles. Statistics South Africa, Bisho

Tyler S (ed) (2006) Communities, livelihoods, and natural resources. Action research and policy change in Asia. Practical Action publishing, International Development Research Centre, Ottawa

Walters C, Holling CS (1990) Large scale management experiments and learning by doing. Ecology 71:2060–2068

Xu J, Ma ET, Tashi D, Fu Y, Lu Z, Melick D (2005) Integrating sacred knowledge for conservation: cultures and landscapes in southwest China. Ecol Soc 10:7

Chapter 4
Learning and Adaptation: The Role of Fisheries Comanagement in Building Resilient Social–Ecological Systems

Daniela C. Kalikoski and Edward H. Allison

4.1 Introduction

This chapter focuses on how robust self-organizations can be formed within fisheries comanagement systems. Over the last 30 years, comanagement has been increasingly advocated as a blueprint solution for small-scale fisheries crisis. Many governments, NGOs, and international and donor organizations are catalyzing projects for implementing fisheries comanagement. On the one hand, the international attention devoted to promoting and supporting comanagement is an important accomplishment; it recognizes that without the help and support of fishers, government can do little to help achieve sustainable, equitable, and resilient fisheries management. On the other hand, as comanagement becomes "mainstream," it risks being regarded as a straightforward technical and organizational process, through which states devolve both rights and responsibilities for the difficult tasks of resource conservation and livelihood improvement. This carries the danger that the widespread occurrence of ill-conceived comanagement systems, which do not account for comanagement's core values, will, through their inevitable failures, leave a legacy of degraded commons and impoverished resource users, thus leading to a backlash against participatory approaches to management (Pinkerton 2003; Wilson et al. 2003). The message is that people-centered, devolved approaches to renewable natural resource management risk being discredited if ecosystem conditions continue to decline.

D.C. Kalikoski
Institute of Humanities and Information Science, Federal University of Rio Grande (FURG), Av Italia Km.08, Rio Grande, RS, Cep: 96201-900, Brazil
e-mail: danielak@furg.br; Daniela.kalikoski@pq.cnpq.br

E.H. Allison
Policy, Economics and Social Sciences, The World-Fish Center, Penang, Malaysia
e-mail: e.allison@cgiar.org

This work derives from a consultancy for the UK Department for International Development and the Food and Agriculture Organization of the United Nations (FAO) Sustainable Fishing Livelihoods Program.

For comanagement systems to be resilient, they need to allow for feedback learning in the face of disturbance that produces a change from which people can learn (Berkes et al. 2003). It is disturbance (e.g., political, ecological) that initiates cycles of adaptive renewal. Thus, the interplay between disturbance, and the capacity to respond to and shape change, is what makes renewal and reorganization possible in the adaptive renewal cycle. Interplay is an important component for building resilience in social–ecological systems (Berkes et al. 2003; Folke et al. 2003; Berkes and Seixas 2005). Adaptive capacity is defined as the ability of actors, in a social–ecological system, to cope with novel situations without losing their options for the future (Walker et al. 2004). Systems with high adaptive capacity are able to reconfigure themselves without significant decline in crucial functions in relation to primary production, hydrological cycles, social relations, and economic prosperity.

Berkes and Jolly (2001) extend the concept of adaptation to embrace the responses of communities of resource users to increase the chances of success/survival in a changing environment. They differentiate two types of responses: coping mechanisms and adaptive strategies. Coping mechanisms are short-term emergency responses to abnormal seasons or years. Adaptive strategies are "ways in which individuals, households and communities change their productive capacity and modify local rules and institutions to secure livelihoods." These concepts are the key for achieving resilient comanagement systems.

This chapter takes a critical look into the past to draw some insights for the future on the factors affecting successes in the field of fisheries comanagement in terms of building robust and resilient organizations. These insights are defined in terms of the degree to which a system builds internal capacity for learning, adapting, innovating, and self-organizing as mechanisms to deal with the pressure from internal and external factors (Carpenter et al. 2001; Olson et al. 2004). Lessons are drawn from a comparison of the commonalities and differences in outcomes of selected comanagement case studies: Lake Malombe and Lake Chiuta (Malawi, Africa); Patos Lagoon and Arraial do Cabo (Brazil), and the Areas for Management and Exploitation of Benthonic Resources (Chile). The chapter ends by discussing the lessons learned from the cases, focusing on the key characteristics of comanagement with reference to their contribution (or not) in enabling the conditions for building resilient social–ecological systems. It closes with a discussion on the policy implications of implementing fisheries comanagement regimes.

4.2 Case Studies

4.2.1 On the Right Track: Comanagement's Positive Outcomes in Chile and Malawian Lake Chiuta

Territorial use rights to fisheries (TURFs) was a comanagement arrangement type that originated in Chile on account of a major crisis in the "loco" fishery that followed a closed season for a 3.5-year period (Castilla and Fernández 1998; Orensanz et al. 2005). Fishermen continued to fish illegally and the economic consequences and social distortions created by traditional administrative management measures (such as closures, quotas) motivated the subsequent search for management alternatives. Three main aspects were key to the establishment of the Chilean TURFs (Orensanz et al. 2005): (1) the presence of historical fishing territories in Chile; (2) the organization of artisanal fishers that lobbied the incorporation of the TURFs into the fisheries legislation; and (3) and informal management experiments (e.g., closures, protection of nursery grounds, removal of predators and competitors, translocations, and manipulation of species upon which loco preys) conducted voluntarily by fishers in some communities. Hilborn et al. (2005) describe this comanagement initiative in Chile as a successful one if compared with the disastrous situation that faced Chilean Benthic resources a decade ago, where TURF systems had proved to be the rights incentives to prevent and restore overfished benthic resources. These authors compared the severe contrast between the status of the stocks within the TURFs and those in open access "historical grounds" and found out that "fishermen are highly protective of the first, while a 'tragedy of the commons' situation prevails in the latter" (Hilborn et al. 2005, p. 52). Other positive aspects of the comanagement include gathering of knowledge about the response of the stocks to the harvest, great improvement in marketing practices, improved product quality and reliability of supplies and, most important, strengthening of fishermen's organizations stemming from shared responsibilities and appropriate incentives (Stotz 1997; Hilborn et al. 2005).

In the case of the African Lake Chiuta in Malawi, comanagement originates from a conflict that grew from the late 1980s to the early 1990s between indigenous fishers and the immigrant *nkacha* operators (Njaya 2002) who were allowed to fish in this lake by local leaders. Because of indigenous fishers request in 1995, comanagement evicted 300 *nkacha* operators from the fishery in this lake (Njaya 2002). These migrant fishers were allowed by local leaders to operate in the lake, conflicting with the interests of indigenous fishers who considered these leaders to have been corrupted by *nkacha* seine owners (Hara et al. 1999; Njaya 2002). Local fishers were opposed to the use rights given to migrant fishers on the basis that migrant's fishers: (1) used gears that destroyed fish habitats and caught juvenile fish; (2) landed larger catches than indigenous fishers whose use gear were fish traps and gill nets; (3) fish prices were down due to the above-mentioned items and had a negative effect on trap and gill net fishers (Hara et al. 1999). Hara et al. (1999) draw attention to how resource users identified the damaging effects of *nkacha* and then approached the fisheries department (FD) for advice, indicating their willingness to take care of their exploited resource. Fishermen created the beach village committees (BVC) and established the following regulations: (1) ban on the *nkacha* seine owners; (2) establishment of a minimum 38 mm mesh size; (3) conflict resolution mechanisms; (4) 100 mm as minimum tackable size of chambo; and (5) beach seiners not allowed on the lake and absence of a closed season (Njaya 2002). These fisheries regulations were approved by the Malawi government and are reviewed and enforced by the BVCs and local leaders. The role of communities is related to enforcement (i.e., expulsion of *nkacha* fishers; ban on use of *nkacha*, *m*esh sizes, no seining, immature fish, prosecution), and controlling access to the fishery. Other management actions are carried out on a consultative, co-operative, and delegated basis (e.g., licensing, resource-monitoring, problem solving, among others) (Hara et al. 1999). Communal property rights are considered but there is no clearly stated interest of the Government so that convergence of interests from both partners is assured (Hara et al. 1999).

4.2.2 The Challenge to Participatory Management: The Struggle to Implement Comanagement in Malawi and Brazil

Fisheries comanagement was introduced in Malawian Lake Malombe as a response to fisheries crisis precipitated by collapse of the more valuable species in the fishery (Hara et al. 2002) The Lake Malombe Participatory Fisheries Management Program (PFMP) started in 1993 as a pilot comanagement project to reverse the conditions of fisheries decline (Njaya 2002). The initial motivation of the PFMP was the decline of the "chambo" fishery (from around 8,300 ton in 1982 to less than 100 ton in 1994) and the failure of the existing management regime based on centralized government control and regulation. Established regulations were supposed to help restore fish habitats, protect juveniles and breeding fish and reduce fishing effort. Although these had a sound biological basis, they were not enough to ensure the successful implementation of regulatory measures. Recognized constraints were associated to budget limitation and enforcement capability. In addition, an increasing defiance and open resistance to compliance of regulations from fishermen (Hara et al. 2002) were observed and incidences of violence in the early 1990s against fisheries inspectors out on patrol duties had become common (Hara et al. 2002; Njaya 2002). The Government had to search for an alternative regime. The regime that seemed to provide the best option and chance for success seemed to be the one which involved some amount of self-regulation by the user communities. Government actors hoped that if this approach was successfully introduced it would satisfy both the government's and the user's objectives of biologically-sustainable exploitation of the resource and continued economic viability. These objectives were expected to be achieved at less cost to government on the assumption that self-regulation and increased acceptance of the regulations by users would result in much less need for outside enforcement of the regulations while ensuring sustainable economic viability of the resource for the fishing communities (FD 1993) (Hara et al. 1999). In addition user participation in resource management became one of the conditions for donor aid as it is believed that this will result in greater accountability and also as part of the general drive to empower the formerly disenfranchised population. Thus donor support for finding solutions to the problem of Lake Malombe came in the form of funding for activities aimed at promoting increased involvement of users in management of the fishery (Hara et al. 1999).

The Lake Malombe participatory program (PFMP) was launched and implemented as a multidonor funded project. The main donors were the German Technical Foundation (GTZ), the United Nations Development Program (UNDP), the Overseas Development Administration (ODA), and the World Bank (WB).

In Brazil, the major step to move toward participatory community-based comanagement was the creation of the Marine Extractive Reserve (MER) within the Brazilian legal framework of National System for Conservation Units (known as SNUC). This policy is significant because it represents the first government-sponsored effort toward comanagement and has enormous potential for conserving coastal areas and securing the livelihoods of coastal populations. The MER in Arraial do Cabo, RJ, was created in 1997 to protect the resident beach seining community and the resources their livelihoods depend on (Pinto da Silva 2002). The rich marine environment nourished by coastal upwelling attracted fishers that have been fishing in the cape for centuries. Local fishers in the region employ relatively sustainable fishing methods and depend on migratory stocks, such as mullets, that are managed inside the boundaries of the MER by a body of traditional rules governing the fishery for generations that became legalized and warranted the creation of the reserve (Pinto da Silva 2004). However, Pinto da Silva shows that these rules implemented by the MER are no longer robust and significant social barriers will need to be overcome to revitalize and fully integrate them into the reserve structure. Ownership patterns among beach seiners have changed dramatically even before the implementation of the MER and are concentrated in the hands of few individuals who own the majority of canoes and nets, thereby controlling the associated norms such as access days to the fishing grounds. This author concludes that the MER has not significantly fortified local management institutions and has overlooked or not been able to deal with these obstacles to participation and empowerment. She also concludes that although at different stages, the reserve has demonstrated some characteristics from the entire spectrum of comanagement arrangements, the role of fishers and State has not been ideal and both sides lack the capacity (funds, training, and experience) to support an effective system for participatory governance (Pinto da Silva 2002).

The Forum of Patos Lagoon was created in 1996 as an institutional response to the crisis of estuarine fisheries resources and the miserable situation that small scale fisheries communities in the estuary of Patos Lagoon were continuously facing (Reis and D'Incao 2000; Kalikoski 2002). The weak fishing season of 1995/1996, which helped trigger local changes in fisheries management, had one of the lowest landing volumes in 50 year. The Forum was an initiative of a group of people, led by the Church, the Fishers Colonies and in partnership with the local branch of the Federal Environmental Agency (IBAMA–CEPERG). The objective was to initiate an action plan to reverse the crisis that artisanal fisheries were facing. There was a general consensus on the part of the different actors related to fisheries management that, to reverse the fisheries crisis, a rearrangement of fisheries management was needed to accomplish a better organization of the sector in relation to management policies (Reis and D'Incao 2000). The representatives of the Forum comanagement arrangement concluded that through a collaborative partnership among communities and governmental and nongovernmental organizations, and using a negotiation-style of decision process, the fisheries could move toward a more productive status (Forum of Patos Lagoon minutes). The Forum has been an attempt to share responsibility and authority as concerns the management of fisheries resources. However, it still lacks the mechanisms for empowering the community and delivering fully the principles affiliated with comanagement. The model adopted by the Forum resembles more that of a stakeholder centered comanagement than a community-centered comanagement (Kalikoski 2002; Kalikoski and Satterfield 2004).

4.3 Deconstructing Fishery Comanagement Arrangements

Later we discuss the relationship among adaptive capacity, context-specific social–ecological systems, institutions, and collective action efforts, and, therefore, why certain comanagement regimes perform better than others. We propose that the adaptive capacity of comanagement systems will be dependent upon how well-prepared these systems are to deal with power imbalances, legitimacy crises, adaptive learning mechanisms, and the threat of erosion in social cohesion.

4.3.1 Power Imbalances

The lack of government commitment to devolving sufficient decision-making power to resource users will affect the incentives to their participation in fisheries management on equal terms with government. Unless power is genuinely shared and territorial and managerial rights are assigned to communities or other stakeholder groups, comanagement risks being captured, co-opted, and misapplied by the power holding actors maintaining the status quo in the fisheries governance (Jentof 2003; Pinkerton 2003). Transferring property rights is a mechanism that empowers users to make collective decisions. A property right potentially confers the power to (1) use or manage a resource or area, (2) the power to sell it or grant it, and (3) the power to take its yield as a harvest, rent, or royalty (Scott 2000). The owner of a fishing boat has all three powers over his/her boat: he/she can run it, sell it, and make a profit from the fish landed by it. The same fisher in his role as participant or occupier in the fishery may have only the third power. The first and second powers may be vested in the State or simply poorly specified and thereby appropriated and controlled by the powerful (Béné 2003). This, in theory, leaves the individual fisher with no incentive to look after the fishery as individual restraint may increase the value of the fish stock, but the individual has no powers to capture this extra value.

Transferring property rights back to the communities was an important instrument to deal with power imbalances in comanagement systems of TURFs in Chile, for instance, but was not successful in the MER of Arraial do Cabo in Brazil. In Chile enacting the allocation of TURFs was allowed only among organized fishing communities exploiting benthonic resources by a legal framework that devolves to fishers the power and gives them the means to govern the resources (Hilborn et al. 2005; Orensanz et al. 2005; Parma et al. 2003; Castilla et al. 1998; Castilla and Fernández 1998; González 1996). Resource use within TURFs is based on the exploitation plan that requires fishers to make projections of stock status that are used by the government to set a quota for the area. The fishers under TURF then decide how that quota is to be caught. In addition to TURFS, the fisheries Act contemplate management instruments that include size limits and seasonal closures and a system of marine protected areas (Orensanz et al. 2005). The request of exclusive TURFs of benthic shellfish is possible if communities meet certain criteria according to the law. Artisanal fishers must be part of a traditional organized community and are generally entitled to operate only in the region where they are registered. Artisanal rights are vested in fishers, not vessels, and are not transferable (Orensanz et al. 2005). As concluded by Castilla and Defeo (2001) the allocation of TURFs, when accompanied by a strong community-based comanagement, ameliorates the weaknesses of enforcement regulations, diminishing information and enforcement costs. Fishing grounds outside of benthic TURFs are open to all fishers registered in the region but exclude industrial fleets and artisanal fishers registered in other regions.

In the MER of Arraial do Cabo, power sharing is limited mainly because social capital has been disrupted within the community and government officials are not prepared to engage in such collaborative arrangement (lack of funds, training, and experience) (Pinto da Silva 2002). The MER was created because of the existence of local-based rules devised through generations that were resilient over time. However, those resilient governing institutions became less robust and have been co-opted by a few fishers controlling the fishery for their own benefit (Pinto da Silva 2002). Power devolution via transferring property rights was not sufficient in Arraial do Cabo because social capital has been eroded before the MER was created. Fishing communities were not homogenous, lack cohesion, and the fishery was locally controlled by some powerful groups within the community – the ones that own the boats, control the market, and ultimately decide resource use norms and rules (Pinto da Silva 2002). No means were given to fishing communities by the MER to deal with the power imbalances that have rather augmented over local governance at the community-level. Conflicts inside the MER also increased because powerful groups boycotted it when realized that their own organizational structure were at risk by this new arrangement. Powerless fishers also felt threatened. They were the employees of the power holders and did not have the means to engage within this comanagement otherwise they would risk losing their means of living. Comanagement in this case augmented power imbalances at the community level despite government's efforts to share its power over the governance of fisheries. Historically fishers' experiences with government have generally been negative; they do not trust officials and complain about corruption and inefficiency within government organization. This view has not improved with the creation of the MER since many feel that the reserve is an added responsibility placed on fishers (Pinto da Silva 2004).

Similar challenges halting adaptive capacity in comanagement systems have been observed in Lake Malombe (Malawi) and in the estuary of the Patos Lagoon (Brazil). In Lake Malombe and Lake Chiuta, a legal framework for comanagement is in place the Fisheries Conservation and Management Act (approved in 1997) still kept the management control in the hands of the government. The new Act conceded community participation, resource ownership, and empowerment of local communities, although transferring property rights to user communities was not part of the Act (Njaya 2002). The Act also stipulates that the local-level management groups will function under the "protection" and "advise" of the FD (Hara et al. (2002). Nevertheless, in Lake Chiuta, fishing communities convinced government of their capabilities and comanagement rapidly evolved toward a co-operative community-based comanagement, and this was partially due to the rapid self-organization capacity of local communities. Self organization in this Malawian lake is definitely a key point for cross-scale management. In Lake Malombe, local communities did not take any action to change their marginalized status quo. Although 31 BVCs were created around Lake Malombe as a two-way channel of communication between user groups and the FD (Njaya 2002), most BVCs lack fishers participation (Njaya 2002). Some argue that BVC's composition was influenced by government and donor agencies that ended up controlling: (1) a beach via the listing of its members and type of gear used; (2) the entry of additional gear owners; (3) the access and use of each beach; (4) right to expel members who do not comply with the agreed management measures (e.g., closed season, gears types, etc); (5) the meetings to discuss problems and management solutions; and (6) representation of BVCs members at higher levels of decision making (Njaya 2002). The BVCs have thus been seen as representing government interest rather than those of the communities (Hara et al. 2002). Comanagement under these circumstances may halt therefore the opportunities to institutional adaptive capacity that is a function of (1) self-organization, (2) nurturing diversity for reorganization and renewal, and (3) combining different types of knowledge for learning (Folke et al. 2003, p. 355). The struggle to devolve power from the government to the fishing communities has been a recurrent problem in the Lake Malombe comanagement project and has been evident since the beginning.

Similar to Malombe, in the case of the Forum of the Patos Lagoon, the struggle is also associated to the control that the government ultimately keeps in the final decisions over the establishment of the management rules. Also, the challenge to shift fisheries governance toward decentralization is related to the different levels of preparedness of people and institutions to adapt and make such a shift. The Forum is composed of 21 institutions involved in small scale fisheries management decisions (Kalikoski 2002). All of them have the right of one vote each and the representatives of fishers (e.g., colonies, associations, etc.) were each given the rights to two votes. Assigning more votes to fishers representatives is an attempt to shift the locus of control to the institutions representing artisanal fishers. Despite this effort, the Forum still lacks the mechanisms for empowering the community and delivering fully the principles affiliated with fisheries comanagement. It has been shown to work via a combination of partial empowerment augmented by the support of elite representatives and in some instance the Forum has been co-opted by the powerful institutions (Kalikoski and Satterfield 2004). In fact, the groups with capacity to adapt in this new arrangement are the most successful in promoting their own interests. Among them are fishers colonies (that seek to keep their power over fishers) and the central government (that has to approve final decisions by law) (Kalikoski 2002; Kalikoski and Satterfield 2004). One impediment of power sharing in the Forum is associated with the illiteracy and socio-economic marginalization of fishers that create low expectations among scientists and officials of the management value of fishers' knowledge (Kalikoski and Vasconcellos 2007). Kalikoski's (2002) analysis has demonstrated that the Forum is an attempt to share responsibility and authority as concerns the management of fisheries resources. However, it still lacks the mechanisms for empowering the community and delivering fully the principles affiliated with fisheries comanagement. Following O'Riordan (2003) "the achievement of pluralist power relationships in a society implies the capacity of empowerment, where all individuals are aware of their ability to recognize what is going on in their name, and have a capability to express their needs and reactions in such a manner as to be respectfully heard." According to the same author, "in many instances, however, pluralism gives way to neo-elitism where coalitions collude to determine what is to be done and how. Empowerment thus becomes possible in different forms of policy space" (O'Riordan 2003), such as in the case of the estuary of Patos Lagoon, where it has been shown to work via a combination of partial empowerment augmented by the support of elite representatives. Such procedures can be helpful, as discussed by O'Riordan (2003) "if genuinely representative groups are present," but as it has been shown here, this is still not the case in the Forum of Patos Lagoon where adaptive mechanisms have been easier to power holders. This challenge is amplified if cohesion among community members is weak as in this case.

4.3.2 Legitimacy Crisis

Achieving legitimacy depends on (1) how well the designed rules within the comanagement fully represent the interests of local fishers as a whole and (2) the recognition that actions taken locally are truly legitimized by the responsible federal agencies (Ostrom 1990; Jentoft 2000; Kalikoski and Satterfield 2004). This is strengthened by the existence of a legal framework that formally recognizes collective rights and emphasizes the importance of a coherent integration between different levels of governance. As argued by Jentoft (2000, p. 142): "...legitimacy should not be anticipated regardless of institutional design of comanagement. Comanagement may perhaps be the best available solution to the legitimacy problem but it may also, in itself, be the source of disappointments and loss of legitimacy. What if decisions resulting from collaborative and communicative processes produce regulatory outcomes that do not fulfill expectations of user-groups?"

This is well illustrated by the challenges faced in the comanagement arrangement in Lake Malombe. Hara et al. (2002) observed that the management objectives set by the comanagement institution were mainly government driven. FD attempted via capacity building to align fishers' objectives to its own, i.e., recovering the status of the fisheries in the Lake. Government also retained ultimate power to regulation's design and implementation (e.g., mesh size, net length, and closed-season restrictions) with the promise that at a later (unspecified) stage, greater input into decisions would be transferred to the fishing communities (Hara et al. 2002). The government's concentration of decision power helped to hinder the legitimacy of this comanagement process and, consequently to its failure (Hara et al. 2002). There is no evidence of rebuilding – neither resilient fishing livelihoods nor resource recovery (Hara et al. 2002). Part of the problem relates to the wrong assumption, based on a consultancy report, that local-based community institutions were inexistent, but they were not (Hara et al. 1999). They have been hijacked by BVCs and conflicted with the existent traditional authority systems held by fishing village headmen (Hara et al. 1999).

A different story happened in Lake Chiuta where comanagement can be considered a remarkably successful model (Hara et al. 1999). It is self-sufficient in terms of time and financial resources (Hara et al. 1999). Communities were the first to identify a crisis and proposed regulatory measures in a legal context. Sustainability of the program is associated with the fact that comanagement program was initiated to chase away *nkacha* fishers and the government was identified as a key partner. It took only 4 months to drive away *nkacha* fishers and 2 years later rules revisions were included in fisheries regulations (Hara et al. 1999). Fishers saw in this new organization an opportunity to challenge the historical power held by the village headmen and to empower and reorganize themselves through this new arrangement. "Village headmen had alienated themselves from the fishers following allegations of corruption and collusion with *nkacha* fishers" (Hara et al. 2002). Although this new organization generated an antagonism between these new and old power bases, through the creation of BVC's fishers took more control on their fisheries and acquired the power basis for their local decisions (Hara et al. 2002). Fishers, in this case, showed cohesion and were empowered by comanagement. They played an active role in this new institution since the beginning. Legitimacy of this comanagement arrangement is identified through the following: (1) improved catches; (2) improved relationship with government; (3) improved compliance to regulations; (4) reduced conflicts with the expulsion of *nkacha* fishers; (5) reduced illegal fishing based on fear of sanctions; (6) reduced costs for government; (7) improvement of natural resources conditions; (8) quick actions that may threaten survival of indigenous fishers (Hara et al. 1999, 2002; Njaya 2002).

A major risk to comanagement legitimacy in Lake Chiuta relates to a mismatch between the scale of the comanagement arrangement and the boundaries of the fishery, which is shared with Mozambican fishers under different management jurisdictions (Hara et al. 1999). A similar problem is observed in the MER of

Arraial do Cabo and in the Forum of Patos Lagoon (industrial fishers exploit the same resources outside the comanagement's jurisdictions boundaries) and poses a major challenge to the adaptive capacity to co-manage migratory resources. Comanagement can fail not because it has not been capable to adapt its governance system. It can fail because of the risk imposed by outsiders that have not engaged in the same arrangement and have not compromised to comply with new established rules. The estuary of the Patos Lagoon area managed by the Forum differs from the boundaries of the ecosystem in which the artisanal and industrial fisheries operate. Consequently, the management priorities defined in the Forum also differ from those of fishers, who see no point in enforcing rules inside the estuary when there is no control of access and exploitation of resources in the ocean by industrial fishing operations. This institutional misfit is a factor affecting the acceptance of the Forum among fishers (Kalikoski 2002).

4.3.3 Adaptive Learning Mechanisms

Adaptive learning involves the ability of comanagement institutions to receive and to respond to environmental feedback, through mechanisms for generation, accumulation and transmission of knowledge, flexibility to change rules accordingly, and a time frame to revise regulations and redesign management systems (Gunderson et al. 1995; Berkes and Folke 1998; Holling et al. 1998). It also measures the ability of institutions to learn how to better implement comanagement, through mechanisms that improve participation of resource users in decisions, or the representation of their interests, increasing trust among participants. Flexible social systems that proceed through learning-by-doing are better adapted for long-term survival than are rigid social systems that have set prescriptions for resource use (Holling et al. 1998).

Institutional learning has been identified in the implementation of the TURF system, which evolved through an elaborate process of institutional feedback. Orensanz et al. (2005) explain that evaluation of the TURF's implementation process was conducted by the Institute for the Promotion and Development of Fisheries (IFOP), with national funding to identify challenges and to provide feedback to the managers. This evaluation process included a survey conducted by IFOP of perceived problems among managers, scientists, consultants and leaders of fishers' organizations, and used the results to develop the agenda of a one-week workshop held in September 1999. According to Orensanz et al. (2005), the workshop, cosponsored by government involved all the participants in the management system, plus an international panel. At the end, the panel produced a consensus report with the following recommendations (Orensanz et al. 2005): (1) the need to expand the TURF to encompass the whole fishery; (2) the need to design simpler process to implement the TURFs; (3) the need to simplify data collection system; and (4) the need to work toward the empowerment of fishers organizations and the recognition that education was needed so that fishers and managers could participate actively in the comanagement arrangements. This resulted in an adaptive comanagement that led to what Orensanz et al. (2005) defined as a simpler process that pays substantial attention to the socio-economical aspects of management. The TURF system expanded at a fast pace (e.g., by 2001, 264 more TURFs had been decreed) as fishers' organizations learned about management successes in other regions. Many new organizations are being formed, prompted by the prospects of claiming a TURF. One identified adaptive process was the shift in the origin of the catch brought about marked economic benefits to the fishers, who are now better positioned to arrange sales. In the past, the catch was sold "on the beach" and individual fishers were unable to make convenient sales and price and sale conditions deteriorated when fishers were driven to operate illegally (O. Avilés, pers. comm). In the case of TURFs, sales are prearranged. The organization decides how much to sell and receives offers from middlemen. Middlemen occasionally send their own divers to verify the quality of the locos in the TURF. Once a price is negotiated, fishers bring the catch to the beach on a prearranged date. Given the quality and predictability of supplies from the TURFs, some organizations are evaluating the possibility of advertising shellfish from TURFs with a "certified origin" label. Another significant development was increased emphasis and investment in vigilance, with many villages patrolling their TURFs. There is positive feedback between the establishment of a TURF and fishers' organization (Payne and Castilla 1994) and according to Orensanz et al. (2005) this is, in itself, a significant plus.

In the estuary of the Patos Lagoon studies demonstrate that fishers' knowledge can provide a valuable set of information about the relationship between the fisher and its local environment, and about the characteristics of practices, tools and techniques that led a more sustainable pattern of resource use in the past (Kalikoski and Vasconcellos 2007). Local knowledge can broaden the knowledge basis needed for management and hence improve institutions that mediate the interaction between communities and their use of the resources. This would play a strong link toward cross-scale management and facilitates institutional learning. Recognizing the value of fishers' knowledge is a precondition for the willingness of institutions to involve fishers in the management process. A reforming and restructuring process, including the revision of rules, is occurring within the Forum at this time showing elements for adaptive institutional learning. Change toward a more inclusive process of rule making has been recently observed and fishers' inputs were used to revise Decrees and Laws. Although, inputs from fishers were taken into account in the revision of regulation (e.g., mesh size, and calendars for catfish, mullet, and croaker), their knowledge was only considered valid following considerable scientific scrutiny.

4.3.4 The Threat of Erosion of Social Cohesion

"When resource users find themselves disembedded from the social bonds that connect them to each other and to their community, the dynamic represented in the tragedy of the commons may result" (Hanna and Jentoft 1996, pp. 35–55). The tragedy of the commons, argued by these authors, is the product of social disruption rather than a natural outcome of individual rational behavior, in this case; and once removed social cohesion cannot be easily reestablished. For example, reestablishing management responsibilities within the local community through the design of comanagement regimes and the inclusion of user-knowledge in resource management is a difficult task in the historical context of marginalization or social exclusion that faces small-scale fisheries worldwide. A comanagement arrangement provides an opportunity for communities to influence their development through their participation in the governance system and their involvement in tailoring better management rules to local circumstances (Ostrom 1990). However, as argued by Jentoft and McCay (1995), comanagement institutions must be designed with social integration in mind, and users must be involved in their creation as social cohesion has been shown to be an important precondition supporting comanagement in other geographical settings (Pinkerton 1989).

Despite government's assumption that there were no customary local-based institutions in Lake Malombe, in fact, they existed and the newly comanagement system established helped to disrupt them (Hara et al. 2002). When the BVCs were created though comanagement as a partnership mechanism between fishers and government, customary traditions were disregarded. Also, the kind of partnership established with this comanagement arrangement was mainly to perform enforcement of regulations that expose BVCs to implement confrontational tasks such as (1) collection of money for licenses and handing it over to FD for issuance of licenses; (2) checking of fishing gears for the legal mesh sizes and that they have been licensed; and (3) carrying out patrols especially during the closed season. This augmented conflicts and animosity between BVCs and fishing communities jeopardizing local relationships in the fishing villages.

> "BVCs also saw themselves as doing the 'dirty' work on behalf of the Department of Fisheries...BVCs felt betrayed by government and that government simply used them while all along it had never really intended to hand over this responsibility in the first place" (Hara et al. 1999, p. 16).

The allocation of TURFs in Chile exemplifies an attempt to shift the governance system toward self-governance by strengthening social cohesion through comanagement. The allocation of TURFs given to a fishing community is dependent upon a formal request from the communities to the government. To be eligible to such request, fishing communities should (1) be legally organized in a form of artisanal fisher's associations, co-operatives, or other form of organization; and (2) present a resource management plan describing the status of benthonic resources in the area and a set of actions to ensure the sustainable management of the fishery (González 1996). The resource management plan must include a schedule of annual harvests and other proposed management measures. The organization is also required to produce annual follow-up reports of management performance, including trends in estimated abundance. TURFs are assigned for 4-year periods, renewable upon compliance with the regulations. Fishers' organizations are required to contract consultants for the preparation of the base-line ecological studies, management plans, and follow-ups. So far, the execution of these studies has been almost entirely subsidized by the state through different agencies and programs with the help and involvement of fishers' communities. Central fisheries authority negotiates the management of TURFs on a one-by-one basis with the individual organizations. The internal arrangements in the organizations that receive a TURF are stipulated in written regulations and include rules that limit the entry of new members, as well as dismissal of old ones because of violation of internal regulations (Orensanz et al. 2005). The rent is distributed among members of the organization (e.g., sailors, divers, owners) and varies among communities. A percentage of the rent is destined to communal needs (school, celebrations, maintenance, vigilance, etc.) and elementary forms of welfare (contribution to widows, elders or sick/injured fishers). Some challenges within the Chilean TURFs comanagement system were identified and include (Parma et al. 2003; Orensanz et al. 2005; Hilborn et al. 2005): (1) a lack of formal coordination for TURF management as negotiation is done between the government and the individual organizations. There is no predefined criteria on the TURF devolution process other than the one proposed by the requiring organizations such as ecological-baseline study and the management and exploitation plan. Fisheries administration does not conduct a previous study to investigate if the TURF claimed will affect and exclude other fishers from their historical fishing grounds. (2) The amount and nature of the information required from the fishermen to get a TURF. (3) The TURF per unit area taxation and the uncertainty once subsidies dry up. (4) The coexistence of TURFs and open-access areas, which makes fishers under TURF also gather as much as possible from "open access" grounds, either to sell it or to enhance the TURFS through translocation, despite the existent but enforceable regulation to avoid this.

Although devolution was associated with strengthening social cohesion by triggering fishers to self-organize at the local level, still, along the Chilean coast, not all communities organized themselves to request a TURF. Some communities have been alienated from the decision making process for so many years that they do not have the capabilities of engaging themselves in management functions without some assistance. The fishing communities that have self-organized guaranteed ownership and decision control over fishing resources. The communities that have not organized themselves became marginalized as they do not have the rights to claim the creation of comanagement regimes. This may be a challenge for implementing comanagement in Chile in the near future. The issue of ownership and property rights in fishing practices plays an important role that may jeopardize collective actions and disrupt efficient and equal rights-based systems.

Similar opportunities and challenges are identified in the case of MER of Arraial do Cabo. Three phases are involved in the case of setting up a MER. First, in the preparation phase, a formal request has to be made to the federal government by the local communities with a description of the setting along with an approximate indication of the area traditionally used by the local community. The formal request should describe also the social, economic, cultural, institutional, and biological importance of the setting in which the reserve will function along with arguments in support of their proposal. A branch of the government that deals with traditional peoples, then carries an interdisciplinary assessment study that evaluates the biological and socioeconomic potential of the proposed reserve, and the limiting factors that act against its creation. Once the proposal is accepted, the coastal/marine area is declared State (public) land and a contract is signed whereby the government gives the community usufruct rights as a concession for a period of 50–60 years. Second, in the implementation phase, a management plan is developed, which defines rules, rights, and responsibilities over resource use, in essence representing a social contract among appropriators. This plan must then be approved by the government and published in the federal register to codify the rights and responsibilities of government and resource appropriators. Although the State maintains ownership of the physical area, the members have rights of access to resources in the MER. These rights cannot be traded or sold and can only be passed on through inheritance, something that makes it an incentive for sustainable resource use. Diegues (2008) further described the process of implementation as follows: "A director is appointed for the MER by ICMBio and he/she plays a crucial role in mobilizing financial and technical resources. The members of the MER have to be organized into a legal entity that will act as an intermediary between the State (ICMBio) and the users of the resource. In most cases, a new association has to be created. A utilization plan for the MER has to be compiled and implemented by the association, and officially approved by the government in a comanagement process. This temporary plan establishes the activities and practices that are permitted in the area. It also defines penalties for those who do not obey the rules. If the association's activities deviate from the utilization plan in a way that causes environmental degradation, the contract can be canceled. Next comes the comanagement plan, which replaces the utilization plan and has to be completed in the first 5 years of the MER's existence. Third, in the consolidation phase, the MER must be self sufficient and be able to depend on funds generated by its members. According to Diegues (2008) at present very few MERs have achieved economic self-sufficiency, and rely mostly on funds provided by the federal government. In the very few cases of self sufficiency, funds are originated from contribution of associated members, from levying a percentage on the fish traded by its members, from fees paid by industrial fishing craft that cross the MER's space and from the operations of commercial harbors that exist within them" (Diegues 2008).

As shown in Arraial do Cabo by Pinto da Silva (2004), this final phase is the most challenging as it requires robust locally derived institutions sustained by long-term community participation and government support. The MER of Arraial do Cabo was created with the intention to formalize existent sustainable fishing methods and local-based informal institutions that have governed fishers in Arraial do Cabo for generations, i.e., "rights of day" and "right of way" system (Pinto da Silva 2002). Although these current traditional institutions were incorporated by the MER that govern the MER in Arraial do Cabo, Pinto da Silva (2002, 2004) argues that they are no longer considered robust. Rather "...institutions have weakened and have been hijacked by a handful of vertically integrated individuals to serve their own interests..." (Pinto da Silva 2004, p. 426) The local-based institutions were already disrupted when the MER was created and this has been overlooked by government despite all evaluation phases needed to set up a MER as showed above. Before, decisions were made in a more collective way by boat owners. Ownership was collectively distributed given that it was impossible for one person to own the entire boat. This scenario has changed before the creation of the Reserve and ownership became controlled by two or three people changing considerably the decision making structure of the past. "...Fishers and nonfishers alike refer to the current seining management system as a 'Mafia', in which the canoes/nets, refrigeration, and marketing systems are controlled by a tightly knit group" (Pinto da Silva 2002, p. 217).

Pinto da Silva's argument is that although fishing practices in Arraial do Cabo remain very similar today when compared with 50 years ago, local-based institutions are not the same. Negative social capital is manifested in the hierarchical structures, which have come to control this fishing activity, while a historical legacy of deep divisions within this gear group also complicates and constrains participation. As a result, the reserve has not significantly fortified local management institutions and has overlooked or not been able to deal with these obstacles to participation and empowerment. A deep analysis capturing the existence/lack/challenges of social cohesion of fishing communities is not requested as a precondition to implement a MER, and this is one important weakness of the legal framework for establishing MERs in Brazil. The appropriation and control of local-based institutions by a few fishers was the main source of social disruption and an important element of the MER of Arraial do Cabo´s failure.

4.4 Reflections on Advances in Comanagement Arrangements: Lessons from Case Studies

The narratives presented in this study illustrate that, while creating comanagement may be relatively easy, the challenge lies in sustaining these initiatives over the long-term, and ensuring that they deliver both efficient and equitable outcomes. All cases evaluated here show that an institutional change and renewal was the first step that had led to the creation of comanagement arrangements in response to the signal or to prevent an imminent (and foreseen) collapse of fisheries resources and, consequently, to the high risk that such impact imposed on fishing livelihoods survival.

A major struggle impacting on the adaptive capacity of these systems is to design comanagement arrangements with social integration in mind that allows self-organization and autonomous control over decisions. The presence of a "traditional community" with a strong connection to the resource base and with a system of local governance is key. Also the existence of a legal framework that legitimizes comanagement at higher levels of decision making will help cross-scale management. When adaptive capabilities were not identified neither from the part of the government nor from the communities, fisheries comanagement in these cases has not been able to cope with challenges that appeared along with its implementation. Imposed self-organization generated did not allow for learning and adaptation. Institutional rigidity associated with a complete disregard of fisher's input into the comanagement system also characterized the challenges of adaptive capacity in these cases where comanagement was particularly driven by administrative and political concerns.

Difference and diversity must be taken into account as well as existing power structures that may distort or constrain participation. If not, comanagement could potentially reinforce inequitable power structures instead of promoting broad-based participatory conservation. A mechanism to ensure an assessment of the existence and characteristics of these institutions should be undertaken before including or excluding customary traditional practices. Information on the state of these institutions is essential to design effective regimes to collaboratively manage natural resources.

Where local customary institutions have not been successfully built upon, weaknesses of formal institutions, lack of trust between communities and government and weak social capital are a key constraint to the adaptive capacity of comanagement systems. The analysis here even suggests that in these cases comanagement might further marginalize the fishing communities that they were initially expected to "empower." Wrong assumptions that local communities were self-organized and robust, combined with the lack of adaptive capacity to adjust the comanagement when it became evident that the local system had been in fact eroded, have contributed to the unsuccessful comanagement outcomes. This complicates the possibility to balance the power and restructure internal collective actions, despite the existence of the legal instrument to do so. The devolution of property rights should be done along with incentives for keeping local-based social–ecological systems. When social cohesion and human rights access have been already lost, this may hamper comanagement systems. Rights are meaningless unless practical mechanisms exist to ensure they are legally exercised. However, if actions taken locally come from eroded traditional systems that encourage power imbalances and jeopardize the livelihoods of the poor, then failures will certainly occur.

Involving fishing communities in management depends on the existence of appropriate institutions that are based on a process of shared governance, "the process of communities creating their own pathways to the future" (O'Riordan 2003). This chapter showed that not all institutions created with the comanagement systems are an attempt toward sharing responsibility and authority over the management of fisheries resources. Some comanagement still lacks the mechanisms for empowering the community and delivering such a model of shared governance of fisheries. This is the risk faced when comanagement devolves responsibilities to communities without devolving to them the power to make decisions on management objectives or wider policy. As discussed by O'Riordan (2003), empowerment is by no means a "clean" concept. Without the appropriate power sharing and representation of fisher's set of knowledge-belief-practice system within comanagement systems, it will be difficult to achieve a highly adaptive comanagement.

If social cohesion exists at the local level and collective choices are exercised within the comanagement, self-organization will happen and allow for the use of the best knowledge available that will lead to positive outcomes.

The adoption of comanagement as a management strategy can take different forms: it can integrate existing local systems into the formal new comanagement institution-building, it can build a whole set of institutional arrangement or it can mix both existing traditional systems while creating new arrangements. The cases demonstrated here illustrate that this decision should be context-based. But incentives to create comanagement arrangements from external sources other than communities should be extra careful to understand the conditions and existence of local level customary systems. Traditional institutions should not be disregarded by the comanagement arrangement. Prior to implementing comanagement, a careful analysis should indeed be conducted with fishers to indicate how these institutions should be linked to new comanagement arrangement in place.

Acknowledgements The authors thank Fikret Berkes and Brian Davy for providing valuable comments in an early version of this manuscript.

References

Béné C (2003) When fishery rhymes with poverty: a first step beyond the old paradigm on poverty in small-scale fisheries. World Dev 31(6):949–975

Berkes F, Folke C (1998) Linking social and ecological systems. Management practices and social mechanisms for building resilience. Cambridge University Press, Cambridge

Berkes F, Jolly D (2001) Adapting to climate change: social–ecological resilience in a Canadian western Arctic community. Conserv Ecol 5(2):18

Berkes F, Seixas C (2005) Building resilience in lagoon social–ecological systems: a local-level perspective. Ecosystems 8:1–8

Berkes F, Colding J, Folke C (eds) (2003) Navigating social–ecological systems: building resilience for complexity and change. Cambridge University Press, Cambridge

Carpenter SR, Walker B, Anderies JM, Abel N (2001) From metaphor to measurement: resilience of what to what? Ecosystems 4:765–781

Castilla JC, Manrıquez P, Alvarado J et al (1998) Artisanal 'caletas' as units of production and co-managers of benthic invertebrates in Chile. In: Jamieson GS, Campbell A (eds) Proc North Pacific symp on invertebrate stock assessment and management. Can Spec Publ Fish Aquat Sci 125: 407–413

Castilla JC, Fernández M (1998) Small-scale benthic fisheries in Chile: on co-management and sustainable use of benthic invertebrates. Ecol Appl 8(Suppl):S124–S132

Castilla JC, Defeo O (2001) Latin American benthic shellfisheries: emphasis on co-management and experimental practices. Rev Fish Biol Fish 11:1–30

Diegues AC (2008) Marine protected areas and artisanal fisheries in Brazil. Samudra Monograph. International Collective in Support of Fishworkers, Chennai, India

Folke C, Colding J, Berkes F (2003) Synthesis: building resilience and adaptive capacity in social–ecological systems. In: Berkes F, Colding J, Folke C (eds) Navigating social–ecological systems: building resilience for complexity and change. Cambridge University Press, Cambridge, pp 352–387

González E (1996) Territorial use rights in Chilean fisheries. Mar Resour Econ 11:211–218

Gunderson LH, Holling CS, Light SS (1995) Barriers and bridges in the renewal of ecosystems and institutions. Columbia University Press, New York

Hanna SS, Jentoft S (1996) Human use of the natural environment: an overview of social and economic dimensions. In: Hanna SS, Folke C, Maler K (eds) Rights to nature: ecological, economic, cultural and political principles of institutions for the environment. Island, Washington, DC, pp 35–55

Hara M, Donda S, Njaya FJ (1999) Fisheries co-management: a review of the theoretical basis and assumptions. South Afr Perspect 77:1–32

Hara M, Donda S, Njaya FJ (2002) Lessons from Malawi's experience with fisheries co-management initiatives. In: Geheb K, Sarch MT (eds) Africa's inland fisheries: the management challenge. Fountain, Kampala, Uganda, pp 31–48

Hilborn R, Orensanz JM, Parma AM (2005) Institutions, incentives and the future of fisheries. Phil Trans R Soc B 360:47–57

Holling CS, Berkes F, Folke C (1998) Science, sustainability and resource management. In: Berkes F, Folke C (eds) Linking social and ecological systems. Management practices and social mechanisms for building resilience. Cambridge University Press, Cambridge, pp 342–362

Jentof S (2003) Co-management: the way forward. In: Wilson DC, Raajær Nielsen J, Degnbol P (eds) The fisheries co-management experience. Accomplishments, challenges and prospects. Kluwer, London, pp 1–13

Jentoft S (2000) Legitimacy and disappointment in fisheries management. Mar Policy 24(2):141–148

Jentoft S, McCay BJ (1995) User participation in fisheries management. Lessons drawn from international experiences. Mar Policy 19:227–246

Kalikoski DC (2002) The forum of the Patos Lagoon: an analysis of comanagement arrangement for conservation of coastal resources in southern Brazil. Ph.D. thesis, University of British Columbia Press, Vancouver

Kalikoski DC, Satterfield T (2004) On crafting a fisheries co-management arrangement in the estuary of Patos Lagoon (Brazil): opportunities and challenges faced through implementation. Mar Policy 28:503–522

Kalikoski DC, Vasconcellos M (2007) The role of fishers' knowledge in the co-management of small-scale fisheries in the Estuary of Patos Lagoon, Southern Brazil. In: Haggan N, Neis B, Baird IG (eds) Fishers' knowledge in fisheries science and management. Blackwell/UNESCO, Oxford

Njaya FJ (2002) Fisheries co-management in Malawi: implementation arrangements for lakes Malombe, Chilwa and Chiuta. In: Geheb K, Sarch MT (eds) Africa's inland fisheries: the management challenge. Fountain, Kampala, Uganda, pp 9–30

Olson P, Folke C, Berkes F (2004) Adaptive co-management for building resilience in social–ecological systems. Environ Manage 34(1):75–90

Orensanz JM, Parma AM, Jerez G et al (2005) What are the key elements for the sustainability of "S-fisheries"? Insights from South America. Bull Mar Sci 76:527–556

O'Riordan T (2003) Deliberative democracy and participatory biodiversity. In: O'Riordan T, Stoill S (eds) Biodiversity, human livelihoods and sustainability: protecting beyond the protected. Cambridge University Press, Cambridge

Ostrom E (1990) Governing the commons. The evolution of institutions for collective action. Cambridge University Press, Cambridge

Parma AM, Orensanz JM, Elias I et al (2003) Diving for shellfish and data: incentives for the participation of fishermen in the monitoring and management of artisanal fisheries around southern South America. In: Newman SJ, Gaughan DJ, Jackson G, Mackie MC, Molony B, St John J, Kaiola P, Newman SJ, Gaughan DJ, Jackson G, Mackie MC, Molony B, St John J, Kaiola P (eds) Towards sustainability of data limited multi-sector fisheries. Australian Society for Fish Biology, Department of Fisheries, Perth, Western Australia, pp 8–29, http://www.fish.wa.gov.au/res/broc/report/asfbproc 2001/ASFBProc2001.PDF

Payne HE, Castilla JC (1994) Socio-biological assessment of common property resource management: small-scale fishing unions in Central Chile. Out Shell 4:10–14

Pinkerton E (2003) Toward specificity in complexity: understanding co-management from a social science perspective. In: Wilson DC, Raajær Nielsen J, Degnbol P (eds) The fisheries co-management experience. Accomplishments, challenges and prospects. Kluwer, London, pp 61–76

Pinkerton E (ed) (1989) Co-operative management of local fisheries: new directions for improved management and community development. University of British Columbia Press, Vancouver

Pinto da Silva P (2002) From common property to co-management: social change and conservation in Brazil's first maritime extractive reserve. Ph.D. dissertation, London School of Economics

Pinto da Silva P (2004) From common property to co-management: lessons from Brazil's first maritime extractive reserve. Mar Policy 28(2004):419–428

Reis EG, D'Incao F (2000) The present status of artisanal fisheries of extreme southern Brazil: an effort towards community based management. Ocean Coast Manage 43:7–18

Scott A (2000) Introducing property in fisheries management. In: Shotton R (ed) Use of property rights in fisheries management. FAO Fisheries Technical Paper No. 404 /1. FAO, Rome, pp 1–13

Stotz W (1997) Las areas de manejo en la ley de pesca y acuicultura: primeras experiencias y evaluación de la utilidad de esta herramienta para el recurso loco. Estud Oceanol (Chile) 16:67–86

Walker B, Holling CS, Carpenter SR, Kinzig A (2004) Resilience, adaptability and transformability in social–ecological systems. Ecol Soc 9(2):5

Wilson DC, Nielsen JR, Dengbol P (eds) (2003) The fisheries co-management. Accomplishments, challenges and prospects. Kluwer, London

Chapter 5
Adaptive Capacity and Adaptation in Swedish Multi-Use Boreal Forests: Sites of Interaction Between Different Land Uses

E. Carina H. Keskitalo

5.1 Introduction

Forests are a resource of considerable importance for many national and local economies; for example, the forest industry in the EU has a turnover of some 400 billion Euros, or some 3% of GDP (Parikka 2004; Hazley 2000). Forestry is of great national and regional importance in Sweden, where it comprises some 12% of the total industrial employment (Swedish Forest Agency 2008; Swedish Forest Industries Federation 2007). The terms 'multi-use forests' and 'sustainable forest management' have been applied generally to take into account different forest uses, that is, uses additional to wood production. In the Swedish case, forest uses beyond forestry proper may include reindeer husbandry, tourism and environmental protection. Local use is significant and includes hunting, fishing and berry picking. It has been proposed that the inclusion of such uses is a priority for adaptive and sustainable forest management (Raison et al. 2001); this would entail management of not only forestry but also cooperation among a multitude of other actors. Potentially, the development of such interaction could also reduce communities' vulnerability to external change.

This study focuses on the multiple uses of forested lands in the municipality of Gällivare, which is situated in far northern Sweden. The use of renewable resources in the area includes forestry, environmental protection on a substantial scale, tourism – particularly winter tourism – and reindeer husbandry. In general, reindeer husbandry is practised in the same areas as forestry, with the land having a range of owners from industry to private individuals and reindeer herders having a right to use the areas. Dual land-use systems thus operate in the same areas. In Sweden, the numbers of stakeholders in these sectors typically vary considerably: in the reindeer husbandry area, which comprises about 40% of Sweden's land mass, some 2,500

E.C.H. Keskitalo
Department of Social and Economic Geography, Umeå University, Umeå 901 87, Sweden
e-mail: Carina.Keskitalo@geography.umu.se

persons are actively engaged in husbandry, whereas private forest owners number some 40,000, with corresponding differences in importance for the GNP (cf. Moen 2008; Keskitalo 2008a). In addition, environmental protection plays a large role in many municipalities located inland or close to mountain ranges, such as Gällivare, and is often seen as removing valuable land from forestry.

Studies in northern Sweden have previously noted that regulative and market measures delimit the potential for adaptation by setting the legislative boundaries of reindeer herding and forestry rights and defining the market environment within which these sectors operate (Keskitalo 2008a). However, the extent to which land uses can exploit the space for adaptation (Berkhout et al. 2006) afforded by legislation depends to some extent on the interactions and interaction mechanisms between stakeholders in each locality. In order to operationally define the way in which different sectors may impact each other at the local level, this chapter investigates the sites of local interaction between different sectors in the case of the multi-use of forest. In particular, the research focuses on whether certain existing cooperation measures and actions on the local level are able to allow the adaptations that actors define as relevant. The chapter thus adds a supplementary case to the debate about the extent to which adaptation can be undertaken purely on a local basis (cf. Næss et al. 2005). Accordingly, this study focuses on the role of interactions between sectors as a factor impacting adaptive capacity.

5.2 Theoretical Framework

Adaptation involves making adjustments in response to change and is regularly considered within the framework of vulnerability. The term 'vulnerability' has been used to describe people's 'risk to be wounded by change' (cf. Kates et al. 1985) in a broad sense; that is, it encompasses the totality of stresses that are seen as affecting a situation. Vulnerability is thus a very broad term, encompassing *exposure* to specific stresses, *sensitivity* to them – for example, due to specific characteristics of the ecosystem – and the *adaptive capacity* to respond to their impacts. Adaptive capacity is to a large extent dependent on the socioeconomic situation – including the political and economic resources – of the actors who must adapt. Within the adaptive capacity of a group or other unit, a number of specific *adaptations* or adaptive strategies may be undertaken depending on both existing traditions and the types of resources available (Smit and Wandel 2006; Smit et al. 2000). The boundary between existing coping responses and strategies developed in response to new stresses can be seen as a floating one in that coping actions may be extended in response to a changed situation and ultimately take on the character of an adaptation (Adger et al. 2004).

Recently, a focus has been placed on the underlying vulnerability ensuing from persistent stresses and situations at large, known as *social vulnerability* (cf. e.g. Adger 2000). This perspective highlights that the responses to any stress or event that occurs – be it climate change or globalisation – are conditioned by the existing resources; as it is these that determine the adaptive capacity, one should focus on the underlying situation rather than the attendant stresses (Smit and Wandel 2006). The relevant resources in this regard include material resources, technology and infrastructure; political capital and entitlements; social, human and financial capital; and wealth (Smit and Pilifosova 2001). The broader systems of management and governance, understood as the sum of public and private (including market-driven) decision-making by which societies are governed (cf. Keohane and Nye 2000) play an important role in conditioning the resource access that determines adaptive capacity and the adaptations that may be undertaken. International conventions, market demands and national legislation and regulation may be seen as imposing an external limit on adaptation by determining the land tenure and demand and pricing systems that affect the local level (Keskitalo 2008a). Brooks (2003) and Berkhout et al. (2006) note that terms such as 'adaptation likelihood' and 'adaptation space' can be used to highlight the space, or opportunity, for adaptation – constrained by legislative or other measures – that is available to actors and thus include the entire range of capacities drawn upon by actors at any given time.

Pressure may here be exerted locally – at least to some extent – in the interaction between land uses that draw upon the same resource and within the governance context in which these interact such that development that supports one use often impinges upon others (so called 'vulnerability transfer'). Literature on local or community vulnerability assessment has argued that the local level, although seen within a multi-level governance context, is that at which adaptation needs to take place (cf. Adger 2001). Much work has also highlighted the possibility for local cooperation and interaction between stakeholders to significantly enhance adaptive capacity (e.g. Wondolleck and Yaffee 2000; Olsson et al. 2007). On the other hand, the governance and legislative context fundamentally institutionalises particular aspects of land use, setting the baseline for interaction; there may also be limitations on the extent to which actors can contribute and take part in cooperation on a voluntary basis given their economic context (cf. Keskitalo 2008a, b; Allard 2006; Brugge and Rotmans 2007). On the regional and local levels, the degree to which local adaptive management is developed, for instance, through cooperation between sectors, may influence the extent to which different land users can act within the scope of external regulative or market limitations. However, if cooperation measures do not effect an actual change in conditions, it is possible that they may also increase conflicts between actors or fail to add to adaptive capacity (cf. Keskitalo et al. 2009).

5.3 Case Study Area and Methodology

Gällivare, situated in Norrbotten county, northern Sweden, is a sparsely populated municipality with significant interests in the renewable land-use sectors of forestry, reindeer husbandry, winter tourism and environmental protection. The municipality encompasses 16,000 km² and has about 19,000 inhabitants, of whom some 8,500 live in the central town of Gällivare. The area is thus very sparsely populated, with few actors in the focal branches and a large proportion of the population residing in a single, central community. The Gällivare area includes parts of the Laponian Area, a UNESCO world heritage site encompassing Stora Sjöfallet and Muddus, two national parks with old pine forest and mires, and at least part of a number of nature reserves. The municipality is home to a large downhill skiing resort, Dundret, and features considerable resource and infrastructure development: the Aitik mine, one of the largest copper mines in northern Europe; a number of smaller mines and water power projects; a rail link; and potential wind power development (Fig. 5.1).

In the context of multi-use forests, area resources include substantial nature values, forest for wood production, lichen and reindeer grazing areas, migration and subsistence conditions for reindeer in general, and forest areas where winter tourism is practised. Resources are used extensively over large areas subject to certain constraints. During the last 30 years or so, the pressures on forest resource use have increased through, among other things, more extensive road infrastructure and logging, which have depleted many areas with forest mature enough to contain lichen

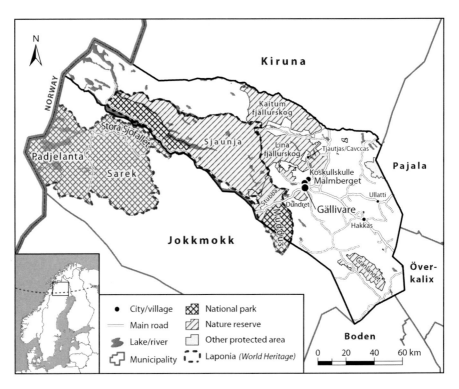

Fig. 5.1 Map of Gällivare municipality with environmental protection areas and indicating neighbouring municipalities. GIS graphics courtesy of Magnus Strömgren

that grows on trees. In addition, the establishment of the Laponian Area – with an eye to protecting both the natural and cultural environment, the latter embracing reindeer husbandry – has made it relevant for a more participative management involving the local community (Rådelius 2002). When the area was designated as a World Heritage Site in 1996, a body known as the Laponia Council was formed that included representation from the county administrative board, the municipalities and Saami representatives. However, conflicts developed early on between a regional (State administrative) focus on environmental protection and local interest in development, and between the proposals of the local municipality for management within a broader partnership and the Saami representatives' emphasis on management within a group with a Saami majority (Nilsson Dahlström 2003; Rådelius 2002; cf. Mijá Ednam 1999).

There is a range of actors in the sectors examined who use forest resources. Almost half of the forested land in Gällivare is owned by the State, including the state forest company Sveaskog, and, to a lesser extent, the State Property Board, which manages non-productive forest. Other owners include the major multinational company SCA (*Svenska Cellulosa Aktiebolaget*), private owners and the Forest Common. The Forest Common constitutes a form of ownership where some of the privately held forest is owned in the form of a historically established, separately administered unit, the revenue from which is often used to support forest management on the private plots. A large proportion of the forested land in the municipality, including slowly regenerating forests near mountainous areas, is protected (Swedish Forest Agency 2000).

Reindeer husbandry is carried on within a system of administrative units (known as *sameby*, lit. 'Saami village', although the term denotes a herding district rather than any unit of habitation). In Sweden – unlike, for instance, Finland, where reindeer herding is a general right – only persons belonging to the indigenous Saami minority may practise reindeer herding, an exception being herding in what are known as concession areas. Reindeer husbandry is practised by a small minority among the Saami, but the livelihood is generally seen as being of cultural importance. Herding operates within a market system where the sale of reindeer meat provides the main income for herders, who act as individual entrepreneurs within a herding unit. While a unit has general user rights for reindeer herding on all land within its area, land is generally owned by forest owners (including the State, companies and private individuals). Many private forest owners live outside Gällivare, for instance, in Stockholm, and their use of and interest in maintenance of their holdings may vary. Reindeer herders may also be private forest owners, this practice being somewhat more common in reindeer herding units that stay in the forest year around than among those who spend part of the year in the mountains. Land is also used by local people under the right of common usage provided for in Swedish law, which allows the public to, among other things, pick mushrooms and berries as well as camp on others' land on the condition that they do not disturb land owners. Hunting (moose and small game) is widely practised among all the population.

The study reported here is mainly on the basis of semi-structured interviews regarding perceived vulnerability, potential and existing adaptations, as well as interaction between the sectors. The interviews were approximately 1–1.5 h in length, conducted primarily during autumn 2008 and in general transcribed in full. The selection of interviewees was on the basis of existing registered companies and administration with their main activity within the municipality in the sectors. In total 27 interviews were conducted, which represent a full selection of the relevant actors as regards the focal sectors and selection parameters with the exception of two winter tourism companies unavailable for interview. The actors interviewed encompassed forestry and forest administration (main land owner Sveaskog, its Model Forest representative, SCA, the Common Property Board, the representative of the two Forest Commons, the Swedish Forest Agency administration, the private forest owners' interest group Norra Skogsägarna and the municipal forest coordinator), chairs of the five main reindeer husbandry organisational units having reindeer grazing in the area, the twelve winter tourism companies, the one environmental protection organisation and the environmental protection administration at the county administrative board. The interviews were inductively coded according to the main factors of interaction between the sectors that were discussed by interviewees. The quotations in the text have been translated from the original Swedish by the author.

5.4 Results

5.4.1 Interaction Between Sectors

5.4.1.1 Reindeer Husbandry's Interaction with Other Sectors

The concepts of vulnerability and adaptive capacity are rather well suited for capturing reindeer herders' concerns regarding the broad range of factors that may impact reindeer husbandry. For instance, one interviewee in reindeer herding noted:

> One talks about wind power development but one doesn't talk about the whole picture ... including all those other things that have an impact: forestry, mining, water power. One cannot look only at a small part: [while] this [any specific use or encroachment] may not be very much ... it impacts all of the Saami village's activities.

Reindeer herding representatives stressed interactions with forestry, a finding consistent with conclusions presented in the literature (cf. Allard 2006; Berg et al. 2008; Keskitalo 2008a, b) and potentially attributable to herders' rights as users of land that is often owned and utilised for logging by other interests. Interviewees indicated the legislative framework as the external limit to acting and cited conflict over the interpretation of legislation relevant to the interaction between forestry and reindeer herding. Both sectors are driven to a large extent by market and profitability requirements, yet the strategies for maximising outcome from the two livelihoods conflict. For instance, reindeer herders noted that the planting of quick-growing

Contorta pine results in very dense forest that is difficult for reindeer to traverse and easy for predators such as bear to hide in. One herder noted that the planting of Contorta could even be seen as having changed the possibilities for land use in the areas – something that would be prohibited according to law:

> We can show that we cannot use these grounds in the same way as we have previously. In our eyes, that is a changed land use ... If Vattenfall [the state water power company] dams mountain areas, they need to pay compensation to those affected. But if forest companies plant Contorta [and the ground cannot be used for herding] ... [why is it that] they do not need to pay?

Forestry is also increasingly using fertilisation in order to increase tree growth, while ground lichen, the main grazing resource for reindeer during winter, thrives on nutrient-poor soil. Regulation mandates that fertilisation cannot be used on areas that are more than 25% lichen covered, but one reindeer herder noted that the assessments of lichen cover may vary widely. He noted on a field visit with a forest manager that:

> He says that it is five, six percent [lichen coverage] and I think it is 30 percent ... I asked him about an area that was entirely covered with lichen ... he said it may be some 25–30 percent [lichen coverage].

In the same herder's assessment, fertilisation may result in the failure of higher percentages of lichen cover to develop where conditions would otherwise be favourable.

Current forest management also includes other features having an impact on reindeer herding: limited clearing or thinning of forest resulting in dense stands with conditions that restrict the development of ground and tree lichen and, over time, changes in ownership or area delineations that result in areas being logged that were previously exempt from logging by agreement. For instance, reindeer herders noted the risk of previously preserved areas being logged that might follow from the State requiring Sveaskog to sell plots to private forest owners in order to stimulate private engagement in forestry. In particular, conflicts exist with large-scale commercial forestry, even though in some cases owners of small forest holdings have sued reindeer herders for damage to forest (e.g. to plantations). However, with regard to private small-scale forestry, many in reindeer husbandry feel that the individual entrepreneurs there must be able to use their land. For instance: 'if a private forest owner applies for a logging permit ... we as a Saami village won't go in and deny this private person permission to log ... that would be wrong'.

The degree of environmental protection in national parks such as Muddus also has an impact on reindeer husbandry. While a certain degree and type of forest management (such as pre-commercial thinning), as opposed to having purely natural forests, support accessibility for reindeer, the more demanding forms of environmental protection prohibit all forest management. Impacts of environmental protection areas may, however, mainly be beneficial, as old spruce forests with tree lichen may be preserved. Herders noted, though, that reindeer herding has not been a selection criterion for which areas should be placed under environmental protection mechanisms.

Interactions also exist between reindeer herding, tourism and local use. These involve snowmobile safaris and local snowmobile driving, as well as hunting, which may disperse or disturb reindeer unless herders are contacted so that these activities can avoid areas where reindeer have gathered. Some reindeer husbandry entrepreneurs are involved in local tourism enterprises and gain income there; however, there exist both cultural and structural obstacles to a larger focus on reindeer-related tourism. One regulative limitation is due to Swedish legislation, which states that the Saami village as an administrative unit cannot be economically involved in activities other than reindeer husbandry and may only engage in subsistence fishing and hunting, and not lease out or economically profit from hunting and fishing rights on its grazing grounds.

5.4.1.2 Interactions Between Forestry and Environmental Protection

Forest managers noted that they face challenges in undertaking forestry when it comes to coordination with other land uses: however, limitations on the traditionally large and nationally important sector of forestry are less restrictive than on users who do not have ownership. In addition, small-scale forest owners in the area are often part of a forest common, meaning that they will receive support for soil treatment, planting, thinning and other management measures. While employment in forestry as such, including administration, has decreased in the area, the economic value of forest production has been maintained through technological development. Thus, for the forestry practised in the area today, the largest impact on wood production may be seen as the extended inclusion of forest land under different forms of environmental protection, as well as the inclusion of environmental values in production (cf. Keskitalo and Lundmark 2010).

Environmental values have been integrated into forestry to a large extent following an amendment to the Forest Code in 1993 whereby environmental and production values were given equal weight. Forest areas have also become subject to extensive environmental protection demands, with the aim, for instance, that the Swedish Forestry Board should be funded to buy private forest areas where key biotopes are found and thereby remove these from commercial exploitation. One interviewee noted, however, that funding for such purposes is limited and that not all key biotopes can be protected in this way. Since the 1993 amendment to the Forest Code, however, and with certain decisions taken in the last decade, many more and also smaller areas than previously have been given protected status. The state forest company Sveaskog, for instance, noted an aim of exempting 20% of its land area from forestry: as a state forest company, demands on Sveaskog for economic competitiveness may be more limited than those on private companies; it also has relatively large land areas and must support a number of different interests in addition to forestry.

Forest companies and owners are also increasingly certified on the terms of the Forest Stewardship Council (FSC) or Programme for the Endorsement of Forest Certification Schemes (PEFC), market-based labelling systems whereby wood harvested according to specific social and environmental criteria may be sold as 'certified'. The market benefits of this system to forest owners, resulting from, among other things, the fact that some wood buyers require certified wood, have led to an institutionalisation of the certification system, whereby, as an interviewee noted, 'the instructions to us [in forest management] are much stricter than the laws are'. In this connection, given the social criteria for certification, it has also become important to show further consideration to reindeer husbandry. However, even given the certification criteria, forestry is practised with increased intensity: much regeneration through planting or seeds takes the form of monocultures of pine and the extent of fertilisation is increasing. Forest managers noted that it may be difficult to go beyond today's aims for environmental protection to fulfil State goals:

> if large areas are set aside, this may make other areas difficult to access. It becomes difficult to expand roads and infrastructure ... quite simply, the area becomes too limited for industry to get the wood it requires.

Similarly, one actor noted that land use pressures are increasing all the time in forestry: 'keeping peace with the Saami is a problem ... since the areas must decrease ... and since you have [so far] avoided the sensitive areas you are getting closer and closer to the sensitive areas'. In addition, local revenue from forestry has fallen drastically over the last decades, given that, among other things, competition on the wood market has resulted in Gällivare, like many other inland areas, no longer having locally based logging entrepreneurs or local sawmills. Instead, harvested wood is to a large extent sold to the large sawmills and pulp and paper plants further south in Norrbotten and the neighbouring county of Västerbotten. As a result, there is substantial interest in retaining existing revenue – and potentially extending this through increased fertilisation and more intensive management – as well as in the municipality having a strong say in State-proposed nature reserve development through larger local management.

Aside from the organisation for administering environmental protection (specifically the county administrative board), the municipality has only one relatively small environmental protection NGO, a local branch of the Swedish Society for Nature Conservation. The organisation noted that because of its small size and the relatively limited coordination between branches within the organisation, as well as attitudes toward environmental protection in the area, its impact may be limited to smaller projects and to endorse letters-to-the-editor with the organisation's name. The representative noted that

> most people in Gällivare use the forest for hunting, fishing, berry picking, and see the forests as a resource. When we talk about nature reserves it ... [often gives the impression] of prohibitions: 'here ... we will not be able to hunt, fish and use these resources'.

Environmental protection is thus often seen as limiting employment in the area by removing forest resources from use in forestry and other local use (cf. Keskitalo and Lundmark 2010).

5.4.1.3 Winter Tourism

The area includes one large downhill skiing resort as well as winter tourism entrepreneurs focused on domestic tourism in Sweden and international snow tourism featuring dog sledding and snowmobile safaris. Many entrepreneurs also to some extent utilise the Saami heritage in their programmes, for instance reindeer safaris or presentations of Saami culture, some of which are organised in cooperation with entrepreneurs in the Saami administrative units. Given the limited tourist industry in Gällivare, some businesses also work across larger areas, including the large neighbouring municipality of Kiruna. Many of the entrepreneurs are focused on the international market and tourists from Germany, Switzerland and elsewhere, noting that such guests are generally interested in experiencing substantial differences vis-à-vis their home countries: very cold temperatures and winters with large amounts of snow. The most important environmental factors are that it is

> clean, no rubbish, quiet, and that you cannot see civilisation. You shouldn't be able to see roads ... there should be great open spaces where you don't meet so many people.

However, interviewees also noted the developed character of these areas: 'There are logging areas and developments ... electric lines and power station cabins ... I can almost feel like we are fooling the tourists ... "it is one of Europe's last wildernesses", and then we walk under power lines'.

While land use by other sectors thus has an impact on the experience resources drawn upon by the tourism industry, tourism is at the same time dependent on easy access to areas through a developed infrastructure. Many noted that tourism has been relatively limited in the area, given limited bus connections with the larger airport in Kiruna, the relatively small airport in Gällivare, and the lack of bus connections to Muddus National Park, which make access difficult. As a result, there exist few tourism activities lasting longer than 1 day or shared facilities where occasional tourists could rent snowmobiles or other equipment or view reindeer in a pen. Some protected areas also restrict access by all-terrain vehicles and snow scooters by law; one person noted, for instance, that gaining a permit for tourism activity in the national park took 2 years and succeeded in the end only because of good relations with the relevant Saami administrative unit. Some tourist companies are also concerned over discussions in the current interim council for the Laponia world heritage site about prohibiting sled dog safaris in Laponia. The interim council includes representation from Jokkmokk and Gällivare municipalities, the county administrative board, the Swedish EPA and Saami administrative units. However, one person noted that 'tourism companies do not have a say in the Laponia Council' and that the reindeer herding sector may thus have some influence on land use in protected areas, with an impact on tourism practices. Many of the tourist entrepreneurs noted, however, that the areas they use are regularly those closer to central towns, resulting in a limited impact on most entrepreneurs from protected area regulations. A larger impact is seen here in forestry, especially clear-cutting, which creates areas that are not attractive for safari tourism.

However, among the sectors studied here, tourism is the one which least emphasises interaction problems. This may relate both to what tourism entrepreneurs described as the relatively limited extent of tourism development in the municipality and to the fact that there thus exist fewer established sectoral conflicts. Tourism is also, however, emphasised by some as the land use sector with the largest potential for expansion, reflected among other things in a large planned expansion at the downhill skiing facility and in the development of a destination company to market Gällivare as a winter tourism location that might compete with more established and better marketed locations in Finland. Tourism entrepreneurs also note regulative limits to the sector, such as costs for permits and frequent changes in regulations on food production, transport and animal husbandry.

5.4.2 Adaptation and Potential Means of Increasing Adaptive Capacity

In response to the impacts identified in the different sectors, adaptations and suggestions for adaptation have emerged in a number of areas. These include inter-sectoral demands, demands on the State for compensation, reorganisation within the sectors and the development of cooperation measures.

In *reindeer herding*, as an adaptation to the general constraints in the situation at large, one of the Saami administrative units in the area is currently undergoing a process of partial restructuring. While reindeer herding is traditionally practised in separate sub-units, known as *siida*, the current reorganisation will include practising reindeer husbandry in one single sub-unit for the entire village. This will be undertaken in order to be able to preserve specific areas for grazing later in the season or year, and to avoid areas with high numbers of predators. In some areas, reindeer herding can thus be seen as close to reorganisation, something assessed as a possibility elsewhere as well (Danell 2005; Moen and Keskitalo in press). However, a differentiation can be discerned in vulnerability among the Saami administrative units depending on the degree of impact from other land use sectors and on whether protected areas exist in the grazing areas herders rely on.

As a result to some extent of the development of forest certification and of Sweden's state investigations into whether to ratify ILO Convention No. 169 (cf. Keskitalo 2008b), new coordination measures have begun to emerge that may support adaptation in reindeer husbandry. Certification has extended the areas for forestry consultation with reindeer herding on logging, among other things, to an area beyond that required by law; in particular, it has increased the coordination requirements for large-scale forestry where the more demanding FSC certification is prominent. In addition, a number of different coordination and dialogue groups have started for reindeer herding–forestry interaction, and there exists coordination between these land users and the county administration (Keskitalo 2008b). One herder noted, however, that the new reindeer management plans, developed during the last few years as a planning tool for coordination with the management plans drawn up in forestry, do not include criteria for areas with tree lichen which would be valuable for reindeer herding. In addition, consultation meeting attendance and field visits may be taxing both in time and costs, which results in problems for small-scale reindeer herding entrepreneurs given that all sectors are expected to bear their own costs.

A number of direct suggestions for potential adaptations include both compensation from the State and forestry for lost grazing resources and clarification of land use rights. Reindeer herders note that a very small fee per produced cubic metre of wood could be used to fund additional feeding of reindeer where shortages of grazing land exist. Herders also noted the potential for management and logging of forest using means other than clear-cutting, such as selective logging, although these greatly differ from the more intensive practices favoured in forestry. Interviewees in forest management noted that changes have been made in accordance with reindeer herders' requests, for instance in the types of machines used for soil treatment and in seasonal shifts of logging. Reindeer herders also noted that attitudes have changed over time with the rise in international market pressures on forestry to adhere to certification principles, with increased initiatives for dialogue, and with younger forest management being more environmentally oriented; however, herders did not consider these sufficient to support grazing requirements.

A focus was also placed on the State's role with regard to grazing resources for the sector. Currently, the State reimburses 50% of costs for periods when grazing is unavailable (grazing emergencies); however, the costs may be hard to bear even outside such emergencies, especially for herders with small herds and limited incomes, who still need to cover pellets, machine costs, fuel and other expenses. With regard to land use rights, one herder also noted that the fact that the State has established reindeer husbandry areas should be taken to indicate that reindeer herding is a preferred land use in the areas, meaning that private forest owners should not be able to sue reindeer herders for grazing animals on their land: 'there should be someone stepping in from the State to say that "these issues cannot be litigated, these are lands with reindeer herding rights"'. In this case, any regulative change would impact the power relations between sectors, and would not be likely to develop as an adaptation on the local level.

In *winter tourism*, the targeted adaptation is development and marketing within the sector rather than competition with other sectors, given the rather modest development of winter tourism thus far. The attempts to develop a Gällivare destination company for marketing the region could, in the opinion of some, be seen as having increased cooperation among entrepreneurs, in turn supporting marketing efforts, cost sharing and the selling of products that may complement individual entrepreneurs' own selections. At present, however, there is some competition for tourists, as the visiting rates are relatively low. Entrepreneurs noted that administration – for instance, applying for permits – is an onerous process which could be supported municipally in order to strengthen entrepreneurship (for instance in the form of a consultant at the municipality) and that improved integration between administration – to some extent seen as unaware of some tourism company requirements – and tourism companies could be developed.

In the case of *forestry*, environmental protection, as well as rationalisation within the sector, illustrates the ongoing restructuring and benefit of larger units, in particular in light of market requirements. Given forestry's status as a major land use outside environmental protection areas, interaction measures for that livelihood have to some extent focused on interactions to appease other land uses such as environmental protection and reindeer herding as an indigenous land use. In the municipality of Gällivare specifically, Sveaskog has taken the initiative to develop a 'Model Forest' area for inter-sectoral cooperation between forestry, reindeer husbandry, mining and environmental protection in order to 'get different land uses together and find an area that could be used ... for testing different mechanisms' (such as forest management and logging techniques). The Model Forest includes cooperation from the Aitik mine, Gällivare municipality, the Swedish University of Agricultural Sciences, the relevant Saami administrative units, a local foundation and Sveaskog. However, the Model Forest is relatively little developed and not part of the international Model Forest network, which acknowledges the status of such cooperation areas; instead, it has evolved separately from this framework in some cooperation with interactive planning methods in Finnish state forests. According to both interviewees at Sveaskog and the municipality, the model forest development is also a response to the decreasing level of employment in forestry and the increasing volume of land placed under environmental protection. Cooperation networks and local management may thus be developed to not only protect but also utilise areas to support local employment in ways acceptable to the many local interests. With specific regard to protection interests, Sveaskog has also developed what it calls an 'Eco Park' as a pure environmental protection set-aside within the Model Forest.

Inter-sectoral cooperation has also been developed in instances other than those discussed above (see Table 5.1. for an overview of impacts, adaptations, and potential adaptation strategies). The inclusion of some of the area in the Laponia world heritage site means that some cooperation takes place through the Laponia Interim Council. Among other things, the group has started developing different time periods for use by tourism and reindeer husbandry; however, actors in tourism, for instance, have noted their absence from the Council. In 2010, the model of the Laponia council administration will be assessed by the government for potential formalisation. The municipality also has had a coordination group for the last 5 years in which the Saami administrative units, a local Saami organisation representing Saami outside reindeer husbandry, and municipal politicians cooperate.

There was, however, no form of cooperation discussed by all actors, even those who are relatively institutionalised. While the model forest development could supply a potential cooperation arena, most interviewees in the selected sectors

Table 5.1 Perceived impacts, adaptations and potential adaptation strategies described by interviewees[a]

Sector	Examples of impact	Adaptation	Potential additional adaptation strategies (some are not only dependent on the focal actor)
Forestry	State environmental protection	Changes in forest management	Further rationalisation
	Certification	Coordination and dialogue groups	
	Increased attention to requirements from reindeer husbandry	Model forest Eco park	
Reindeer husbandry	Forest management	Internal reorganisation of herding	Compensation from State and forestry
	Logging	Reindeer management plans	Potential clarification of land use rights
	Disturbance from tourism	Work in forestry dialogue groups and coordination	Further rationalisation
Winter tourism	Infrastructure limitations and visibility	Development of destination company	Decreased administration
	Use constraints from Saami and Laponia organisation	Cooperation among entrepreneurs	Improved coordination with municipal actors
	Forest management		Simplification of permit procedures and requirements

[a]Given its status as a State requirement and the limited local organisation specifically for it, environmental protection has more of the nature of a framework that impacts other areas and is not included in the table

were not aware of its development; interviewees more directly involved with the development also noted that there had been some concern over the appropriate placing of the Eco Park in light of environmental protection demands, which may have impacted progress on the model forest. With regard to cooperation in general, many of the interviewees noted that cooperation that works well, including forestry–reindeer herding coordination under legislative or certification requirements in certain cases, does so because of personal contacts, often as a result of their local character and long-term relationships. For instance, one person in tourism noted:

> If I call a reindeer herder ... and ask, 'we have planned to go with the dogs to that valley, are you there with the reindeer?' If they are, I take the other side of the mountain ... On the other hand, if I ... contact the Laponia Council and want to drive somewhere then they say I can't because it would disturb reindeer herding.

Similarly, reindeer herders note that positive interactions with forestry are often due to positive personal relations.

5.5 Conclusion: Sites of Interaction Between Land Uses?

Overall, land uses in the case study area constitute a patchwork of practices that need to interact locally within the context of larger-scale requirements, among these State regulative limitations and the market framework in which the actors need to compete. Entitlement to resources differs with both regulative demands and ownership rights, which to some degree reflect the historical importance of different sectors to the State and for local employment. These policies and regulations constitute the external limitation to changes in local situations and institutions, where a number of cooperation measures exist but are relatively little developed or are perceived by interviewees as having a limited role. For instance, while model forests have often been seen as potential laboratories for trying out different policy and management options, along the lines of adaptive management, most interviewees are so far unaware of or ascribe limited importance to the development. As the model forest development is in its early stages, its potential for local adaptive management is difficult to assess. The different measures for cooperation have also to some extent been instituted with a basis in diverging interests, with Saami interests with a large interest in coordination being relatively prominent in the Laponia Interim Council as well as in municipal cooperation, while forestry interests (vis-à-vis especially environmental protection) are more dominant in the model forest and eco-park developments. Given this background, there has also been some concern regarding the inclusion of sectors in the Laponia Interim Council, as well as some concern over the appropriate placing of the Eco Park in light of environmental protection demands.

There thus exist diverging concerns and coordination requirements among the different actors, corresponding to some extent to an early assessment in the case of the Laponia area (Rådelius 2002). Rådelius noted that given their reliance on user rights over large land areas for herding, reindeer husbandry may be the actor most dependent on coordination within Laponia; other actors may show less dependence on resource sharing beyond existing State regulation and administration for their continued use (Rådelius 2002). The present interviews indicate that small-scale tourism, especially in a developing stage, may require cooperation in order to ensure access to areal resources. The requirement of permits for the use of national parks may also create interaction between sectors (such as for tourism in relation to reindeer husbandry). Forestry, on the other hand, is typically a land owner and less dependent on coordination (however, forestry is comparatively less dominating in this case than in many other areas in Norrbotten, cf. Keskitalo 2008a).

While there exist potential areas for interaction and positive examples in terms of personal relations can be identified, impacts to some extent result from the framework of rights that are seen as limiting or enabling, depending on the possibilities the framework accords to the specific sector. Especially for reindeer husbandry, indicated potential adaptations to some extent target actions that cannot be undertaken purely on the basis of local cooperation. While reindeer husbandry most strongly perceives problems in the regulative framework in its interactions with forestry, tourism – as a relatively undeveloped sector in the municipality – has so far experienced limited institutionalised conflicts and impact from other sectors. Whereas tourism has a potential for expansion, forestry and reindeer husbandry may to some extent be seen as developed to the extent that any further progress may require reorganisation (such as in the case of reindeer husbandry) or be dependent on changing the system for current land use in other respects (such as attempts to increase outcome from existing forestry methods in a context of potential further protection) (cf. Moen and Keskitalo in press). This institutionalised character of the reindeer herding– forestry interface has been noted in studies of the legal framework and historical literature (Allard 2006; Moen and Keskitalo in press) and in case studies (Keskitalo 2008b), potentially indicating that the relation between forestry and reindeer husbandry may be seen as a 'persistent problem' characterised by

> on the one hand the complexity of the interactions of broad societal trends and physical (natural) processes [...] and on the other hand by the involvement of many stakeholders with different but plausible perspectives, which leads to problems of management and governance (Brugge and Rotmans 2007).

To some extent, a transfer of vulnerability may be seen as taking place between these two sectors, and, to a more limited extent, environmental protection, where areas for environmental protection are seen as limiting forestry practices.

The potential for cooperation measures to extend adaptive capacity may here depend both on their possibility to attain coherence in terms of interests as well as funding for participation, the latter being a relevant consideration in light of interviewees' concerns over limited time and funding for interaction. There may also exist a need for interventions beyond the local situation and at governance framework level in order to change prerequisites for actors to operate; some recent extension in cooperation can for instance be related to discussions over ILO Convention No. 169 as well as certification, where existing international processes may impact the possibilities for different local groups and land uses.

Acknowledgements This work has been funded by FORMAS, the Swedish Research Council for Environment, Agricultural Sciences and Spatial Planning. The author is grateful for comments from two anonymous referees.

References

Adger WN (2000) Social and ecological resilience: are they related? Prog Hum Geogr 24 (3):347–364
Adger WN (2001) Scales of governance and environmental justice for adaptation and mitigation to climate change. J Int Dev 13:921–931
Adger WN, Brooks N, Bentham G, Agnew M, Eriksen S (2004) New indicators of vulnerability and adaptive capacity. Technical Report No. 7, Tyndall Centre for Climate Change Research, Manchester, UK
Allard C (2006) Two sides of the coin – rights and duties: the interface between environmental law and Saami law based on a comparison with Aoteoaroa/New Zealand and Canada. Doctoral thesis, Luleå University of Technology, Luleå

Berg A, Östlund L, Moen J, Olofsson J (2008) A century of logging and forestry in a reindeer herding area in northern Sweden. For Ecol Manage 256:1009–1020

Berkhout F, Hertin J, Gann DM (2006) Learning to adapt: organizational adaptation to climate change impacts. Clim Change 78:135–156

Brooks N (2003) Vulnerability, risk and adaptation. A conceptual framework. Tyndall Centre Working Paper No. 28, Tyndall Centre for Climate Change Research, Manchester, UK

Danell Ö (2005) Renskötselns robusthet – behov av nytt synsätt för att tydliggöra rennäringens förutsättningar och hållbarhet i dess socioekologiska sammanhang. Rangifer Report No. 10 (2005):39–49

Hazley CJ (2000) Forest-based and related industries of the European Union – industrial districts, clusters and agglomerations. ETLA, The Research Institute of the Finnish Economy, Taloustieto Oy

Kates RW, Ausubel JH, Berberian M (eds) (1985) Climate impact assessment: studies of the impact of climate and society. Wiley, Chichester

Keohane RO, Nye JS Jr (2000) Governance in a globalizing world. In: Nye JS, Donahue JD (eds) Governance in a globalizing world. Brookings Institution Press, Washington, DC

Keskitalo ECH (2008a) Climate change and globalization in the Arctic. Earthscan, London

Keskitalo ECH (2008) "Konflikter mellan rennäring och skogsbruk i Sverige". In: Sandström C, Hovik S, Falleth EI (eds) Omstridd natur. Trender & utmaningar i nordisk naturförvaltning. Borea, Umeå, pp 248–268 (in Swedish)

Keskitalo ECH, Lundmark L (2010) "The controversy over protected areas and forest-sector employment in Norrbotten, Sweden: forest stakeholder perceptions and statistics". Soc Nat Resour 23(2):146–164

Keskitalo ECH, Sandström C, Tysiachniouk M, Johansson J (2009) Local consequences of applying international norms: differences in the application of forest certification in northern Sweden, northern Finland, and northwest Russia. Ecol Soc 14(2):1. http://www.ecologyandsociety.org/vol14/iss2/art1/

Mijá Ednam (1999) Samebyarnas Laponiaprogram. Gällivare, Miljá Ednam

Moen J (2008) Climate change: effects on the ecological basis for reindeer husbandry in Sweden. Ambio 37(4):304–311

Moen J, Keskitalo ECH (in press) Interlocking panarchies in multi-use boreal forests in Sweden. Ecology and Society, 21 p

Næss LO, Bang G, Eriksen S, Vevatne J (2005) Institutional adaptation to climate change: flood responses at the municipal level in Norway. Glob Environ Change 15:125–138

Nilsson Dahlström Å (2003) Negotiating wilderness in a cultural landscape: predators and Saami reindeer herding in the Laponian world heritage area. Acta Universitatis Upsaliensis, Uppsala University, Uppsala

Olsson P, Folke C, Galaz V, Hahn T, Schultz L (2007) Enhancing the fit through adaptive comanagement: creating and maintaining bridging functions for matching scales in the Kristianstads Vattenrike Biosphere Reserve Sweden. Ecol Soc 12:28. <http://www.ecologyandsociety.org/vol12/iss1/art28/>

Parikka M (2004) Global biomass fuel resources. Biomass Bioenergy 27(6):613–620

Rådelius C (2002) Självstyre eller samförvaltning? Problem och möjligheter utifrån en studie av världsarvet Laponia. Licenciate thesis, Institution for Industrial Economy and Social Sciences, Luleå University of Technology, Luleå

Raison RJ, Brown AG, Flinn DW (eds) (2001) Criteria and indicators for sustainable forest management. IUFRO Research Series 7. CABI Publishing, Wallingford

Smit B, Pilifosova O (2001) Adaptation to climate change in the context of sustainable development and equity. In: McCarthy JJ, Canziani OF, Leary NA, Dokken DJ, White KS (eds) Climate change 2001: impacts, adaptation, and vulnerability, contribution of working group II to the third assessment report of the intergovernmental panel on climate change, published for the intergovernmental panel on climate change. Cambridge University Press, Cambridge

Smit B, Wandel J (2006) Adaptation, adaptive capacity and vulnerability. Glob Environ Change 16(3):282–292
Smit B, Burton I, Klein RJT, Wandel J (2000) An anatomy of adaptation to climate change and variability. Clim Change 45:223–251
Swedish Forest Agency (2000) Forest 2000 county programme (in Swedish). http://www.skogsstyrelsen.se/episerver4/templates/SNormalPage.aspx?id=18798&epslanguage=SV Accessed 10 Dec 2008
Swedish Forest Agency (2008) Swedish statistical yearbook of forestry. Swedish Forest Agency, Jönköping
Swedish Forest Industries Federation (2007) About forest industry (in Swedish). http://www.skogsindustrierna.se/litiuminformation/site/page.asp?Page=10&IncPage=4148&IncPage2=232&Destination2=226&Destination=227 Accessed 10 Dec 2008
van der Brugge R, Rotmans J (2007) Towards transition management of European water resources. J Water Resour Manage 21:249–267
Wondolleck JM, Yaffee SL (2000) Making collaboration work: lessons from innovation in natural resource management. Island, Washington, DC

Chapter 6
From the Inside Out: A Multi-scale Analysis of Adaptive Capacity in a Northern Community and the Governance Implications

Sonia Wesche and Derek R. Armitage

6.1 Introduction

Northern latitudes are experiencing significant impacts as a result of global climate change (ACIA 2004, 2005). Expanding resource development in the region further exacerbates processes of environmental change. Although northern systems are adapted to a dynamically changing environment (Robards and Alessa 2004), the increasing scope, intensity, and variability of change may inhibit the inherent capacity of social–ecological systems to cope with resulting impacts (Duerden 2004; Adger and Vincent 2005).

In the context of historically produced socio-political and economic inequities, small northern Aboriginal communities experience vulnerability to a myriad of emergent challenges and changes. O'Brien and Leichenko (2000) have referred to similar experiences elsewhere as "double exposure," where communities must simultaneously respond to the multi-faceted effects of globalization and climate change. Arguably, northern Aboriginal communities face additional constraints due to existing conditions and the legacy of colonial influences. They continue to experience significant external influences causing rapid changes to their culture, to livelihood options, and broader economic systems, and to the environment around them, which has long been a primary source of sustenance. As such, it may be more correct to call this "multiple exposures".

S. Wesche
Geography and Environmental Studies, Wilfrid Laurier University, Waterloo, Canada;
Current address: National Aboriginal Health Organization - Métis Centre, 220 Laurier Ave. West, Suite 1200, Ottawa, ON K1P 5Z9, Canada
e-mail: swesche@naho.ca

D.R. Armitage
Department of Geography and Environmental Studies, Wilfrid Laurier University, Waterloo, ON, Canada N2L 3C5
e-mail: darmitage@wlu.ca

Building adaptive capacity in the context of multiple exposures demands careful attention to its determinants (local and nonlocal) and the manner in which adaptive capacity is linked to emerging governance regimes in the north. We thus consider the key elements of adaptive capacity and explicitly connect endogenous determinants, enabling factors, and strategies in a way that generates a uniquely integrative perspective. In the northern Aboriginal context, the relational aspects of adaptive capacity are particularly important. Relative isolation, close kinship ties, and a survival-based history encourage inter-dependence. Yet, the nature of these dimensions is shifting rapidly with external influences, the influx of non-Aboriginal populations, and the "opening up" of the North for resource development. Formative institutions (societal conventions, norms, and rules) of these societies have changed dramatically as a result.

This chapter draws on research in Fort Resolution (Deninu Kue), Northwest Territories (NWT), a small Dene and Métis community on the south shore of Great Slave Lake. We have two objectives. The first is to examine the social and institutional dimensions of adaptive capacity from a multi-level perspective, and take into account both endogenous and exogenous influences. Secondly, we connect the analysis of adaptive capacity with implications for emerging institutional and governance processes that may foster adaptive capacity in northern communities experiencing significant social and ecological change.

6.2 Context and Methods

Located in the NWT and intersecting with the southern shore of Great Slave Lake, the Slave River Delta is an important node in a series of delta ecosystems of the Mackenzie Basin (Fig. 6.1). Members of the adjacent community of Fort Resolution have long depended on the delta's high biological productivity and diversity for food (e.g., fish, moose, caribou, beavers, muskrats) and other resources (e.g., pelts, timber). Recent and predicted changes in climate coupled with industrial development and river regulation cause variation in the hydrological regime that drives the delta system, thus constraining local livelihoods (Wolfe et al. 2007). Major sociocultural and economic shifts over the past four decades combined with increasing political clout and ongoing land claim negotiations add complexity and alter the adaptive capacity of the system.

Of the approximately 485 residents in Fort Resolution, over 90% are Aboriginal, mostly Dene and Métis (Statistics Canada 2007). The community is made up of interconnected, kin-based groups, most of whom share a joint cultural history. Many individual family groups lived in small hamlets (most of which are now abandoned) or seasonally used the land along one of the three river systems within the traditional territory: Little Buffalo River, the Slave River and Delta (including Fort Resolution), and Taltson River (where the main settlement was Rocher River). As Dene increasingly settled in Fort Resolution during the mid to late 1900s because of shifts in government policy and economic opportunities, and new legislation extended

6 From the Inside Out: A Multi-scale Analysis of Adaptive Capacity

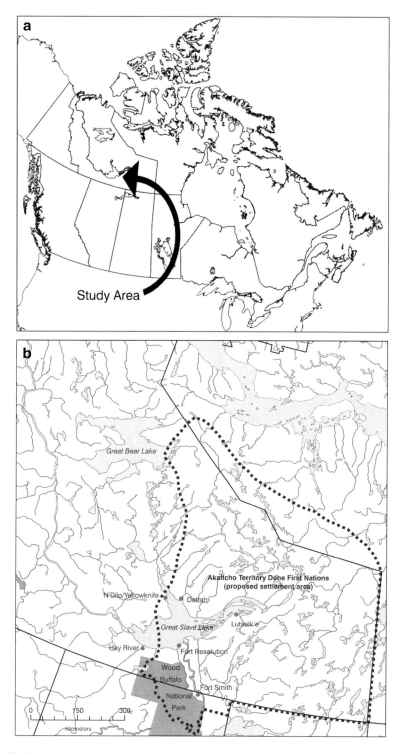

Fig. 6.1 (Continued)

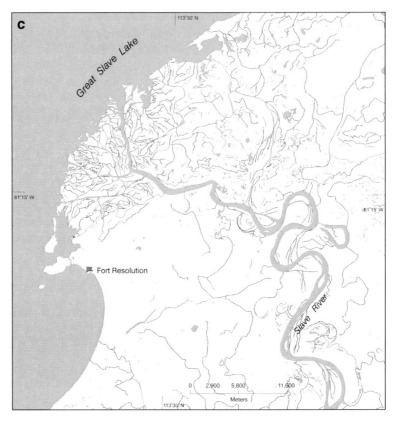

Fig. 6.1 (**a**) The case study area, (**b**) Traditional territory of the Akaitcho Dene (Treaty of 1900 land), (**c**) The Slave River Delta and community of Fort Resolution

treaty status[1] to many who were not previously eligible, the demographic character shifted from being Métis-dominated to Dene-dominated. The Métis make up approximately 35–40% of the current population.

Whether Dene or Métis, most residents have knowledge about, and strong intergenerational ties to, the surrounding traditional land-use area, now referred to as Deninu Kue First Nation (DKFN) traditional territory. Changes in the biophysical environment – particularly climatic and hydrological – are impacting local livelihoods. Local observations indicate shifts in climatic variables such as temperature, seasons, precipitation, snow pack, winds, and storms. Primary trends include warmer winters, more variable transition seasons, and more variable and unpredictable weather. Changes have been observed in the following hydrological system variables: river discharge, water levels, ice freeze up, ice thickness and quality, ice break

---

[1]Individuals with treaty status are members of a First Nation that has a treaty relationship with the Crown. Treaty rights in Akaitcho Territory include education, health care, hunting and fishing, annuities, and land rights (still under negotiation), among others.

up, flood frequency and extent, and water quality. Important trends include increased variability in flow, reduced water levels and flooding, and reduced ice quality. These changes cause challenges for travel safety, among others. There are also many noted shifts relating to wildlife, including fish, migratory birds, ptarmigans, muskrats, beavers, other fur species, caribou, moose, bison, and black bears. Observed trends include changes in migration patterns, population size, and health, which alter availability and hunter access for harvesting. Furthermore, warmer winters result in animal pelts not reaching "prime," translating to lower income for trappers. Other observations include changes in vegetation and fire patterns. Although a mix of livelihood impacts are noted for most of the above variables, the large majority are negative (Wesche 2009).

Residents are also dealing with changing socio-cultural conditions. Most of the current population has spent much of their lives in settlements rather than on the land, resulting in challenges to the transfer of traditional skills and knowledge to the younger generations and a shift in both the range of land-based activities and the number of current land users. For example, rather than camping out on the land for extended periods during specific seasons, it is now common for land-users to take day-trips for hunting, trapping, or fishing, thus concentrating resource-use in areas closer to town. Social change has also been driven by fluctuations in individual access to the market economy during the past decades. Declines in the viability of the fur trade, combined with past eras of abundant alternate employment in the resource sector (e.g., mining, forestry, and the commercial fishery) and current opportunities with mining and government organizations have reduced reliance on local natural resources for food and economic well-being for large segments of the population. Regardless of these changes, resource harvesting activities remain an important component of social, cultural, and economic systems in Fort Resolution.

Kinship is strongly linked to place of origin within the traditional territory, which determines many people's connection to place and knowledge of the land. Relationships and networks based on family origins shape local social structures. Despite some degree of community integration, relationships and allegiances between same-origin families still remain important and affect everyday operation of the community in multiple ways. Personal relationships influence the political landscape, where power swings back and forth between two dominant family groups, each of a similar size. These allegiances influence political and economic decision-making at the local level (e.g., in terms of membership on the First Nation or Métis Council), and they also play out in the school yard where children from different families come into conflict because they know they are not "supposed" to get along. In this way the personal becomes political, and vice versa. With the current political structure and electoral procedure supporting the perpetuation of these family groups, individuals from smaller families have limited influence over community direction.

Beyond the local scale, governance regimes are in transition. Along with three other First Nations, DKFN is involved in land claim negotiations regarding Akaitcho Territory (Treaty 8), thus finding itself in the context of evolving self-government arrangements that may include co-jurisdiction of lands and resources. Additionally, the Métis are seeking a separate contract agreement with the Government of Canada to consolidate their Aboriginal rights. These changing relationships provide both challenges and opportunities for adaptive capacity.

This paper is based on integrative and collaborative research to engage local residents and regional actors in discussions about adaptive capacity at multiple levels. The concept evolved from local leaders' interest in documenting human experiences of environmental change, and community input was sought at all project stages. Data collection took place between 2004 and 2007 during seven visits to the community of Fort Resolution in different seasons. This study draws upon a mix of predominantly qualitative data from 33 interviews with elders and subsistence harvesters, two workshops, five focus groups, a questionnaire administered to 104 heads of household, and 19 key informant interviews with leaders involved in resource management at local and regional levels (see Wesche 2009). Questionnaire data were analyzed via descriptive statistical procedures in SPSS® (version 15.0) software, and textual transcriptions and notes from interviews, workshops, and focus groups were analyzed using QSR NVivo® (version 7), a qualitative research software. Semi-structured interviews and the questionnaire provided information on patterns of change and adaptation, as well as cross-level linkages among governance departments, and the social and organizational networks and linkages beyond the community. Focus groups and workshop discussions provided information and context for the consideration of adaptation issues and strategies. Because of space limitations, this chapter offers a synthesis of relevant items relating to adaptive capacity; for full details on collected data, see Wesche 2009.

6.3 Adaptive Capacity at the Local Level: Endogenous Determinants

The adaptive capacity of a system is defined by the attributes that allow it to prepare for, adjust effectively to, and recover from existing or anticipated disturbance while maintaining critical structures and functions (Walker et al. 2002; Adger 2003a; Smit and Pilifosova 2003; Olsson et al. 2004). Adaptive capacity is socially and geographically differentiated, with localized processes often dictating how specific groups respond to exposure. The vulnerability, international development, and community development literatures recognize the importance of identifying and strengthening existing endowments or assets (Kretzmann and McKnight 1993; Beck and Nesmith 2001) held by households and communities, and building on strategies that are already being implemented in response to changing conditions (Agrawal 2008).

These endowments may take a variety of forms. While a system's operational context (e.g., requirements and issues of a technical, financial, social, institutional, or political nature) plays an important role in determining adaptive capacity, other strategic and largely informal attributes such as power, scale of operation, knowledge valuation, culture, and community characteristics are often more influential in localized contexts where common resources are shared (Armitage 2005). Thus, in such settings adaptive capacity relies largely on the state of social and institutional relationships, the nature of social interactions regarding overlapping or conflicting interests, and the means by which these factors permit positive collective action (Adger 2003a; Armitage 2005). Levels of connectedness, trust, and reciprocity are of particular importance; however, these dimensions remain difficult to measure.

Adaptive capacity is not isolated to individual levels of organization. Rather, the ability to effectively respond to extreme or variable conditions is influenced in part by the broader enabling environment, the availability of resources at higher scales of organization, and access to those resources through cross-scale linkages (Yohe and Tol 2002; Brooks 2003; Smit and Pilifosova 2003; Adger et al. 2005; Smit and Wandel 2006). The nature of this broader social context may differentially facilitate or hinder access to adaptation resources and opportunities for specific groups depending on social attributes (e.g., race, class, gender; Cutter 1995; Pelling and High 2005; IPCC 2007).

Different combinations of the following features of communities and regions are recognized as strong determinants of adaptive capacity: economic wealth, technology, information, knowledge and skills, infrastructure, institutions, social capital and networks, and equity (Smit et al. 2001; Smit and Pilifosova 2003; IPCC 2007). While these factors are all applicable and important in the case of Fort Resolution and other similar northern communities and situations, the crucial role played by social relationships in facilitating or hindering adaptation both at the household level and collectively through joint action has only received more recent recognition (Adger 2003a, b; Armitage 2005; Pelling and High 2005; Adger et al. 2007).

The primary endogenous determinants of adaptive capacity as identified by residents in Fort Resolution are summarized in Table 6.1 (Wesche 2009). As in many small communities, there are multiple location-specific determinants of adaptive capacity. These determinants vary in space and time (see Smit et al. 2001) and act to either constrain or enhance the ability of individuals, households, and the community as a whole to adapt (Kelly and Adger 2000). The critical determinants of adaptive capacity documented in Fort Resolution include (1) the ability to draw on diverse knowledge and skills; (2) levels of access to specific resources (particularly financial) and new technology to enable travel on the land, improve travel safety, and increase economic options; (3) the extent to which local institutions (governance organizations and social customs) provide resources and support; (4) the nature of social networks in providing access to food, harvesting, and economic opportunities; and (5) the extent to which individuals and households have equitable access to resources and jobs.

Determinants of adaptive capacity related to the community's operational context – knowledge and skills, access to resources and technology, and institutional support – are more evident because of their tangible nature. However, underlying community characteristics including social networks and the level of equity among residents also play a significant role in shaping the capacity of individuals and households to adapt, and in influencing whether or not existing adaptive capacity remains latent or becomes engaged (Kelly and Adger 2000). In the Fort Resolution

Table 6.1 Determinants of adaptive capacity at individual and household levels

Determinant	Application	Examples
Knowledge and skills	• The transfer of traditional knowledge to youth and its continued use improves livelihood choice, travel safety, harvesting flexibility, and connection to culture and identity • Western knowledge, skills, and technology enable opportunities for training and career development in the wage economy, and improve the ability of individuals to connect with institutions and resources beyond the local level • Drawing on both traditional and western knowledge increases livelihood flexibility	• Knowledge of river currents and ice break up patterns improves risk assessment when traveling by snowmobile on spring ice • Knowledgeable land users are better able to alter the timing and methods used for harvesting • Access to the Internet and knowledge about how to use it allows local business owners to market directly to potential visitors (e.g., accommodations, tour operations, handiwork sales) • Innovative individuals with diverse skills act as 'go-to guys' for wide-ranging requests, such as tour guiding, research assistance, provision of traditional food, manual labor, and small business partnerships
Access to resources and technology	• Land users increasingly require capital input (e.g., for harvesting equipment) • Technology improves harvesting flexibility and travel safety under uncertain conditions • Money and resources facilitate involvement in new ventures	• Owners of boats or snowmobiles can travel on the land, while those who cannot afford such investments are limited to harvesting in town or along roadways • GPS devices facilitate way-finding and satellite tele-phones enable communica-tion when stranded. • Owners of diverse traveling and camping equipment can engage in camp-building contracts for resource developers or initiate tourism operations
Institutional support	• Local organizations provide access to resources and employment opportunities • Traditional customs dictate that residents be willing to help others when needed	• Local governance organizations offer periodic access to specific resources (e.g., financial pay-outs from industry Impact Benefit Agreements, harvesting equipment) and contract employment (e.g., trail clearing, forestry work, shuttle driving) • DKFN and Métis Local provide honoraria for sitting on committees and for participation at community meetings • Residents come together to provide emotional support and food in times of family crisis
Social networks	• Sharing of food and harvesting equipment among relatives and friends improves access for vulnerable populations	• Some hunters provide meat to relatives who lack means to hunt in exchange for the service of drying meat

(*continued*)

Table 6.1 (continued)

Determinant	Application	Examples
	• Social networks increase flexibility to combine resources for new ventures	• Boat owners invite relatives or friends to join them on hunting and fishing excursions
		• Two households can combine knowledge, networks, and assets to pursue economic ventures (e.g., camp-building contracts for resource developers)
Equity	• Inequitable access to resources and jobs increases social divisions	• Tensions result from the periodic allocation of managerial contracts (e.g., for forestry, maintenance work, etc.) to the same individual or family group
		• Many Métis are frustrated by exclusion from DKFN-sponsored contracts

situation, adaptive capacity clearly depends not only on the availability of resources, but also on whether those resources are accessible by group members at all social strata (including the marginalized and vulnerable). In Fort Resolution, the Métis and those with Rocher River (an abandoned hamlet on the Taltson River) origins often feel underrepresented within the elected leadership, that their environmental and social interests are not well supported, and that they are marginalized from accessing important livelihood resources.

6.4 Enablers of Adaptive Capacity: Scaling Up

Assessments of adaptive capacity are specific to culture, place (Adger 2003a), and scale (Adger and Vincent 2005; Vincent 2007). At the same time, systems do not function in temporal and spatial isolation. It is essential to consider processes and interactions across scales in terms of space, time, and social organization (Gunderson et al. 1995; Gunderson and Holling 2002; Dietz et al. 2003; Cash et al. 2006). While social linkages are shown to play a critical role at the local scale, adaptive capacity also relies on a broader enabling context created by linkages across scales – those at regional, territorial, national, and even global levels. Furthermore, while a local focus is critical for adaptation, complementary vulnerability reduction measures may be required at multiple scales (Naess et al. 2005). In light of the opportunities and constraints to adaptive capacity in Fort Resolution, we outline the connections to higher-level governance and political dimensions, and the manner in which they enable or limit the development of adaptive capacity.

While there are multiple exogenous determinants of adaptive capacity that may interact with endogenous conditions, we emphasize four in particular: (1) the effects of government-based support programs, and their role in creating a social safety net; (2) the impacts of existing and potential economic transitions on people and livelihoods in the region; (3) Aboriginal–Aboriginal and Aboriginal–State relations; and (4) emerging self-government arrangements in Akaitcho Territory.

6.4.1 Government Support Programs

Despite the road connection, Fort Resolution is relatively isolated and limited local wage opportunities exist. Moreover, land users have seen fur prices decline while costs for harvesting equipment increase. As noted above, additional financial pressures include increased outlay for the purchase of specialized equipment to mitigate safety concerns due to changing climatic conditions. These local economic constraints highlight the importance of social assistance programs, the allocation of resources and jobs through local governance organizations, and harvester support programs. While each of these support programs provides essential resources for adaptation for both harvesters and nonharvesters, harvester support programs are of particular interest as they have important links to some of the key endogenous determinants of adaptive capacity relating to cultural continuity and economic diversification.

In terms of supporting basic needs, federal resources devolved to and administered by the Government of the Northwest Territories (GNWT) are applied to a number of programs (e.g., income support, family allowance, pension) that act as a formalized social safety net. Although the number of beneficiaries in Fort Resolution has declined from 111 to 52 individuals from 1996 to 2006 as incomes have increased, the unemployment rate still hovers around 18% (Northwest Territories Bureau of Statistics 2007). As in many other small northern communities, social assistance continues to provide a significant portion of total income. Although the subsistence economy may account for one-quarter to one-half of total economic value in northern communities, its value is not reflected in economic accounts (Warren and Egginton 2008). The flexibility to harvest provides a supplementary source of revenue, which can be supported through fixed incomes such as those described above. For example, a number of able-bodied pensioners rely on monthly income to support land use from which they derive subsistence, so there exists to some degree a transfer of financial resources into harvesting outputs.

In the absence of sufficient wage opportunities and in light of the interest in preserving culture and maintaining livelihood choices for northern residents, harvester support programs are particularly important for enabling continued land use and subsistence activities. The GNWT has implemented a complement of programs to support continuity within the territory's traditional economy. The Genuine Mackenzie Valley Fur Program has a targeted marketing strategy to promote the sale of high quality authentic NWT-harvested fur, while sub-programs (Guaranteed Advances, Prime Fur Bonus Program, and Grubsteak Program) have been established to reduce individual risks to trappers. They provide financial support to encourage the harvest of high quality pelts, and help to stabilize fur market fluctuations through guaranteed minimum pelt prices. Subsidies for equipment purchase are also periodically available through local institutions. Funds from the GNWT Community Harvester Assistance Program are distributed to defray capital and operating costs for harvesting activities, and DKFN also periodically provides hunting shells to its members. Another complementary suite of GNWT programs – Take a Kid Trapping, Trapper Recognition, and Trapper Workshops – aim to promote and recognize trapper skills, and counteract the degradation of traditional knowledge and skills in the NWT (Rossouw 2004; Government of the Northwest Territories 2008). Culture camps run at the community level through the First Nation or the school also help to reinforce these skills. In combination, these resources help to support the viability of trapping activities for individual residents.

Individuals in Fort Resolution are not always positive about these support programs as they are limited in their ability to provide desired resources and are difficult for some to access. However, despite some limitations, there is a general recognition that such programs are linked to important enabling resources that contribute to the capacity of harvesters and their family networks to adapt to changing environmental conditions. Improving access to the means (equipment, financial flexibility, knowledge) for traveling and harvesting on the land can improve food security for households and their social networks, and increase opportunities for economic diversification. Land users also highlighted how territorial harvester support programs encourage the continued use and transfer of traditional knowledge, a key resource for adaptation.

6.4.2 Economic Transition

The north has experienced a relatively rapid economic transition over the last half century. This involves the combined effects of a declining fur trade, a shift toward extractive, nonrenewable industries, and a government settlement policy that encouraged a bush-settlement transition, peaking in the mid-1900s. However, many adults who spent their lives on the land had few skills to take advantage of wage employment and suffered from a loss of identity and way of life. Alcohol (and later other drugs) became a coping mechanism, and its use expanded rapidly in the 1970s after a new road was completed, linking Fort Resolution to the towns of Pine Point (a company town attached to the lead–zinc mine by the same name) and Hay River. As a result, Fort Resolution residents widely acknowledge a generational fracture in the transfer of traditional knowledge and connection among people.

The social consequences of the bush-settlement transition, increased access to drugs and alcohol, and rising disparities in material wealth stemming from selective employment at Pine Point (1964–1988) were exposed through a period of social unrest from the 1970s to the early 1990s. During this time the community suffered several tragic suicides and a rise in domestic and community violence which strained social networks. Since that time, a number of residents have begun the process of healing and re-building healthier and more prosperous lives. While the active unrest of previous decades has mostly subsided, continuing drug use has shifted in part toward harder drugs, and bootlegging (local sales) of alcohol and drugs by a handful of residents reinforces interpersonal tensions.

The rise and fall of employment opportunities has been challenging for many residents. Since the decline of the Great Slave Lake fishery in the early 1970s (Mackenzie River Basin Board 2003) and closures of the local sawmill and Pine Point mine in the 1980s, most of the remaining local wage work is in government offices. Some individuals maintain more than one part-time job, whereas others are unable to break into the market, resulting in feelings of inequitable treatment (Workshop Participants 2005). Some residents have also taken advantage of opportunities in the burgeoning diamond mine industry in other regions of the NWT, although a lack of professional qualifications and the existence of criminal records limit participation. Many families have resorted to drawing on a combination of occasional or seasonal employment, trapping income, and welfare to meet basic needs.

There is a range of possible or likely developments in Akaitcho Territory in the foreseeable future, including hydro-electric development on Taltson and Slave Rivers, mining, oil and gas, logging, and tourism, some of which are already under discussion or in the early stages of development. There is also growing concern about the downstream effects on water quantity and quality caused by the operation and planned expansion of Alberta Oil Sands activities (Campbell and Spitzer 2007). While community leaders feel they have some control over – and experience adapting to – developments within their territory, issues like transboundary pollution add heightened concern. For some residents, these developments offer both opportunities for wage employment and novel business ventures (e.g., building and maintaining worker camps, producing specialized industrial products). For others, the lack of diverse knowledge and skill sets, limited access to resources, social exclusion, and inequitable employment opportunities may severely challenge their capacity to adapt.

6.4.3 Aboriginal and State interrelationships

Enabling environments, consisting of social structures that help society function at different levels (Klein and Smith 2003) are required to support collective adaptation. In the case of Fort Resolution, and in keeping with the federal government's fiduciary responsibilities, implementation of most programs (e.g., social support, harvester support, employment training) is top–down (see Fig. 6.2); funding and services are allocated directly from the federal government or by way of the territorial government.

The limits of this historically-determined approach are well-documented, with recognition of the need for a dramatic shift in Aboriginal–State relations to improve equitability and sustainability (RCAP 1996). Akaitcho communities possess many of the fundamental building blocks to design and develop institutions and programs adapted to the modern realities of settlement life and flexible enough to accommodate increasing social–ecological change. However, the historical

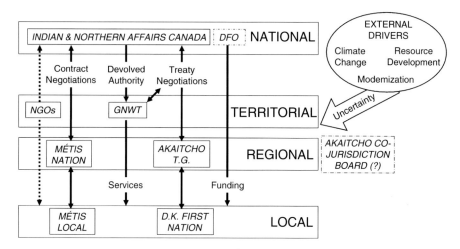

Fig. 6.2 Primary organizational relationships across levels and drivers of change (The co-jurisdiction board indicated here is one possible model). *DFO* Department of Fisheries and Oceans, *NGOs* Non-Governmental Organizations, *GNWT* Government of the Northwest Territories, *Akaitcho T.G.* Akaitcho Territory Government, *D.K. First Nation* Deninu Kue First Nation

relationship with the Government of Canada has evolved as a paternalistic one, where Aboriginal communities are on the receiving end of programs, policies, and resources. Many Fort Resolution residents feel that such relationships have generally been divisive (at least in the past), and as a result are disinclined to recognize the potential opportunities of nonAboriginal organizations to support adaptation. Continued mistrust of government agencies results in local organizations protecting their own interests, with limited willingness to engage with external facilitators to help develop an effective and viable adaptation planning process. Paradoxically, a lack of trust also incites tension among local organizations due to concerns about the inequitable allocation of resources and opportunities by government authorities.

In this context, regional Aboriginal institutions (Akaitcho Territory Government and Métis Nation) play an increasingly important role in mediating relationships between local governments and those at territorial and federal levels. In the case of the Dene, the Akaitcho Territory Government (ATG) exists to represent the collective interests of its constituent First Nations, and coordinate transitional efforts towards self-governance. The ATG Lands and Environment Division provides a regional forum to enhance local to regional knowledge transfer on relevant issues and facilitate the standardization of policies and procedures for joint decision-making in the territory (Boucher 2006a). Conversely, this additional layer of bureaucracy can act as an information barrier if communication channels break down, or as a funding sink due to elevated administrative costs.

At the same time, some competitive and protectionist stances have developed among the four First Nations within Akaitcho Territory, stemming in large part from the perception that inclusiveness will infringe upon local rights and benefits. Each community is working mostly for its own benefit, rather than the benefit of the broader region. While there is substantial political leverage when Aboriginal governments work together, such efforts are often stymied by the perceived "divide and conquer" approach taken by the Government of Canada and industry. For example, resource companies have been known to approach communities independently to try and broker a deal, insisting that local benefits will be diluted if other communities in the region become involved. Local governments on the receiving end are caught in a dilemma, feeling pressured to choose between desired economic benefits and their loyalty and ties with surrounding local governments and residents. In Akaitcho Territory, one such instance led to Fort Resolution's exclusion from an Impact Benefit Agreement with a mining operation located on the far side of Akaitcho Territory (Boucher 2006b). A combination of strong Aboriginal leadership and solid reciprocal relationships between and among communities is crucial to ensure that all relevant parties are included in important land use negotiations. Improved regulation will also likely emerge as communities move toward self-governance, which will help to standardize procedures. Movements towards these outcomes must be both desired and enabled at the local level.

6.4.4 The Road to Self-Governance

Dene and Métis have been participating independently in negotiations over self-governance and land rights since the 1990 breakdown of the Dene–Métis comprehensive claim (Fumoleau 2004). These negotiations offer optimism for enhanced involvement of local authorities and residents in decisions regarding the traditional territory, and for incorporating mechanisms to strengthen adaptive capacity and support adaptation options.

In the case of Akaitcho Territory, the first stages of this process have already been implemented. An Interim Measures Agreement (IMA) was signed in June 2001, authorizing Akaitcho First Nations to pre-screen water and land use applications and permit requests (Indian and Northern Affairs Canada 2001). More recently, local environmental officers worked with elders and land users in each of the Akaitcho Dene First Nations (DFNs) to outline lands of specific cultural, spiritual, and environmental interest in their joint traditional territory. The result was a federally-sanctioned Interim Land Withdrawal in 2007, safeguarding approximately 62,000[2] km from mineral staking, sale, or lease while negotiations continue (Parks Canada 2007). Pending ratification of the Treaty 8 Final Agreement, it is anticipated that renewable resource management decision-making for Dene Title Lands will be transferred to the Akaitcho DFNs, and for the remaining "Co-jurisdiction Lands"[2] to

---

[2]The Akaitcho Dene First Nations are pursuing a "co-existence" agreement, which emphasizes the recognition of pre-existing rights determined through the original treaty, as an alternative to the widely used comprehensive claims process. The likely outcome is an administrative division of Akaitcho territory into Dene Title Lands (under full authority of the DFN) and Co-jurisdiction Lands (where authority is shared).

a joint management board with representation from the federal government, the GNWT and Akaitcho DFNs. Although many issues remain, community expectations are that such shifts in governance will build social networks (locally and regionally), help to integrate diverse information and knowledge sources, support mutual learning, and improve decision processes and outcomes.

The evolution of regional governance arrangements has local complications, however. For example, despite their shared history, differential legal treatment has created divisions between the Dene and Métis. The Métis tend to be strongly affiliated with the territory adjacent to Fort Resolution where they had historically founded small hamlets, including those at Little Buffalo River and the mouth of the Slave Delta. Many Dene are attached to other more isolated parts of the traditional territory. Consequently, individuals and family groups are differentially concerned about specific territorial zones, which contrasts with the centralized management of the territory at both community (DKFN) and regional (Akaitcho Territory) levels. While the territory is administered as a single block, it is not necessarily perceived as such by those who use it. Furthermore, the Métis do not have any legal land rights at present, leaving land use planning and decision-making largely in the hands of the First Nation. Dene and Métis positions regarding the Pine Point Pilot Project – a preliminary bulk extraction of one million tons of lead and zinc at the site of a decommissioned mine between Fort Resolution and Hay River – offer an illustrative example. While K'atlodeeche First Nation (Hay River Reserve) and DKFN (Fort Resolution) have both signed agreements with Tamerlane Ventures indicating their support for the project, the Fort Resolution Métis Council remains opposed because of concerns about environmental impacts and the lack of direct benefits to its members (Fort Resolution Métis Council 2007; Bickford 2008).

Both the Dene and Métis have entered into land and rights negotiations with an awareness of the power dynamics at play between them and the federal and territorial governments, and with the recognition that they are functioning within a western framework rather than one they themselves devised. Like other First Nation groups (e.g., Nadasdy 2003), they envision the process as the only realistic way at this point to preserve their territory and way of life. However, this choice comes with costs to the very way of life that they are trying to protect. It moves people off the land and into offices or industrial sites, and the financial influx for wage earners reinforces social and economic divisions at the local level (see Sect. 6.4.2). However, on balance, negotiating such claims and enhancing control over decision-making in their traditional territory are perceived by residents to offer benefits over the long term.

For the Dene, the treaty negotiation process has taken on an adaptive approach, which bodes well for the development of a more flexible and functional Final Agreement. There is a concerted effort underway for community negotiators to learn from the experiences of existing land claim and co-management arrangements, to apply positive aspects to current negotiations, and make alterations where necessary (Boucher 2006b, pers. interview). For example, the First Nations in Akaitcho Territory (including in Fort Resolution) have chosen to pursue "Co-Existence" under their Treaty Land Entitlement negotiations (Akaitcho

Territory Government 1995) rather than the more common Comprehensive Claim process. While the outcome of this model is uncertain, the concept is being developed in this case with some of the following principles in mind: respect for and accommodation of Dene sovereignty and political autonomy, collective rights, a separate Dene legal system, free and informed consent in dealings with the Government of Canada, and the exercising of Dene jurisdiction (Akaitcho Territory Government 1995).

The application of adaptive management principles to the negotiation process is indeed unique. The intention is to proceed with a negotiation–implementation cycle in order to learn, determine what works and what doesn't, and develop capacity at each stage. The first phase of decision-making devolution began with the 2001 IMA. It recognizes that the Akaitcho DFNs have "their own internal processes for determining the use of lands and water" (Akaitcho Territory Dene First Nations et al. 2001, p. 2) and outlines the conditions for developing a pre-screening process for land and water use applications. As such, each community has a chance to first implement its own process, evaluate its ability to implement, and subsequently address outstanding issues within a smaller scale process before implementation of the entire Final Agreement. This iterative learning process allows negotiators to assess the effectiveness of IMA policies and either modify or transfer them to other areas (e.g., lands management), and use the IMA as a springboard to extend discussion to items that would otherwise be left off the agenda (e.g., jurisdiction over prospecting permits). Community leaders anticipate positive structural outcomes from this process, as residents collectively gain more control over their territory and affairs. At the same time, collaboration among local governance organizations is essential to ensure that benefits extend both to First Nation members and to other residents within each community, requiring focused efforts to build community cohesion around collective assets.

More recently, a policy change at the national level has added a layer of uncertainty to the emergence of self-government, with subsequent implications for adaptive capacity. Specifically, through devolution the Government of Canada seeks to enable the territories to become more self-sufficient, with greater local control and accountability over decisions that affect the north, especially with regards to natural resource development (Government of Canada et al. 2004). While regional Aboriginal Governments in the NWT[3] are at the table as part of this government-to-government-to-government process (with national and territorial governments), Akaitcho Territory Government lacks active involvement in the

---

[3]As of 2000, the Aboriginal Summit, comprised of a caucus of Northwest Territories regional Aboriginal government leaders, represented collective Aboriginal interests in negotiations with the federal and territorial governments on devolution and resource revenue sharing. After the 2007 folding of the Summit, individual regional governments have represented their own interests. Akaitcho Territory Government and the Deh Cho First Nations maintain observer status while continuing to negotiate land claim or treaty agreements (Indian and Northern Affairs Canada 2003; Government of Canada et al. 2004; Irlbacher-Fox and Mills 2007; Indian and Northern Affairs Canada 2008).

talks due to its unresolved land claim. Key Aboriginal concerns about possible outcomes include potential adverse impacts on recognized and unrecognized Aboriginal rights, adequate resource revenue sharing with Aboriginal governments, and meaningful Aboriginal participation in implementation (Irlbacher-Fox and Mills 2007). Understandably, ATG's concerns are heightened because of its limited input into the process and lack of knowledge about how federal-territorial devolution may affect ongoing negotiations. Local leaders in Fort Resolution also worry that increased GNWT authority undermines their nation-to-nation relationship with the Government of Canada, adding another layer of bureaucracy and diverting funds which could otherwise be transferred for use at the local level.

6.5 Strengthening Adaptive Capacity in a Northern Social–Ecological System

While all communities have some degree of adaptive capacity, it is generally not specifically developed to combat environmental change (Handmer 2003). Adaptation measures can be targeted to deal with a specific pressure (Downing 2003); however, building adaptive capacity reflects the idea of developing resilience within a system to deal with a range of shocks and stresses. While it is important to note the elements in society that undermine adaptive capacity – in other words, what makes people vulnerable (Handmer 2003) – a sole focus on negative aspects and barriers threatens to re-pathologize Aboriginal people, ignoring their many inherent strengths and capabilities. As mentioned previously, it is important to identify and build on existing assets (Kretzmann and McKnight 1993; Beck and Nesmith 2001) to foster a system's ability to withstand change (Walker et al. 2004).

In the context of the endogenous and exogenous dimensions of adaptive capacity highlighted above (see Fig. 6.3), we outline a number of opportunities to address these challenges and foster adaptability. While the opportunities outlined here are not exhaustive, they reflect the insights and recommendations of Fort Resolution residents and others who are active regionally. In this regard, they provide a road map for policy, decision-making, and institutional development.

6.5.1 Building Social and Cultural Capital

Issues around social dynamics lie at the base of many environmental and resource management problems. Here we define social and cultural capital as the actual or potential resources (material or nonmaterial) that stem from one's social networks and connection to a cultural milieu. Strong social bonds among community members and connection with culture and land-based traditions are important components of adaptive capacity in Fort Resolution. Culture is important in shaping values and

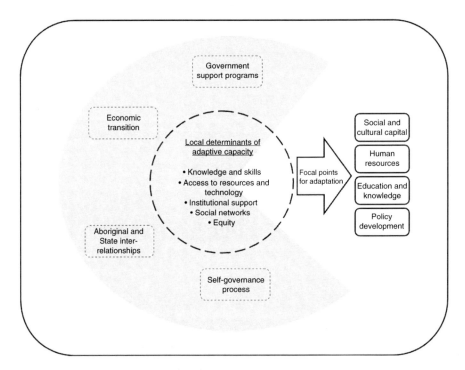

Fig. 6.3 Overview of adaptive capacity issues in Fort Resolution

attitudes, and associated norms "embody place-specific memory of change, feedbacks, and repercussions" (Robards and Alessa 2004, p. 416). For example, the Dene have incorporated lessons about adaptive capacity into their legends and laws (Newton et al. 2005), which are fundamentally relationship-based. Dene Laws offer a guiding framework for re-developing the connections and ways of being that are important for individual and collective wellbeing.

Aboriginal Peoples have emphasized the importance of revitalizing cultural ties (Smith 1999; McGregor 2004). "Connecting is about establishing good relations" (Smith 1999, p. 149), with multiple outcomes for community wellbeing. In Fort Resolution, potential exists for building on shared community identity and history to instill collective stewardship values and encourage people to work together. Despite internal factionalism, local residents continue to feel connected at a deep level (Wesche 2009), as evidenced by the strong support provided to one another in crisis situations. A focus on "building the home fire" (Boucher 2006b, pers. interview) is essential, requiring individual healing and cultivating the strengths of existing relationships.

Existing mechanisms that support cultural development in Fort Resolution include curriculum-based Chipewyan language classes and culture days at Deninu School, Chipewyan Language Centre activities (e.g., sewing, hide tanning), and "culture camps." These activities encourage the transfer of traditional knowledge and language to community members through experiential learning.

In the north, traditional culture and livelihoods rely heavily on local level kinship bonds and other social ties, and these continue to influence local and regional politics. While participation in external networks may be limited, and social norms are shifting, residents still maintain significant shared values and understandings. For example, Fort Resolution participants remarked repeatedly on the need for protecting water quality and the health and numbers of traditional food species, and there was general agreement about linkages between healthy land and healthy people. The development of community-level collaborative efforts may be best stimulated through an initial focus on such areas of common interest.

Regarding adaptive capacity, while technical or financial solutions may improve short-term coping capacity, the implementation and persistence of long-term adaptive strategies require attention to the structural underpinnings of the social system. Individual healing and the rebuilding of relationships is seen to start in the home, where many residents are already choosing healthier lifestyles and providing safe learning environments for their children. At the same time, forces external to the household can also contribute to restructuring social ties. For example, desire exists for a re-empowerment of family units through the redistribution of authority for collective decision-making at the local level (Boucher 2006a, pers. interview). In addition, governing bodies at all levels can directly support the development or preservation of social capital to achieve program objectives or deliver services. They can also offer indirect support by establishing favorable enabling conditions for social network development and by increasing their sensitivity to policy-social capital interactions (Government of Canada 2005).

6.5.2 Improving Human Resources

Locals indicate capacity-building needs in such areas as access to financial and human resources, expertise in governance and resource management, and sustainable employment opportunities. Preparing local leaders requires incorporating traditional values within the education system, and the re-empowerment of family units. Efforts to build local human capacity may also draw from external resources. Investing in building a supportive and dynamic learning environment that offers a range of opportunities will attract both external and local employees. Deninu School has achieved success in this regard as a result of initiatives undertaken by internal managers. This community- and relationship-focused leadership has resulted in longer employee retention, more effective engagement in community events and committees, and stronger employee-community member relationships.

Despite an out-migration trend for those seeking educational and employment opportunities, there is still a strong sense in Fort Resolution of cultural identity being tied to place (Wesche 2009). This sense of place and attachment can serve as a positive foundation for long-term planning. Prioritizing relationship- and capacity-building activities with long-term residents would likely provide lasting benefits to the broader community, as they have an inherent interest in community sustainability and may be more easily mobilized to develop and implement plans. Furthermore, if economic opportunities are made available locally (e.g., through resource development), this may encourage some previous residents to return, injecting new human resources and ideas.

6.5.3 Education and Knowledge Transfer

At individual and household scales, land user adaptation to changing environmental conditions requires specialized knowledge and skills. Since youth are not receiving a traditional upbringing, culture camps on the land are one method that locals use to transfer traditional knowledge through experiential activities with elders. Land users stress the importance of knowing the land that one is traveling on, and of having sufficient foundational skills to survive should a problem arise.

At a collective level, northerners have a major stake in research and policy outcomes, requiring improved capacity for engaging in these processes (Abele 2006) through action at multiple levels. Improving research capacity involves targeted efforts in educating northern youth, educating community leaders about effective governance, creating new centralized research facilities (Graham and Fortier 2005; e.g., the proposed Great Slave Lake research center (Boucher 2006b)), and developing regional research planning, coordination and education centers (e.g., the new Yukon Research Centre of Excellence initiative focused on climate change adaptation (Council of the Federation 2007)). Researcher-community collaboration is also increasingly identified as a significant northern research component (Krupnik and Jolly 2002; Armitage and Clark 2005; Graham and Fortier 2005), helping to improve contextual applicability, policy-relevance (Wolfe et al. 2007), knowledge exchange, and local level capacity-building.

In the north, technology can be particularly helpful at bridging distances and distributing information, for example, through interactive CD-ROMs on Arctic climate change (Fox 2004) and resource development (Willett and Janes 2005), and publications outlining aboriginal experiences of environmental change (e.g., Sherry and Vuntut Gwitchin First Nation 1999; Krupnik and Jolly 2002; Nickels et al. 2006; Riewe and Oakes 2006; SnowChange 2008). The internet has enabled new types of knowledge transfer and access to on-line courses, facilitating education within the community, and providing options for those coping with multiple responsibilities (e.g., families, jobs, care-giving).

At the territorial level, events like the NWT Climate Change Leadership Summit and annual Science in the Changing North conference are essential for exchanging environmental knowledge across communities and organizations. This is further reinforced by multidisciplinary, northern-focused research and knowledge transfer projects funded by organizations such as ArcticNet and the International Polar Year. Community leaders noted that improvements in knowledge retention and exchange may be facilitated through the creation of a centralized environmental database for Akaitcho Territory, whose access could be mediated by contributing parties. Small governance organizations are vulnerable to the loss of specialized staff who acquire specialized knowledge about the subject matter and human, financial, and material resources in a particular field. A centralized database would provide permanent records of environmental research data, results and contacts for consultation and insurance against such losses.

6.5.4 Policy Development for Adaptation

Effective policy response is an important mechanism for strengthening adaptive capacity and expanding adaptation options (Smit and Pilifosova 2003). While true, the north faces a recognized challenge in developing and implementing policies that address the diverse needs of northerners at multiple levels of organization, where governing bodies may have incongruent or sometimes competing jurisdictional objectives. The lack of a NWT adaptation strategy has resulted in erratic consideration and application of responses to change at community and territorial levels to date. Self-governance negotiations may provide opportunities to build a more adaptive approach to decision-making (as described above). For example, the collaborative allocation of authority and a forum (e.g., a co-jurisdiction board) where more equitable relationships can be brokered offer the possibility to encourage fair treatment and mutual support among member bodies. These structures provide a venue for (re)building trust among individuals and groups, learning together, and developing the flexibility to act rapidly and in unison when required.

At the local level, existing projects and interests can be built on or extended to improve adaptive capacity. For example, DKFN initiated a Land and Resources Management Plan in 2002, which was subsequently derailed by lack of funding (FREWC 2003, 2004; Simon 2006). Revitalizing this effort as part of a broader community planning process would provide a structured framework to guide environmental- and resource-related decision-making

Federal and territorial governments must also engage actively in adaptation planning and implementation, as they hold the necessary resources and staying power to "do the heavy lifting" (Ritchie 2006). While federal climate change policies are still in flux, it is recognized by the GNWT as a serious concern for the environment and northern livelihoods (Government of the Northwest Territories 2007). The GNWT's response strategies are generally mitigation-focused, although adaptation has recently garnered more attention. The territorial greenhouse gas strategy (released in 2001, updated in 2007) supports the development of a Climate Change Network (Government of the Northwest Territories 2007) and an impacts and adaptation plan. The pervasiveness of climate change and the linked nature of planning for multiple exposures to change encourage integration or "mainstreaming" within other government programs (Smit and Wandel 2006; Adger et al. 2007; Ford et al. 2007). For example, the Aboriginal Aquatic Resources and Ocean

Management (AAROM) program in the Great Slave Lake watershed recognizes that climate change impacts all environmental aspects under its mandate, and must be considered in all decisions and actions (Giroux 2006). Such programs can bring a range of stakeholders to the table and encourage inter-community collaboration around common issues (Akaitcho Territory Government 2006).

6.6 Conclusions

Adaptive capacity to social–ecological change is an outcome of many interconnected factors operating at different scales (Fig. 6.3). In northern, Aboriginal communities, the socio-cultural and political implications of colonialism require that adaptive capacity be assessed holistically. Here we have generated an integrative perspective on adaptive capacity in a small northern community that highlights key endogenous and exogenous determinants, and links them to adaptation strategies. We discuss the implications for emerging institutional and governance processes.

While financial and technical inputs are important for developing adaptive capacity in Fort Resolution, especially to facilitate short-term coping responses, access to resources and benefits must be more equitably distributed. Residents also highlighted the need to address underlying social and institutional structures to ensure that improvements are sustainable over the long term. Community members consistently emphasized the importance of social and cultural ties as either facilitators or barriers to local-level adaptation. They also stressed the need for a holistic approach to education and skill development for northern residents that incorporates both traditional and "western" components, offering a solid foundation for thriving in a complex and changing world.

At the same time, strengthening adaptive capacity necessarily occurs within a broader enabling context shaped by economic transition, specific government programs, relationships between and among Aboriginal and State entities, and Aboriginal self-governance aspirations. Mainstreaming adaptation throughout government programs at territorial and federal levels, therefore, is fundamental to supporting environmental change planning and capacity building at the local scale. Other useful activities to foster adaptation include building social and cultural capital, developing human resources, and improving education and knowledge transfer.

Relationships in the north are evolving. Aboriginal groups are asserting their collective rights, and the benefits of stakeholder participation in economic development and decision-making are increasingly recognized. The implications for adaptive capacity are significant. In this regard, experiences in Fort Resolution and the wider Akaitcho Territory are relevant for other northern communities – in Canada and elsewhere in the circumpolar north – where vulnerabilities are many and the challenges complex.

Acknowledgements We are very grateful to community members in Fort Resolution for sharing their time and knowledge with us. Thanks also to Pam Schaus for producing the study area maps. This research would not have been possible without funding from multiple sources, including the Social Sciences and Humanities Research Council, the Northern Scientific Training Program, the Canadian Polar Commission, the Oceans Management Research Network, Natural Resources Canada's Climate Change Impacts and Adaptation Program, and the Natural Sciences and Engineering Research Council's Northern Research Chair Program.

References

Abele F (2006) Policy research in the north: a discussion paper. Walter and Duncan Gordon Foundation, Toronto
ACIA (2004) Impacts of a warming Arctic: Arctic climate impact assessment. Cambridge University Press, Cambridge
ACIA (2005) Arctic climate impact assessment. Cambridge University Press, Cambridge
Adger WN (2003a) Social aspects of adaptive capacity. In: Smith JB, Klein RJT, Huq S (eds) Climate change, adaptive capacity and development. Imperial College, London, pp 29–49
Adger WN (2003b) Social capital, collective action, and adaptation to climate change. Econ Geogr 79(4):387–404
Adger WN, Agrawala S, Mirza MMQ, Conde C, O'Brien K, Pulhin J, Pulwarty R, Smit B, Takahashi K (2007) Assessment of adaptation practices, options, constraints and capacity. In: Parry ML, Canziani OF, Palutikof JP, Van der Linden PJ, Hanson CE (eds) Climate change 2007: impacts, adaptation and vulnerability. Contribution of working group II to the Fourth Assessment Report of the Intergovernmental Panel on Climate Change. Cambridge University Press, Cambridge, pp 717–743
Adger WN, Arnell NW, Tompkins EL (2005) Successful adaptation to climate change across scales. Glob Environ Change 15:77–86
Adger WN, Vincent K (2005) Uncertainty in adaptive capacity. C R Geosci 337(4):399–410
Agrawal A, Agrawal A (2008) The role of local institutions in adaptation to climate change. IFRI working paper #W08I-3. School of Natural Resources and Environment, University of Michigan, Ann Arbor
Akaitcho Territory Dene First Nations, Government of Canada, and Government of the Northwest Territories (2001) Interim measures agreement (Policy document). Lutsel K'e, Akaitcho Territory
Akaitcho Territory Government (1995) In the spirit and intent of treaty 8: co-existence in Akaitcho Territory (Policy document). Akaitcho Territory Government, Fort Resolution
Akaitcho Territory Government (2006) Akaitcho communities protect and monitor own waters. Tucho spirit. pp 1–4
Armitage DR (2005) Adaptive capacity and community-based natural resource management. Environ Manage 35(6):703–715
Armitage DR, Clark D (2005) Issues, priorities and research directions for oceans management in Canada's North. In: Berkes F, Huebert R, Fast H, Manseau M, Diduck A (eds) Breaking ice: renewable resource and ocean management in the Canadian North. Arctic Institute of North America and University of Calgary Press, Calgary, pp 337–362
Beck T, Nesmith C (2001) Building on poor people's capacities: the case of common property resources in India and West Africa. World Dev 29(1):119–133
Bickford P (2008) Tamerlane test mine over another hurdle. Northern News Services May 19. Retrieved 30 May 2008, from www.nnsl.com/northern-news-services/stories/papers/may19_08tt.html
Boucher P (2006a) Personal interview, April 3. Fort Resolution

Boucher P (2006b) Personal interview, May 7. Fort Resolution
Brooks N (2003) Vulnerability, risk and adaptation: a conceptual framework. Working Paper No. 38. Tyndall Centre for Climate Change Research, Norwich
Campbell D, Spitzer A (2007) High and dry. Up here: explore Canada's far north. Retrieved 11 Feb 2008, from www.uphere.ca/node/141
Cash DW, Adger WN, Berkes F, Garden P, Lebel L, Olsson P, Pritchard L, Young OR (2006) Scale and cross-scale dynamics: governance and information in a multilevel world. Ecol Soc 11(2) 8. [online] URL: www.ecologyandsociety.org/vol11/iss12/art18
Council of the Federation (2007) Climate change: leading practices by provincial and territorial governments in Canada. Retrieved 27 Jan 2008, from www.gnb.ca/cf/index-e.asp
Cutter SL (1995) The forgotten casualties: women, children, and environmental change. Glob Environ Change 5(3):181–194
Dietz T, Ostrom E, Stern PC (2003) The struggle to govern the commons. Science 302 (5652):1907–1912
Downing TE (2003) Lessons from famine early warning and food security for understanding adaptation to climate change: toward a vulnerability/adaptation science? In: Smith JB, Klein RJT, Huq S (eds) Climate change, adaptive capacity and development. Imperial College, London, pp 71–100
Duerden F (2004) Translating climate change impacts at the community level. Arctic 57 (2):204–212
Ford JD, Pearce T, Smit B, Wandel J, Allurut M, Shappa K, Ittusujurat H, Qrunnut K (2007) Reducing vulnerability to climate change in the Arctic: the case of Nunavut, Canada. Arctic 60 (2):150–166
Fort Resolution Métis Council (2007) Letter to Mackenzie Valley Environmental Impact Review Board regarding Tamerlane venture. Retrieved 1 May 2008, from www.mveirb.nt.ca/upload/project_document/1194384269_final%20comments%20from%20FRMC.pdf
Fox SL (2004) When the weather is Uggianaqtuq: Inuit observations of environmental change [Digital media]. National Snow and Ice Data Centre, Boulder
FREWC (2003) Fort Resolution Environmental Working Committee final report 2002–2003. Fort Resolution, NT: Deninu Kue First Nation
FREWC. (2004). Fort Resolution Environmental Working Committee final report 2003–2004. Fort Resolution, NT: Deninu Kue First Nation
Fumoleau R (2004) As long as this land shall last: a history of Treaty 8 and Treaty 11, 1870–1939. University of Calgary Press, Calgary
Giroux D (2006) Personal interview, November 23. Fort Resolution
Government of Canada (2005) Social capital as a public policy tool: Project report. Policy Research Initiative, Ottawa
Government of Canada, Government of the Northwest Territories, and Aboriginal Summit (2004) Northwest territories lands and resources devolution Framework Agreement. Retrieved 1 May 2008, from nwt-tno.inac-ainc.gc.ca/pdf/dv/FWA-March04_e.pdf
Government of the Northwest Territories (2007) NWT greenhouse gas strategy, 2007–2011. Environment and Natural Resources, Yellowknife
Government of the Northwest Territories (2008) Furs, agriculture and fisheries. Retrieved 8 Sept 2008, from www.iti.gov.nt.ca/fursagriculturefisheries/
Graham J, Fortier E (2005). From opportunity to action: a progress report on Canada's renewal of northern research (Report submitted by the Institute On Governance to the Planning Committee for the Dialogue on Northern Research). Natural Sciences and Engineering Research Council of Canada, Ottawa
Gunderson L, Holling CS (eds) (2002) Panarchy: understanding transformations in human and natural systems. Island Press, Washington
Gunderson L, Holling CS, Light S (eds) (1995) Barriers and bridges to the renewal of ecosystems and institutions. Columbia University Press, New York

Handmer J (2003) Adaptive capacity: what does it mean in the context of natural hazards? In: Smith JB, Klein RJT, Huq S (eds) Climate change, adaptive capacity and development. Imperial College, London, pp 51–69

Indian and Northern Affairs Canada (2001) Government of Canada, the Northwest Territories and Akaitcho First Nations sign Interim Measures Agreement. Retrieved 21 Mar 2008, from www.ainc-inac.gc.ca/nr/prs/m-a2001/2-01162_e.html

Indian and Northern Affairs Canada (2003) Devolution: NWT lands and resources devolution. Retrieved 1 May 2008, from dsp-psd.pwgsc.gc.ca/Collection/R2-276-2003E.pdf

IPCC (2007) Climate change 2007: impacts, adaptation and vulnerability. Contribution of Working Group II to the Fourth Assessment Report of the Intergovernmental Panel on Climate Change. In: Parry M, Canziani O, Palutikof J, van der Linden P, Hanson C (eds) Cambridge University Press, Cambridge, UK and New York, NY, pp 976

Irlbacher-Fox S, Mills SJ (2007) Devolution and resource revenue sharing in the Canadian North: achieving fairness across generations. Retrieved 1 May 2008, from www.gordonfn.org/resfiles/Forum_DiscussionPaper.pdf

Kelly PM, Adger WN (2000) Theory and practice in assessing vulnerability to climate change and facilitating adaptation. Clim Change 47(4):325–352

Klein RJT, Smith JB (2003) Enhancing the capacity of developing countries to adapt to climate change: a policy relevant research agenda. In: Smith JB, Klein RJT, Huq S (eds) Climate change, adaptive capacity and development. Imperial College, London, pp 317–334

Kretzmann JP, McKnight JL (1993) Building communities from the inside out: a path toward finding and mobilizing a community's assets. ACTA, Chicago

Krupnik I, Jolly D (eds) (2002) The earth is faster now: indigenous observations of Arctic environmental change. Arctic Research Consortium of the United States, Fairbanks

Mackenzie River Basin Board (2003) Mackenzie River Basin: State of the aquatic ecosystem report 2003. Mackenzie River Basin Board Secretariat, Fort Smith

McGregor D (2004) Coming full circle: indigenous knowledge, environment and our future. Am Indian Q 28(3&4):385–410

Nadasdy P (2003) Hunters and bureaucrats: power, knowledge, and aboriginal-state relations in the Southwest Yukon. UBC Press, Vancouver

Naess LO, Bang G, Eriksen S, Vevatne J (2005) Institutional adaptation to climate change: flood responses at the municipal level in Norway. Glob Environ Change 15:125–138

Newton J, Paci C, Ogden A (2005) Climate change and natural hazards in northern Canada: integrating indigenous perspectives with government policy. Mitig Adapt Strateg Glob Change 10:541–571

Nickels S, Furgal C, Buell M, Moquin H, Nickels S, Furgal C, Buell M, Moquin H (2006) Unikkaaqatigiit – putting the human face on climate change: perspectives from Inuit in Canada. Inuit Tapiriit Kanatami, Nasivvik Centre for Inuit Health and Changing Environments, Ajunnginiq Centre (NAHO), Ottawa

Northwest Territories Bureau of Statistics (2007). Fort resolution – statistical profile. NWT community profiles. Retrieved 22 Nov 2007, from www.stats.gov.nt.ca/Profile/Profile.html

O'Brien K, Leichenko R (2000) Double exposure: assessing the impacts of climate change within the context of economic globalization. Glob Environ Change 10(3):221–232

Olsson P, Folke C, Hahn T (2004) Socio-ecological transformation for ecosystem management: the development of adaptive co-management of a wetland landscape in southern Sweden. Ecol Soc 9(4)2. [online] URL: www.ecologyandsociety.org/vol9/iss4/art2

Parks Canada (2007) Government of Canada takes landmark action to conserve Canada's North. Retrieved 21 Mar 2008, from news.gc.ca/web/view/en/index.jsp?articleid=362739

Pelling M, High C (2005) Understanding adaptation: what can social capital offer assessments of adaptive capacity? Glob Environ Change 15:308–319

RCAP (1996) Report of the Royal Commission on Aboriginal Peoples. Communications Group, Ottawa

Riewe R, Oakes J (eds) (2006) Climate change: linking traditional and scientific knowledge. Aboriginal Issues, Winnipeg

Ritchie D (2006) Personal interview, April 25. Yellowknife
Robards M, Alessa L (2004) Timescapes of community resilience and vulnerability in the circumpolar North. Arctic 57(4):415–427
Rossouw F (2004) Personal interview, August 5. Yellowknife
Sherry E, Nation Vuntut Gwitchin First (1999) The land still speaks: Gwitchin words about life in Dempster country. Aasman Design, Whitehorse
Simon P (2006) Personal interview, May 10. Fort Resolution
Smit B, Pilifosova O (2003) From adaptation to adaptive capacity and vulnerability reduction. In: Smith JB, Klein RJT, Huq S (eds) Climate change, adaptive capacity and development. Imperial College, London, pp 9–28
Smit B, Pilifosova O, Burton I, Challenger B, Huq S, Klein RJT, Yohe G (2001) Adaptation to climate change in the context of sustainable development and equity. In: McCarthy JJ, Canziani OF, Leary NA, Dokken DJ, White KS (eds) Climate change 2001: impacts, adaptation, and vulnerability. Contribution of Working Group II to the Third Assessment Report of the Intergovernmental Panel on Climate Change. Cambridge University Press, Cambridge, pp 877–912
Smit B, Wandel J (2006) Adaptation, adaptive capacity and vulnerability. Glob Environ Change 16:282–292
Smith LT (1999) Decolonizing methodologies. University of Otago Press, Dunedin
SnowChange (2008) Northern indigenous views on climate change and ecology. Retrieved 6 Feb 2008, from www.SnowChange.Org
Statistics Canada (2007) Fort resolution. 2006 community profiles. Retrieved 6 June 2008, from www12.statcan.ca/english/census06/data/profiles/community/Index.cfm
Vincent K (2007) Uncertainty in adaptive capacity and the importance of scale. Glob Environ Change 17:2–24
Walker B, Carpenter S, Anderies J, Abel N, Cumming G, Janssen M, Lebel L, Norberg J, Peterson GD, Pritchard R (2002) Resilience management in social–ecological systems: a working hypothesis for a participatory approach. Conserv Ecol 6(1):14
Walker B, Holling CS, Carpenter S, Kinzig A (2004) Resilience, adaptability and transformability in social–ecological systems. Ecol Soc 9(2)5. [online] URL: www.ecologyandsociety.org/vol9/iss2/art5
Warren FJ, Egginton PA (2008) Background information: concepts, overviews and approaches. In: Lemmen DS, Warren FJ, Lacroix J, Bush E (eds) From impacts to adaptation: Canada in a changing climate 2007. Government of Canada, Ottawa, pp 27–56
Wesche S (2009) Responding to change in a northern aboriginal community (Fort Resolution, NWT, Canada): linking social and ecological perspectives. Unpublished PhD, Wilfrid Laurier University, Waterloo
Willett M, Janes E (2005) Your land, your future: engaging youth in resource development and sustainability issues [CD-Rom Set]. Cranberry Consulting, Yellowknife
Wolfe BB, Armitage DR, Wesche S, Brock BE, Sokal MA, Clogg-Wright KP, Mongeon CL, Adam ME, Hall RI, Edwards TWD (2007) From isotopes to TK interviews: towards interdisciplinary research in Fort Resolution and the Slave River Delta, Northwest Territories. Arctic 60(1):75–87
Workshop Participants (2005) DKFN-GEWEX climate days workshop. 11–12 July. Fort Resolution, NT
Yohe G, Tol RSJ (2002) Indicators for social and economic coping capacity – moving toward a working definition of adaptive capacity. Glob Environ Change 12(1):25–40

Chapter 7
Vulnerability and Adaptive Capacity in Arctic Communities

Robin Sydneysmith, Mark Andrachuk, Barry Smit, and Grete K. Hovelsrud

7.1 Introduction

Arctic communities face challenges related to changing environmental conditions, including climate change, and their interconnections with dynamic global economic, political, and social systems. Fishermen in Finnmark County, Norway have a viable cod industry because of available stocks and their ability to export their products to markets in the south; yet they are susceptible to changing ocean conditions affecting fish stocks and to shifts in markets and institutional arrangements. Subsistence harvesting by Inuit of seals, whales, and other species is highly influenced by changing habitat conditions and by government regulations and international agreements. Efforts to address climate and other changes in the Arctic involve individuals, communities, and governance institutions that, as the examples above allude, are influenced by local, regional, and global forces. Climate change is exacerbating many existing challenges that Arctic communities face related to health and a variety of social issues, wildlife harvesting and animal husbandry, community and transportation infrastructure, and to resource extraction and other competing land uses. Understanding the ways that climate change affects communities and identifying adaptation options requires a broad assessment of the interrelated stresses that communities are facing.

Interest in the ways that people in the Arctic are affected by climate change and how they can adapt has been echoed by researchers, policy makers, and local and indigenous communities (NRI 2002; Government of Nunavut 2003; AHDR 2004;

R. Sydneysmith
Department of Sociology, University of British Columbia, 6303 North West Marine Drive, Vancouver, BC, Canada V6T 1Z1
e-mail: robin.sydneysmith@ubc.ca

M. Andrachuk and B. Smit
Department of Geography, University of Guelph, Guelph, ON, Canada

G.K. Hovelsrud
Center for International Climate and Environmental Research-Oslo, CICERO, Oslo, Norway

Denmark Ministry of Environment 2004; Kofinas et al. 2005; IPY 2005; Watt-Cloutier et al. 2005). Considerable information is now available on climate change and its physical impacts in the Arctic (McCarthy and Martello 2005; ACIA 2005; IPCC 2007; Furgal and Prowse 2008) including evidence from local observations (Berkes and Jolly 2001; Krupnik and Jolly 2002; Ford et al. 2006; Nickels et al. 2006; Tyler et al. 2007; West and Hovelsrud 2008; West and Hovelsrud 2009, forthcoming). Documentation of how Arctic *communities* are affected and how they might adapt is an emerging but still underdeveloped area of research. The project, Community Adaptation and Vulnerability in Arctic Regions (CAVIAR), comprises a consortium of researchers from across the Arctic under the auspices of the International Polar Year 2007–2008 (IPY 2007–2008) designed to address these research and policy needs.

The chapter begins by outlining the CAVIAR project including the key concepts and terms that define the common analytical framework of the project. This framework provides the context and structure for comparing insights into the nature of vulnerability, including the social–ecological processes that shape adaptation and adaptive capacity. In the following sections, empirical material from CAVIAR cases is used to demonstrate commonalities among the types of challenges faced by Arctic communities and the ways that people have been adapting to these challenges. Despite the differences in location and livelihoods, there are key, recurring vulnerabilities that are being experienced by Arctic communities, including the changing conditions to which people are exposed and sensitive, and the types of adaptive responses undertaken. The final section provides some insights into the nature of adaptive capacity in Arctic communities and factors that may constrain or facilitate adaptation to future changes.

7.1.1 The Community Adaptation and Vulnerability in Arctic Regions Project

The CAVIAR project was founded on a conviction that understanding the ways that communities in the Arctic interact with climatic *and* nonclimatic conditions is essential for identifying areas of existing and potential future vulnerabilities. The project seeks to document how communities experience changing conditions, are affected by them, and adapt to them, as a basis for assessing approaches and prospects for adapting to future changes. CAVIAR aims to generate results that are policy relevant at multiple scales of Arctic governance, such that it may contribute to policy and planning for adaptations to existing and future challenges and risks. The international research endeavor includes 26 case studies being carried out in all eight Arctic countries (see Fig. 7.1). Communities range from inland forestry-dependent communities to Sámi reindeer herders, from remote Inuit communities, to larger "gateway" cities. The process by which case study communities were selected is detailed in Smit et al. (2008). The cases stand alone, but by

7 Vulnerability and Adaptive Capacity in Arctic Communities

Fig. 7.1 CAVIAR case study communities

following a common framework also provide the opportunity for comparison and integration of insights across the Arctic. There are three elements of CAVIAR research: (1) a common framework for all case studies that conceptualize vulnerability and adaptive capacity, (2) an orientation towards community engagement and participatory approaches that actively involve local people in the research process, and (3) comparison of case study findings based on the common framework.

The main components of the CAVIAR framework are summarized in Fig. 7.2. CAVIAR research into the vulnerability of Arctic communities, particularly in light of climate change, involves the assessment of existing exposure-sensitivities, existing adaptive strategies, future exposure-sensitivity, and future adaptive capacity (Ford and Smit 2004; Smit and Wandel 2006; Smit et al. 2008). Vulnerability is understood here as the manner and degree to which a community is susceptible to conditions that directly or indirectly affect its well-being or sustainability. Exposure-sensitivities relate to the susceptibility of people or livelihoods to a stimulus or stress

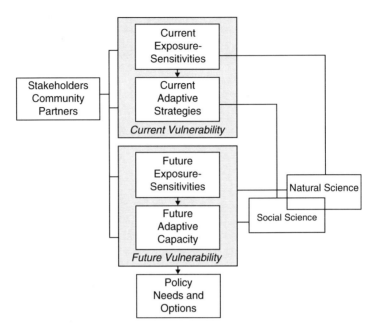

Fig. 7.2 The CAVIAR framework (from Smit et al. 2008)

and are rooted in the community's physical location, social and economic situation, governance, and political systems (Smit and Wandel 2006). There is an explicit understanding within the CAVIAR framework of exposure-sensitivities as dynamic across both temporal and spatial scales. Adaptive strategies are actions, plans, or policies taken by individuals, groups, or governance institutions in response to or anticipation of risks or opportunities (Kasperson and Kasperson 2001; Yohe and Tol 2002; Smit and Pilifosova 2003). Adaptive capacity is a reflection of an individual, group, or community's ability to develop and follow through with adaptive strategies. This conceptualization of adaptive capacity is similar to resilience, in that it attempts to capture a community's ability to adjust to or recover from harm or stress arising from cumulative changes or catastrophic events (Folke et al. 2002; Walker et al. 2002). Both the exposure-sensitivity and adaptive capacity elements of the framework implicitly recognize the central role of institutions, especially governance institutions. Institutional processes influence exposure-sensitivities insofar as they affect the ways in which people live and depend on resources and how people perceive and understand environmental change. Governance institutions are critical to adaptation and adaptive capacity as they provide the necessary frameworks for decision making and action that facilitate or constrain adaptive responses and provide pathways for adaptive capacity to be realized (Agrawal 2008).

Identifying and documenting information under each element of the CAVIAR framework require local input and involvement, as well as information from other sources (Smit et al. 2008; Keskitalo 2004; Pearce et al. 2009a). Community engagement in the case studies ensures that the research is locally relevant and focused on finding solutions to complex problems of governance and adaptation to environmental change (Duerden and Kuhn 1998; Usher 2000; Berkes et al. 2005; Turner et al. 2003a, b). Information about exposure-sensitivities and adaptive strategies is gained from the experiences and insights of people who are living in Arctic communities. It is expected that meaningful engagement of local people in CAVIAR research will also contribute to adaptive capacity via a community's enhanced understanding of environmental change and increased involvement in governance (Pearce et al. 2009a).

The common analytical framework that guides the CAVIAR research process is detailed in Smit et al. (2008). A principal aim of this coordinated approach is to permit comparison of multiple case studies to identify patterns or common features in the dimensions of vulnerability and adaptive capacity. This chapter draws on results emerging from a selection of CAVIAR case studies to explore common features or contexts of community vulnerability and adaptive capacity.

7.2 Vulnerability Contexts

A host of social, economic, and political changes are taking place in the Arctic, many of which are linked to or compounded by climatic and related environmental change. These interacting conditions threaten and/or present opportunities to communities in complex ways. Adaptations are mostly undertaken in response to past or current threats being experienced at the community level and not in anticipation of future threats. Particular adaptive strategies reflect the application of existing resources and capabilities to solve or manage specific problems that threaten individual or community wellbeing. The following sections are framed around four broad "vulnerability contexts" (Table 7.1), which reflect patterns or similarities in types of exposure-sensitivities and which lead to a variety of associated adaptive strategies in communities across the Arctic. Although the four vulnerability contexts are treated separately below, they are seldom exclusive of one another since each community experiences a mix of exposure-sensitivities and employs a suite of adaptive responses. Adaptive strategies are often in response to more than one "type" of exposure-sensitivity (O'Brien et al. 2004), while multiple exposure-sensitivities may be addressed, in whole or in part, by one adaptive strategy.

7.2.1 Local Culture and Society

Many Arctic communities are experiencing significant cultural and social changes, particularly in light of the influences of "southern" culture and resultant social changes which create stresses and prompt adaptations. Indigenous cultures in the Arctic are well known for their ability to absorb stresses and adapt to change

Table 7.1 Vulnerability contexts and their attributes

Vulnerability contexts	Attributes
Local culture and society	Cultural and social changes are ongoing, driven by "southern" influences and historical legacies, which influence the nature of exposure-sensitivities and the form of adaptive responses.
	Media, government, technology, tourists, activists, researchers and others play a role in cultural and social changes.
Subsistence-related livelihoods	In communities where harvesting of resources for local consumption is an important part of households and livelihoods.
	Sensitivity to changes in resource availability because of both physical conditions and institutional arrangements.
	Implications for income, food security, and culture.
	Remoteness and cost of living influence nature of exposure-sensitivities and adaptive capacity.
Market-related enterprises	In communities where natural resources are exploited for external markets.
	Sensitivity to demand and prices in markets, as well as changes in environmental conditions and institutional arrangements.
	Remoteness and transport costs influence imports and exports.
Community infrastructure	Buildings and roads in some communities at risk due to local biophysical conditions.
	Sensitivity to biophysical processes and institutional arrangements that influence access to resources for managing risks.

(Csonka and Schweitzer 2004). Over the last century, the pace and scope of change has been unprecedented, creating conditions which often are difficult to manage as familiar, time-tested livelihood practices are altered or replaced. The influence of central governments and governance, new technologies, tourists, media, activists, and researchers all play a role in driving changes in local culture and society. These changes in turn lead to new types of exposure-sensitivities as people and communities adjust to new types of livelihoods and new social settings often accompanied by new or changing governance regimes (Young et al. 2008; Keskitalo 2008).

For Inuit, as well as other indigenous groups in Arctic North America, changes undergone in the past century mark a profound transformation from the semi-nomadic and largely subsistence ways of life which prevailed not more than a few generations ago to settlement in permanent communities and the absorption of many technical, economic, and social elements of western society. The legacy of external influences, including the roles of governments in changing patterns of occupancy, subsistence, and education, continues with "southern" culture, economic expectations, and social norms available through television and other media and widely adopted, particularly by youth. In many respects, Inuit today straddle two worlds: on the one hand seeking to take advantage of a variety of new social and economic conditions thrust upon them and, on the other hand, attempting to maintain traditions and practices essential to both physical and cultural survival. For example, traditional livelihoods have benefited from the adoption of "new" technology such as firearms and motorized transport. But people and communities are also vulnerable to the potential loss or erosion of indigenous language and traditional ecological knowledge which is still critical to survival. The availability of imported food and supplies from the south reduces the threat of hunger but increases exposure to the negative health effects of substituting traditional country foods for processed southern food (Ford 2010; Furgal and Prowse 2008).

The erosion of traditional livelihood activities, cultural practices, and institutional relationships results not only in spontaneous cultural adaptations to new ways, but also in deliberate adaptive strategies to manage rapid social change. Throughout the Arctic, indigenous groups and communities identify the maintenance of indigenous language and the passing on of traditional ecological knowledge and practices to younger generations as important strategies for countering the distractions and sometimes negative influence of the infiltration of global society. In Russia and Finland there are sharply contrasting perspectives among reindeer herders on the nature of the interaction of indigenous and nonindigenous traditional ecological knowledge. In Russia, a common perspective is that cultural mixing between indigenous pastoralists and ethnic Russians is diluting traditional ecological knowledge about managing and maintaining healthy reindeer herds and as a result may be limiting the range of adaptation options considered by herders. In Finland, on the other hand, Sámi and non-Sámi herders share concerns about the challenges of reindeer husbandry and have found ways to fuse traditional and new knowledge to better respond to challenging environmental and economic conditions. Elders in Fort Resolution, Northwest Territories, and the hamlet of Kugluktuk, Nunavut in Canada report declining youth participation and interest in land-based activities and increased participation in the wage economy outside of the area (e.g., mining employment). From a household economy point of view, particularly among those more accepting of southern values, paid employment is an attractive option to some of the relative hardships of subsistence activities. However, unemployment remains high and has been linked to an increase in a variety of social problems. (Prno et al. 2009, forthcoming). Adaptive strategies include programs to encourage youth to stay in school to enhance their capacity to cope with rapidly changing social and economic conditions. Simultaneously, community and territorial leaders seek to maintain traditional knowledge, skills, and language. Against this backdrop of social and cultural changes, farther reaching institutional changes are emerging through on-going self-government negotiations and outcomes of land claim processes.

It is not only indigenous communities that are vulnerable to the transformation of local culture and society. The Norwegian community of Kjøllefjord is struggling to maintain its identity as a traditional fishing village built around the centuries old Norwegian cod fishery. The local fishing fleet in Kjøllefjord has been declining for several decades, although it has recently stabilized and currently employs around 30% of the municipal workforce (West and Hovelsrud 2008). Environmental and economic conditions are affecting the profitability of the cod fishery but the community is also exposed and sensitive to other changing conditions. For example, the fishing workforce is challenged as older fishers approach retirement and younger members of the community choose to pursue education and other opportunities in urban centers to the south rather than take up a career in commercial fishing. Such demographic and labor-force shifts, which are in themselves connected to changing social and cultural values, are challenging the social and economic viability of the community. At the same time, these changes open the door for new or modified livelihood opportunities and strategies to adjust and cope with change, such as efforts to connect tourism with fisheries (see Sect. 3.2).

7.2.2 Subsistence-Related Livelihoods

Renewable resources are of critical importance to most Arctic communities, either as part of subsistence or as commodities produced for external markets and in some cases for both. Many communities which depend on the use of resources as part of household livelihoods and food systems are sensitive to changes in resource availability. Resource availability is vulnerable to changes in biophysical and environmental conditions as well as to new or changing institutional arrangements, and social conditions. Subsistence activities are also sensitive to the high cost of living and related aspects of remoteness in Arctic communities which affects both the cost of supplies and access to markets (e.g., furs). The vulnerability of subsistence related livelihoods to these and other interacting conditions has implications for household livelihoods in terms of both income and food security. There are also broader implications for cultural importance of hunting and fishing and other subsistence oriented livelihood activities (Table 7.1).

For example, in most Inuit communities, the harvest of wildlife – fish, whales, caribou, waterfowl – is a key component of household food supply important for healthy diets, community well being, and culture. The practice and role of wildlife harvests are being affected by changing environmental conditions, institutional arrangements, and social and cultural expectations and norms. These different exposure-sensitivities give rise to specific but interdependent adaptive strategies. Changing climatic conditions for example, have triggered changes in the location, timing, and abundance of some species and have affected access to wildlife through changes in ice, snow, and permafrost conditions. People in indigenous communities in North America are adapting to changes in migratory patterns and abundance by shifting the timing and location of hunts and/or substituting one species for another (e.g., Hovelsrud et al. 2008). Where environmental conditions make hunting grounds difficult or dangerous to access, routes may be altered or abandoned entirely. In Igloolik, for example outboard powered boats are replacing snowmobiles as the primary mode of transport to key harvesting areas as ice conditions become more dangerous and less predictable (Ford et al. 2006).

Institutional arrangements that govern or otherwise restrict and control the harvest of wildlife in the Canadian Arctic are on-going and in a state of flux presenting another set of changing conditions to which subsistence livelihoods must adapt. Local management systems (e.g., hunters and trappers associations) are increasingly linked into larger, multi-stakeholder comanagement arrangements with higher levels of government and (in some cases) international interests. Devolution of federal powers to territorial governments, land claims settlements, and aboriginal self-governance processes also exemplify institutional and governance changes that directly or indirectly affect subsistence livelihoods. Increasingly, indigenous groups have roles in decisions related to the quantity of wildlife that can be harvested, who has access to harvesting grounds, when particular species can be harvested, as well as monitoring of wildlife and programs for supporting harvesters. In the Inuvialuit Settlement Region, for example, community-based Hunter and Trapper Committees (HTCs) work with territorial and federal government agencies to develop recommendations for annual harvesting quotas. The government agencies hold decision-making authority on quotas, but HTC's dispense harvesting tags to individuals in their communities. Comanagement regimes influence exposure-sensitivities insofar as they work across larger temporal and spatial scales than decisions made by individual harvesters. Consequently, they may facilitate and/or constrain options for communities and household subsistence but provide a regionally based governance strategy for controlling and managing stocks (e.g., limits or controls on harvests, protection of calving grounds) which helps to ensure long term viability of subsistence resources.

Societal and cultural changes in Arctic North America have created a variety of conditions which increase the exposure-sensitivities for subsistence related livelihood activities. For the most part, these are not recent changes but the ongoing legacy of European contact and the continued infiltration of "southern" or outside cultural, social, economic, and technological influences. In general the modernization of Inuit life has increased the financial cost of hunting through increased dependence on imported supplies and equipment which in turn depends on employment income. These conditions are of course not new and interact with other elements of Inuit life in complex ways. For example, some communities report declining wildlife resources in close proximity to permanent settlements requiring increased travel to hunting grounds, which increases fuel costs at the same time that it increases exposure to environmental risks some of which are increasing because of climate change (Prno et al. 2009, forthcoming). Alaska provides insights into the cultural practices of subsistence hunting, fishing, and food sharing as a means of addressing vulnerabilities and adapting to changing conditions similar to those being experienced in the Canadian Arctic (Kofinas et al 2005). Communities in rural Alaska (as well as other areas of the Arctic) manage the burden of increased hunting costs and potential risks by depending on highly active and successful harvesters. Instead of all community members attempting to meet household needs individually, communities cooperate with other harvesters for greater efficiency. Community members provide in-kind contributions of fuel, equipment use, or other forms of exchange to so called "super-hunters" (Kofinas et al 2005). In return for the support they receive, these high producing harvesters, attempt to harvest a surplus of game and/or fish which can be shared through community and intercommunity food sharing networks.

In Russia, Norway, Sweden, and Finland, exposure-sensitivities of livelihoods and food security are largely related to the pastoral practices of indigenous Sami reindeer herders. Reindeer husbandry has elements of both subsistence livelihood activity and market related enterprise (Sect. 2.3). Methods are highly adaptive to annual climate variability and weather, but there is evidence that changing environmental conditions are challenging these communities. In recent years herders have increasingly faced higher costs and loss of livestock due to warmer winters, more frequent freeze-thaw cycles, and heavier, wetter snowfalls that make foraging more difficult and expose herds to higher risks of disease. Adaptive strategies draw upon traditional knowledge and techniques to manage these challenging conditions, such as moving the herd to different foraging grounds. One contemporary response to adverse foraging conditions in Sweden is to bring supplemental feed in by truck, but this option is expensive and not available to all herders (Keskitalo 2008). Market conditions also influence resource management decisions which may enhance or constrain adaptive strategies undertaken in response to environmental conditions (Tyler et al. 2007).

Subsistence related activities, especially in North America, are still an important dimension of community life and contribute, often substantially, to household livelihoods. The infiltration of southern influences, opportunities, and institutions interacts with subsistence practices in complex ways, enhancing some activities but also adding new burdens and costs. Many subsistence activities are vulnerable to changing environmental conditions including climate change, but subsistence is also an important outlet for adaptations to other stressors or changes affecting life in Arctic communities.

7.2.3 Market-Related Enterprises

For reasons of history and geography, natural resource use and local economies are generally more oriented to market related enterprise in the Arctic regions of Scandinavia and northwestern Russia than in North America. Markets and population centers are more proximate, transportation infrastructure more accessible, and trade linkages well established (e.g., Norway: fishing, oil and gas; Finland, Sweden: forestry, reindeer husbandry). Communities along the remote northern coast of Norway, for example, have strong market ties to southern centers and the broader European Union. Some trading relationships such as the stock fish (dried cod) trade between the Lofoten Islands and Italy are centuries old.

The municipality of Norrbotten in northern Sweden provides an example of a relatively remote, culturally diverse region which exemplifies the interplay of exposure-sensitivities arising from social and economic forces (at multiple scales) and changing environmental conditions related to climate change. Examples can be taken for instance from Gällivare municipality in which natural resources such as forestry, reindeer husbandry, and mining have been traditionally important. These sectors are susceptible to pressures of globalization (e.g., increasing foreign competition, volatile markets, and changing technology) on the local economy and society in much the same way as other Arctic communities in the study. First, the region is an economic periphery; hence beyond employment, most of the economic benefits of primary industries (forestry, mining) are exported along with the resource (Marchak 1995). Second, the region is susceptible to social demographic trends that see increasing numbers of young people leaving the region for southern urban centers (Persson and Ceccato 2001). Governance changes that took place mostly in the 1990s were intended to increase local autonomy and decision making especially with respect to local economic development by placing more responsibility in the hands of local and regional authorities; nevertheless, up to the end of the century the region was largely a recipient of state support and the public sector remained the largest single employer (Persson and Ceccato 2001). This broad context of exposure-sensitivities is being aggravated by climate change which is having its most notable effect on the forestry and reindeer husbandry sectors. Forestry is susceptible to disruption of harvesting schedules and transportation resulting from the effects of longer, warmer autumns and increasing winter thaws which translate directly into increased costs. Reindeer husbandry is also exposed to the impacts of climate change on transportation although the sensitivity is different due mostly to scale. Reindeer husbandry is particularly susceptible to conditions which (a) affect animal health (e.g., limits to winter grazing caused by winter thaw-freeze events and heavier snowfalls, reduced water availability from natural springs in summer, and increased insect pests which stress animals), and (b) cause disruption and increased difficulty during key migration times in spring and autumn when animals need to be moved between summer and winter feeding grounds (Keskitalo 2008). Current adaptive strategies such as importing forage supplements are costly and not available to all herders. Other adaptive strategies are more political in nature, involving Sami indigenous rights and resource governance issues including state interventions. Forestry strategies to contend with and adapt to long term changes in growing season – a potential opportunity to realize higher productivity – rest more with the decision making and planning capacity of industry and government rather than with local communities. As such governance institutions and market conditions are and will continue to play a key role in determining adaptive strategies and future adaptive capacity of forestry in these regions (Keskitalo 2008).

Enterprises built around tourism play an increasingly important role in many communities. Tourism helps counter declining contributions from the fishing sector in northern Norway or to supplement household income in Inuit communities in Arctic Canada and among Sami reindeer herders in Sweden. Tourism enterprises are vulnerable to a host of conditions ranging from external market forces and governance related constraints to the environmental impacts of climate change. Trophy hunting of the polar bear in northern Canada is an example of the often complex linkages between traditional, subsistence based activities, market related enterprises, and external cultural values and governance regimes. Currently, polar bears provide a lucrative source of income for Inuit guide outfitters catering to the U.S. and European markets. But this particular enterprise, itself an adaptation to past changes to Inuit life, is especially susceptible to foreign and international governance regimes. The recent listing of the polar bear as 'threatened' under the U.S. Endangered Species Act prevents hunters from bringing any polar bear parts into the U.S. (Platt 2009) which will substantially reduce income for certain Inuit communities. Should the polar bear be subsequently listed by CITES (the Convention on International Trade in Endangered Species of Wild Fauna and Flora) it would become illegal to transport polar bear parts to any of the convention's 175 signatory nations, all but eliminating this particular source of income for certain Inuit communities in the Canadian Arctic.

Institutional changes led by international governance regimes and changing social values have had a different effect in Húsavík. This community in the north east of Iceland has long depended on commercial whaling and fishing as the foundation of local culture and economy. Whaling in Iceland is subject to changes in international governance regimes embodied in the International Whaling Commission. The exposure-sensitivity of the community to global scale governance institutions is set against a backdrop of other internal and external stressors including changes that have taken place in the Icelandic fisheries management system involving privatization through Individual Transferable Quotas (ITQ) of formerly common property resources. Fishing has seen erratic growth in production in recent decades at the same time that it has declined as a proportion of GDP and in importance to community livelihoods (Eythórsson 2000; Iceland Ministry of Fisheries and Agriculture 2009). A global moratorium in 1985 brought an end to commercial whaling and changes in social values and pressure from domestic and international environmental organizations superseded the end of whaling for consumptive use in Iceland by 1990 (Donovan 1989; Einarsson 2009). The community of Húsavík, however, has responded creatively over the past two decades and developed a successful whale watching sector which has subsequently formed the basis of a bourgeoning local tourism industry. The biophysical resource remains but the enterprises which whales support have changed. The adaptive strategies undertaken in Húsavík, particularly with respect to the whale watching industry, and also through adaptation in the fishery (e.g., changing fleet structure and conversions of former fishing and whale hunting boats into whale watching vessels) to respond to multiple pressures have been relatively effective (Einarsson 2009). The financial crisis that hit Iceland in 2008 has put the adaptive capacity of the community and nation to an unprecedented test and, in the short term at least, any concerns about the potential impacts of climate change may pale by comparison.

While connections to external markets and cultures engender exposure-sensitivities for some communities, these connections also provide a means of acquiring information and technology useful to the formulation of adaptive strategies. Norwegian fisheries-based communities, such as Kjøllefjord in Finnmark County and villages in the Lofoten Islands, have long histories of managing climatic variation and its ongoing impacts on local resources (particularly the influence of the Gulf Stream on cod and other marine species). Over the last several decades, stocks have fluctuated or declined as fish move to new spawning and feeding grounds in response to fishing pressure and to changes and variation in ocean conditions. Governance of fisheries has likewise changed in an effort to manage multiple stresses on the fishery. Fishermen have adopted new types of fishing vessels, navigation equipment, and fishing equipment. Efficiency measures in processing and shipping fish are ongoing to help compete in the global market place. Flexibility remains a key attribute of market related enterprises in order to navigate multiple, ongoing changes in market dynamics, resource availability, internal and external governance regimes, and environmental change. Flexibility to react quickly to changing conditions, to make use of multiple types of knowledge and technology, flexibility to pursue new opportunities or new markets, and flexibility to think about and plan for uncertainty are increasingly important hallmarks of current adaptive strategies and future adaptive capacity (West and Hovelsrud 2009, forthcoming).

7.2.4 Community Infrastructure

Across the Arctic there is a broad range of exposure-sensitivities and adaptations related to infrastructure and the built environment. Types of exposure-sensitivities are influenced by landscape features the physical processes at work, and the types of infrastructure involved. Communities in northern Norway, Sweden, and Finland are generally well connected via transportation (air, road, rail, or ferry), energy, and communications networks, disruption of which creates vulnerabilities for community livelihoods and local enterprises that are dependent on the flow of supplies into the community and to the export of local goods produced for external markets. In many communities in Greenland and Arctic Canada, livelihoods are vulnerable to interruptions of the flow of goods, services, and clientele which may only be able to reach communities via infrequent barge service, ground transport that may only be available on winter ice roads or other seasonal routes, or by expensive air transport. Generally, the more remote a community is the more important subsistence activities are to household and community adaptive strategies.

In addition to the exposure and sensitivity of transportation or 'external' infrastructure linking them to the south, whether in Scandinavia, Greenland, or Canada, communities are also subject to the impacts of various biophysical processes and landscape hazards on 'local' infrastructure. Many of these processes are being compounded or accelerated by climatic changes taking place in the Arctic.

The community of Tuktoyaktuk is literally perched on the north coast of Canada. Residential and community buildings have been erected on a narrow peninsula that is highly exposed to the weather, waves, and ice of the Arctic Ocean. The shoreline is especially susceptible to erosion due to the prevalence of gravel and ice in the peninsula upon which the community is built. Adaptive strategies have included some shore protection and relocation of buildings that were at immediate risk. There are indications that the combined effects of reduced sea ice cover and increased storm activity brought about by climate change will increase rates of erosion (Johnson et al. 2003; Manson and Solomon 2007). There are concerns that the community may eventually have to be moved but the urgency of when this might occur is influenced more by external political and economic factors than the immediate physical threat. At present, the municipal government is focused on reinforcing existing shoreline protection measures. Whereas physical adaptations take place within the community, the capacity to follow through with many adaptive strategies is dependent on political will and funding that must be accessed from territorial and federal governments. The case of Tuktoyaktuk illustrates that adaptive strategies are sometimes more a function of institutional processes and governance than a direct response to local or physical need.

In the case of Clyde River, Nunavut, community safety, local roads, building foundations, and water lines are at risk from a variety of landscape hazards associated with hydrology, melting permafrost, and slope stability. Adaptive strategies such as upgrades to waste water management, thermo-siphons, and adjustable building foundations (e.g., engineered pilings) are in use but not universal. Hazard mapping (Fig. 7.3) has identified zones of low, medium, and high risk areas for infrastructure in the community. Increased knowledge and understanding of local landscape conditions emerging from the combined efforts of community members and researchers is facilitating efforts to manage landscape risks as they unfold and will enhance the capacity of the community to plan and prepare for increased hazards in the future.

The ability of communities to take action based on scientific knowledge and monitoring of environmental change or based on direct experience with landscape hazards and infrastructure damage is constrained by the lack of resources to carry out remedial or adaptive actions. Repair, upgrade, or replacement of infrastructure typically involves the infusion of external resources and is dependent on applicable governance regimes and the associated political or institutional will to undertake specific actions and adaptations. Risks to infrastructure associated with permafrost degradation and erosion are predictable risks to significant investments. As with the case of Tuktoyaktuk, whether or not strategic, proactive adaptation takes place lies not with the physical or technical need but within the decision making processes and resource allocations of often remote governance institutions.

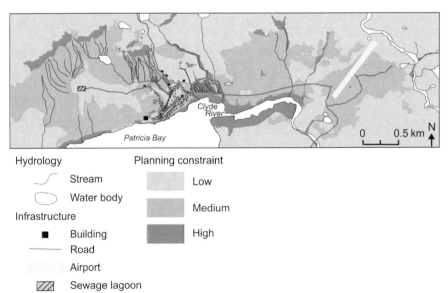

Fig. 7.3 Composite landscape constraint map of Clyde River (Trevor Bell 2009)

7.3 Adaptive Capacity

Adaptive capacity embodies the ability of a system to cope, recover, or adapt to any hazardous conditions (Smit and Pilifosova 2003; Smit and Wandel 2006) without losing options for the future (Folke et al. 2002). Adaptive capacity also relates to the ability of a system to expand its coping range in the process of responding to hazardous conditions (Adger 2006). Social factors which contribute to the adaptive capacity of a social–ecological system include society's access to resources as well as the ability of members of a community to come together and act collectively when faced with threats (Adger et al. 2004). In the context of community, adaptive capacity refers to the totality of local resources and capabilities plus external linkages and networks that may be used or accessed to respond to or cope with change or stress. Definitions of the variables or dimensions of adaptive capacity vary but include financial capital, equipment, technology, and communication networks, human capital including appropriate skills and knowledge, social resources, including social capital and related networks and relationships, and institutional capacity (see Matthews and Sydneysmith this volume) such as governance processes and structures that support decision making. Institutions, especially governance institutions, in effect make adaptive capacity real (Adger 2003) by providing both a context and a process through which adaptations can take place. Institutions shape how risks are perceived and responded to at both individual and collective levels, they play a central role in how local resources may be activated and mediate potential external interventions (Agrawal 2008).

Adaptive capacity is context (some would say "hazard") specific and varies often considerably across time and space (Brooks 2003; Yohe et al. 2003). It is best understood in terms of the various features and conditions through which it is shaped (Smit and Wandel 2006) including not only how it is constituted but also the processes that translate into adaptation (Brooks 2003). The factors determining such processes depend on the nature of the "systems that are adapting" including, for example, questions of scale (i.e., households, communities vs. nation states) (Brooks 2003; West and Hovelsrud 2008). The interaction of different systems across different scales is also an important condition of adaptive capacity. So while the primary focus may be on local conditions that affect adaptive capacity, consideration of broader social, political, and economic forces must also form part of the analysis (Brooks 2003; Vincent 2007).

7.3.1 Local Capacity in a Global Context

Adaptive strategies are examples of adaptation in practice exemplified primarily by local responses, plans, needs, or (re)actions implemented at the community scale. Although adaptation occurs locally, the capacity to adapt is the product of conditions, resources, and other variables that flow from multiple scales. The CAVIAR case studies illustrate the multiscale nature of adaptive capacity and have identified the interplay between the local and the global as an important dynamic in the adaptive capacity of Arctic communities. Local leadership and knowledge, social networks, economic resources (e.g., human, financial, equipment), and local institutions are components of capacity which operate at a local scale. But adaptive capacity is also influenced by forces and events such as policy decisions and market fluctuations that originate or take place thousands of miles away. Adaptive capacity in Arctic communities may be enhanced by new knowledge, ideas and resources, or technology from southern sources, but there may also be negative effects. This section highlights several dimensions of community adaptive capacity in the context of global forces and influence, including the changing nature of traditional ecological knowledge, the limitations of local capacity, and the importance of linkages to external resources and higher order governance institutions.

The socio-cultural transition that has taken place with indigenous communities throughout the circumpolar north has altered the character and context of their adaptive capacity. Cultures are clearly not static but in the Arctic, as with indigenous peoples in many parts of the world, there is a relatively sharp divide between the nomadic hunters characteristic of Inuit society prior to contact with Europeans and contemporary Inuit communities. Notwithstanding the infiltration of institutions and practices from the south, traditional ecological knowledge (TEK) remains prominent in discussions of livelihoods and the survival of indigenous peoples in North America and Europe. In the context of managing and adapting to the rapid pace of climate change in the Arctic, TEK is an important element of local adaptive capacity. Traditional ecological knowledge provides a framework through which peoples with a long history in a place interpret, manage, and make use of the biophysical world around them. In the past, TEK was the foundation of household and community capacity to cope with environmental variation and manage the natural resources upon which their livelihoods depend. One concern about the influence of western science and other outside social and economic forces is that TEK is gradually displaced or at the very least, altered and its value to ensuring people's survival diminished. From this perspective, the erosion of TEK is a loss of adaptive capacity. From another point of view, the transformation of TEK through its interaction with "outside" forces is a *source* of future adaptive capacity.

Regardless of whether the future of TEK is cast in pessimistic or optimistic terms, two points emerge from the CAVIAR case studies. First, TEK is an important feature of northern identities and is widely viewed in Arctic communities as a defining characteristic of being an indigenous person (e.g., Notzke 1994; Cruikshank 2005). The social and cultural benefits that result from the maintenance, or in some cases the rebuilding, of TEK enhance adaptive capacity in both direct and indirect ways. For example, TEK is essential to being able to obtain food from the land and sea which has direct benefits for household livelihoods and community wellbeing. Less obvious, TEK provides a practical link to cultural values, traditions, and meanings that act to counter negative social and economic conditions such as unemployment/idleness and substance abuse. Many communities identify the maintenance of Inuit, Cree, Sámi, and other indigenous languages as a key element of helping to ensure the survival of traditional ecological knowledge and its continued contribution to the capacity of indigenous communities in the Arctic. Language anchors culture, allowing it to shift or swing with changing conditions while maintaining vital connections with history and place.

Second, given that TEK is, by definition, experiential (Notzke 1994), new forms of knowledge and understanding are continually absorbed into the lexicon of local knowledge and capacity, whether introduced by outsiders in the form of western science and technology or brought about through exposure to new patterns of climate and environmental conditions. Traditional ecological knowledge is thus an evolving understanding of change, defined in terms of how it incorporates and *not* how it excludes external or "nontraditional" sources of "knowledge". Guns and snowmobiles, Geographic Information Systems (GIS), and radio equipment are examples of new technology and knowledge that have been readily adopted into traditional livelihoods. Such technology adds new dimensions and opportunities to the pursuit of livelihoods altering the application and thereby the nature of TEK in its contemporary setting. The evolution of contemporary TEK is not limited to the embrace of modern technology and gadgets. The inclusion of TEK in governance regimes, such as the various comanagement agreements which have been developed in northern Canada, institutionalizes the linkage between TEK and western science based models of environmental management.

The incorporation of technology, resources, and services from "external sources" into local indigenous community life enhances adaptive capacity to the extent that it reduces certain risks. The increasingly widespread use of GPS by Inuit in Canada is a recent example of technology that reduces the risk of being out on the land. Originally adopted for convenience and safety, GPS is increasingly important in the context of climate change as weather patterns, ice formation, and other "natural" signals traditionally used for navigation become less predictable. Similarly, the increased availability of foods from the south has altered food security and enhanced the capacity of communities to avoid the threat of starvation or extreme hunger when wildlife is scarce. On the other hand, of course, there are well documented negative health effects associated with increasing dependence on highly processed foods (Furgal 2008). The relationship between adaptive capacity and traditional ecological knowledge is complex although experiences from the CAVIAR project indicate that capacity is generally enhanced by drawing on both traditional and other sources of knowledge.

While arctic communities are adaptive in many respects, they face considerable forces of stress and change. The combined effects of climate change and general social and economic transformations are accelerated by the forces of globalization and stretch community resources and capacity. Adapting to or managing environmental change, specifically climate change, is not necessarily a top priority for local leaders often already overcommitted to other concerns. The stock of people and resources available are typically absorbed in managing day to day community or municipality issues such as care of the elderly, education of youth, employment and economic opportunities, and maintenance of infrastructure. There is a scarcity of capacity to take deliberate or planned action, especially in smaller communities.

But there is considerable knowledge and concern about environmental change. Enhancing adaptive capacity thus hinges on the articulation of local resources with external sources of support which largely arrive through the mechanisms and institutions of regional and national governance. For example, the town of Ivalo in the Inari region of Finland has strengthened its capacity to predict, manage, and respond to flooding of the Ivalo River as a result of closer ties between local officials and higher level agencies. These ties have been strengthened in the wake of a major flood in 2005. In the past institutional blockages and cultural differences between "more traditional" local leaders and the "technocratic bureaucracy" in Helsinki represented a major limitation on coordinated emergency response in the past. Removal of such barriers and improved cooperation and communication between the different levels of government reinforces the point that the linkages between local institutions (including local leadership) and external governance regimes are an important dimension of adaptive capacity.

7.3.2 Flexibility and Diversity

Flexibility and diversity of institutions, livelihood practices, economic activities, and other social processes are important, linked dimensions of adaptive capacity in Arctic communities. Arctic residents often cite their experiences with harsh environmental conditions and remoteness from major governance and economic centers as evidence of their self-reliance and ability to cope with difficult conditions and change. The knowledge and skills accumulated and passed down between generations include understanding of the need to be flexible in response to adverse conditions and unforeseen hazards or opportunities. Survival may depend on the capacity to shift activities, take a different route, adopt a new technology or revert to a technique or process used in the past. Diversity is also important in the sense of having a diversity of options, a choice in the course of action or response to stress and, subsequently, the ability to be flexible. The factors that enable and/or constrain flexibility and diversity, whether within a particular livelihood or among various, perhaps competing livelihood strategies have yet to be clearly defined, although interesting examples have emerged from some CAVIAR case studies that illustrate these important dimensions of adaptive capacity.

Inuit hunters in North America for example, readily substitute one species of game for another or shift entirely from reliance on terrestrial game to fishing or marine mammals to ensure an adequate supply of food for their households and community. They have both flexibility in terms of the skills, knowledge, and willingness to change their hunting practices, and they have options as to what, where, and when to hunt or fish. The need to make these choices, to alter and adapt hunting practices or travel routes, is increasing as climate change impacts (a) availability of game, for example, through changes in the patterns/timing of migration and (b) access to hunting grounds, for example, through changes in snow and ice conditions and the declining predictability of weather. Many such decisions and adaptations are predicated on traditional ecological knowledge, however, as Arctic communities begin (or in some instances continue) to respond and adapt to the changing climate, TEK may be insufficient to help guide effective adaptations unless TEK itself adapts (Wenzel 1999). Will such knowledge be able to absorb and make sense of new conditions such as new wildlife migration patterns, less predictable ice and weather conditions, altered fishing grounds, or more challenging extreme events? Or will cultural models of weather and climate, (Kempton et al. 1995) especially of variability and cycles, trap communities in assumptions based on past experience and the belief that things will return to "normal" at some point in the future. Across the Atlantic, Sámi reindeer herders in Finland offer a slightly different perspective on the theme of flexibility, TEK, and adaptive capacity. The Sámi identify shifts in how "old" and "new" knowledge are used together and how the application of "old" knowledge has changed to incorporate the "new".

In contemporary times, introduction of new technologies and wage employment have added to the suite of choices and opportunities for community members to generate a livelihood from multiple sources. But it is unclear as to whether or not this sort of change represents an increase in the flexibility of livelihoods and, subsequently, an increase in adaptive capacity. On the one hand, the opportunity for paid work reduces dependence on subsistence activities and its associated risks. On the other hand, as youth spend their time in schools and adults engage in paid employment, they have less time to be out on the land (or ice) and fewer opportunities to make use of the diversity of resources that were available for their ancestors.

Along the northern coast of Norway, flexibility and diversity take on a different meaning in the context of adaptive capacity as compared to Inuit communities in Canada. The linkages between subsistence activities and livelihoods are less prevalent if not absent altogether. In the Norwegian context, household income and local economies have for centuries been built around the cod fishery, and local knowledge and identity are inextricably linked to such activities. In recent decades, however, the fishery has been in decline along much of the coast as a result of over fishing, international competition and, most recently in relation to climate change. Fishermen from the Lofoten Islands and in communities such as Kjøllefjord along the Finnmark coast report that flexibility is a critical component of being able to make a living from the sea. Flexibility includes the ability to respond to changes in resource availability, to invest in new technology, and to adjust fishing practices or activities to comply with changing regulations. Fishermen report that in the past the versatility and flexibility of their industry enabled them to adapt to variations in climate and cyclical shifts in the resource-base. Today environmental variability remains but the challenges fishing communities face are compounded by factors which constrain or limit traditional forms of flexibility. Increasingly, the diversity of livelihoods and/or local economies is a critical dimension of community adaptive capacity. Again, the town of Kjøllefjord is illustrative. Here local leaders have championed diversification by embracing the construction of a wind farm on the hillside above the town; they have pursued investment in the renewal of the agricultural sector, and supported the development of coastal cultural, and tourism initiatives connected to traditional fisheries. These efforts to diversify the local economy are being pursued to fill the void created by outmigration and declining employment in land-based fish processing, and to create opportunities in other arenas that attract people to stay or even to migrate to the community (West and Hovelsrud 2009, forthcoming). It is within the human and social resources of people that many of the elements of adaptive capacity lie. The capacity of communities to adapt to changing conditions will be maintained or even enhanced to the extent that they are able to process the implications of change and maintain flexible livelihoods built around diverse activities and sources of sustenance and income.

7.4 Conclusions

This chapter has drawn upon case studies in the Arctic as part of the international IPY 2007–2008 CAVIAR project. The CAVIAR project offers a framework to conceptualize and investigate local conditions that contribute to defining vulnerability and adaptive capacity from community and scientific points of view. The CAVIAR framework provides a common language and set of concepts that are being deployed in the field by researchers in eight countries to facilitate efforts to compare and synthesize experience from a diverse set of social and ecological circumstances and disciplinary approaches. Project investigators are involved to varying degrees in participatory approaches and community engagement which aims to link research to community issues. The preceding discussion is thus a snap shot of ongoing processes of shared learning and research, some of which, in addition to providing new knowledge of the processes of vulnerability and adaptation, might enhance the adaptive capacity of the communities with which we work. The chapter has focused on a discussion of key emerging findings with respect to community vulnerability and adaptive capacity in Arctic regions.

Arctic communities are vulnerable to changing conditions in several ways. Local culture and social processes in Arctic communities have experienced significant changes because of the influence of social, political, and economic forces from the south. These pressures continue and in some respects are increasing in the context of globalization and climate change. In places where subsistence activities continue to be an important part of household livelihoods, communities are vulnerable to environmental conditions and institutional arrangements which may limit or affect the reliability and/or their access to resources vital for food security. Communities dependent on market-related enterprises through the exploitation of natural resources for external markets are sensitive to fluctuations in the resource availability, price, demand, and competition from other sources. High transportation costs related to remoteness increases their exposure to negative market pressure. Arctic communities have limited but critical local infrastructure upon which they depend, some of which is especially vulnerable to biophysical processes and increasingly influenced by climate change. The vulnerability of these communities is often exacerbated by institutional arrangements that limit access to resources or otherwise constrain options and opportunities for adaptation.

The ability of communities to deal with these ongoing changes is related to a wide range of factors some of which emerge from local conditions, knowledge, and capacity, while others stem from the influence of institutional arrangements and governance regimes operating at local, regional, national, and global scales. Some of these factors enhance adaptive capacity in complex and interactive ways such as the role of traditional ecological knowledge in conjunction with western science and technology. Similarly, the flexibility and diversity of Arctic livelihoods and local economies is a key dimension of adaptive capacity in communities throughout the circumpolar region. As climate change and economic development increase pressures on key Arctic natural resources, local capacity on its own will likely be overwhelmed. Governance and the institutions through which it is delivered will be critical to successful adaption and survival of Arctic communities. Adaptive capacity will likely be high where local institutions are strong, have broad community support and good linkages to external governance institutions with an interest in and commitment to Arctic community well-being.

Acknowledgements The CAVIAR project is an international collaboration and would not be possible without the willingness of researchers involved in the project to share their research, experience, and insight from working in the wide range of Arctic communities engaged in the project; this includes not only principal investigators but also the many graduate students and research assistants so often the unsung heroes of fieldwork and, perhaps most especially, the many residents and members of the communities themselves who give frequently and repeatedly of their time as research partners, guides, informants and hosts, and especially the without whom a project of this breadth would not be possible. Individual case study researchers have contributed substantively to the empirical data, ideas, and analysis presented in this chapter. In particular, we thank and acknowledge the invaluable contributions of (in alphabetical order) Elena Alexandrova, Helene Amundsen, Derek Armitage, Trevor Bell, Ryan Brown, Tatiana Bulgakova, Shauna BurnSilver, Halvor Dannevig, Frank Duerden, Níels Einarsson, James Ford, Carina Keskitalo, Gary Kofinas, Ralph Matthews, Tristan Pearce, Jason Prno, Stine Rybråten, Anna Stammler-Gossmann, Monica Tennberg, Terhi Vuojala-Magga, Sonia Wesche, and Jennifer West. Any errors or omissions are, of course, the responsibility of the authors.

References

ACIA (2005) Arctic climate impact assessment. Cambridge University Press, Cambridge, p 1042
Ader WN, Brooks N, Bentham G, Agnew M, Eriksen S (2004) New indicators of vulnerability and adaptive capacity. Tyndall Centre for Climate Change Research. Technical Report 7
Adger WN (2003) Social capital, collective action, and adaptation to climate change. Econ Geogr 79(4):387–404
Adger WN (2006) Vulnerability. Glob Environ Change 16:268–281
Adger WN, Agrawal A (2008) The role of local institutions in adaptation to climate change. IFRI Working Paper #W08I-3, International Forest Resources and Institutions Program, School of Natural Resources and Environment, University of Michigan
Arctic Human Development Report (2004) Arctic Human Development Report. Stefansson Arctic Institute, p 242

Berkes F, Jolly D (2001) Adapting to climate change: social–ecological resilience in a Canadian western Arctic community. Conserv Ecol 5:18

Berkes F, Huebert R, Fast H, Manseau M, Diduck A (2005) Breaking ice: renewable resource and ocean management in the Canadian North. University of Calgary Press, Calgary

Brooks N (2003) Vulnerability, risk and adaptation: a conceptual framework. Tyndall Centre for Climate Change Research Working Paper 38

Cruikshank J (2005) Do glacier listen? Local knowledge, colonial encounters, and social imagination. UBC, Vancouver

Csonka Y, Schweitzer P (2004) Societies and cultures: change and persistence. In: Einarsson N et al (eds) Arctic human development report. Stefansson Arctic Institute, Akureyri, pp 45–68

Denmark Ministry of Environment (2004) Adapting to the climate of the future. Environmental essays 29. Ministry of Environment and Environmental Protection Agency, Copenhagen

Donovan GP (1989) The comprehensive assessment of whale stocks: the early years. Reports of the International Whaling Commission, Special issue 11: 1989, i–iv, pp 210. http://www.iwcoffice.org/publications/pubpre1999.htm#report. Accessed 30 Jul 2009

Duerden F, Kuhn RG (1998) Scale, context, and application of traditional knowledge of the Canadian North. Polar Rec 34(188):31–38

Einarsson N (2009) From good to eat to good to watch: whale watching, adaptation and change in Icelandic fishing communities. Polar Res 28:129–138

Eythórsson E (2000) A decade of ITQ-management in Icelandic fisheries – consolidation without consensus. Finnmark Research Centre Alta Norway. Paper presented at the eighth IASCP conference; constituting the commons: crafting sustainable commons in the new millennium, Bloomington, Indiana, USA, May 31–June 4 2000. http://dlc.dlib.indiana.edu/archive/00000251/00/eythorssone1041500.pdf. Downloaded 30 Jul 2009

Folke C, Carpenter S, Elmqvist T, Gunderson L, Holling CS, Walker B, Bengtsson J, Berkes F, Colding J, Danell K, Falkenmark M, Gordon L, Kasperson R, Kautsky N, Kinzig A, Levin S, Göran Mäler K, Moberg F, Ohlsson L, Olsson P, Ostrom E, Reid W, Rockström J, Savenije H, Svedin U (2002) Resilience and sustainable development: building adaptive capacity in a world of transformations. Scientific background paper on resilience for the process of the world summit on sustainable development on behalf of The Environmental Advisory Council to the Swedish Government, Kasperson, JX, Kasperson RE (2001) International workshop on vulnerability and global environmental change. Risk and vulnerability programme report 2001–01, Stockholm Environment Institute, Stockholm, Sweden

Ford JD (2010) Vulnerability of Inuit food systems to food insecurity as a consequence of climate change: a case study from Igloolik, Nunavut. Reg Env Change

Ford J, Smit B (2004) A framework for assessing the vulnerability of communities in the Canadian Arctic to risks associated with climate change. Arctic 57:389–400

Ford JD, MacDonald J, Smit B, Wandel J (2006) Vulnerability to climate change in Igloolik, Nunavut: what we can learn from the past and present. Polar Rec 42:1–12

Furgal C (2008) Climate change health vulnerabilities in the North. In: Seguin J (ed) Human health in a changing climate: a Canadian assessment of vulnerabilities and adaptive capacity. Health Canada, Ottawa, pp 303–366

Furgal C, Prowse TD (2008) Northern Canada. In: Lemmen DS, Warren FJ, Lacroix J, Bush E (eds) From impacts to adaptation: Canada in a changing climate 2007. Government of Canada, Ottawa, pp 57–118

Government of Nunavut (2003) Nunavut Climate Change Strategy. Government of Nunavut, Iqaluit

Hovelsrud GK, McKenna M, Huntington HH (2008) Marine mammal harvests and other interactions with humans. Arctic marine mammals and climate change. Ecol Appl 18(2):135–147

Iceland Ministry of Fisheries and Agriculture (2009) Information Centre of the Icelandic Ministry of Fisheries and Agriculture. Official government website. http://www.fisheries.is/. Accessed 30 Jul 2009

IPCC (2007) Climate change 2007: the physical science basis: contribution of working group I to the fourth assessment report of the intergovernmental panel on climate change. Cambridge University Press, Cambridge, p 996

IPY (2005) International polar year themes. IPY: International Polar Year. http://www.ipy.org/development/themes.htm. Accessed 28 May 2009

Johnson K, Solomon S, Berry D, Graham P (2003) Erosion progression and adaptation strategy in a northern coastal community. In: Eighth international conference on Permafrost

Kasperson JX, Kasperson RE (2001) Global environmental risk. United Nations University Press, New York, NY

Kempton W, Boster JS, Hartley JA (1995) Environmental values in American culture. MIT, Cambridge, MA

Keskitalo ECH (2004) A framework for multi-level stakeholder studies in response to global change. Local Environ 9(5):425–435

Keskitalo ECH (2008) Climate change and globalization in the Arctic: an integrated approach to vulnerability assessment. Earthscan, Sterling, VA, p 254

Kofinas G, Forbes B, Beach H, Berkes F, Berman M, Chapin T (2005) ICARP II – science plan 10: a research plan for the study of rapid change, resilience and vulnerability, in social–ecological systems of the Arctic. In: Proceedings from the second international conference on arctic research planning (ICARP II), Copenhagen, Denmark, 10–12 November 2005

Krupnik I, Jolly D (2002) The earth is faster now: indigenous observations of climate change. Arctic Research Consortium of the United States, Fairbanks, AK, p 384

Manson GK, Solomon SM (2007) Past and future forcing of Beaufort Sea coastal change. Atmos Ocean 45:107–122

Marchak MP (1995) Logging the Globe. Montreal. McGill-Queens University Press

Matthews R, Sydneysmith R (2010) Adaptive capacity as a dynamic institutional process: conceptual perspectives and their application. In: Armitage D, Plummer R (eds.) Adaptive Capacity and Environmental Governance. Springer, Heidelberg

McCarthy JJ, Martello ML (2005) Climate change in the context of multiple stressors and resilience. In: Arctic climate impact assessment (ACIA 2005). Cambridge: Cambridge University Press, pp 945–988

Nickels S, Furgal C, Buell M, Moquin H (2006) Unikkaaqatigiit – putting the human face on climate change: perspectives from Inuit in Canada. Joint publication of Inuit Tapiriit Kanatami, Nasivvik Centre for Inuit Health and Changing Environments at Universite Laval and the Ajunnginiq Centre at the National Aboriginal Health Organization, Ottawa

Notzke C (1994) Aboriginal peoples and natural resource in Canada. Captus University Press, York

Nunavut Research Institute (NRI) (2002) Gap analysis of Nunavut climate change research. Nunavut Research Institute, Iqaluit

O'Brien K, Leichenko R, Kelkar U, Venema H, Aandahl G, Tompkins H, Javed A, Bhadwal S, Barg S, Nygaard L, West J (2004) Mapping vulnerability to multiple stressors: climate change and globalization in India. Glob Environ Change 14:303–313

Pearce TD, Ford JD, Laidler GJ, Smit B, Duerden F, Allarut M, Andrachuk M, Baryluk S, Dialla A, Elee P, Goose A, Ikummaq T, Joamie E, Kataoyak F, Loring E, Meakin S, Nickels S, Shappa K, Shirley J, Wandel J (2009a) Community collaboration and climate change research in the Canadian Arctic. Polar Res 28:10–27

Persson LO, Ceccato VA (2001) Economic development and policy in Norra Norrland – the Swedish periphery. Report prepared for the peripheral regions workshop, Montreal, Canada, 11–13 October 2001. http://www.ucs.inrs.ca/a/inc/regionsduS/persson.PDF. Accessed 30 Jul 2009

Platt J (2009) Will Canada ban polar bear trophy hunting? Sci Am. http://www.scientificamerican.com/blog/60-second-science/post.cfm?id=will-canada-ban-polar-bear-trophy-h-2009-04-17

Prno J, Wandel J, Bradshaw B, Smit B, Tozer L (2010 forthcoming) Community vulnerability to climate change in the context of other risks in Kugluktuk, NU, Pol Res (re-submitted September 2010)

Smit B, Pilifosova O (2003) From adaptation to adaptive capacity and vulnerability reduction. In: Smith JB, Klein RJT, Huq S (eds) Climate change, adaptive capacity and development. Imperial College Press, London, pp 9–28

Smit B, Wandel J (2006) Adaptation, adaptive capacity and vulnerability. Glob Environ Change 16:282–292

Smit B, Hovelsrud G, Wandel J (2008). CAVIAR: Community adaptation and vulnerability in Arctic regions, University of Guelph, Department of Geography, Occasional Paper No. 28

Turner BL II, Kasperson RE, Matson PA, McCarthy JJ, Corell RW, Christensen L, Eckley N, Kasperson JX, Luers A, Martello ML, Polsky C, Pulsipher A, Schiller A (2003a) A framework for vulnerability analysis in sustainability science. Proc Natl Acad Sci USA 100(3):8074–8079

Turner BL II, Matson PA, McCarthy JJ, Corell RW, Christensen L, Eckley N, Hovelsrud-Broda GK, Kasperson JX, Kasperson RE, Luers A, Martello ML, Mathiesen S, Naylor R, Polsky C, Pulsipher A, Schiller A, Selin H, Tyler N (2003b) Illustrating the coupled human–environment system for vulnerability analysis: three case studies. Proc Natl Acad Sci USA 100 (3):8080–8085

Tyler NJC, Turi JM, Sundset MA, Strøm Bull K, Sara MN, Reinert E, Oskal N, Nellemann C, McCarthy JJ, Mathiesen SD, Martello ML, Magga OH, Hovelsrud GK, Hanssen-Bauer I, Eira NI, Eira IMG, Corell RW (2007) Saami reindeer pastoralism under climate change: applying a generalized framework for vulnerability studies to a sub-arctic social–ecological system. Glob Environ Change 17:191–206

Usher PJ (2000) Traditional ecological knowledge in environmental assessment and management. Arctic 53(2):183–193

Vincent K (2007) Uncertainty in adaptive capacity and the importance of scale. Glob Environ Change 17:12–24

Walker B, Carpenter S, Anderies J Abel N, Cumming G, Janssen M, Lebel L, Norberg J, Peterson GD, Pritchard R (2002) Resilience management in social-ecological systems: a working hypothesis for a participatory approach. Conserv Ecol 6(14). http://www.consecol org/vol6/iss1/art14

Watt-Cloutier S, Fenge T, Crowley P (2005) Responding to global climate change: the perspective of the Inuit circumpolar conference on the Arctic climate impact assessment. Inuit Circumpolar Conference. http://www.inuitcircumpolar.com/index.php?ID=267&Lang=En. Accessed 27 May 2009

Wenzel GW (1999) Traditional ecological knowledge and Inuit: reflections on TEK research and ethics. Arctic 52:113–124

West JJ, Hovelsrud GK (2008) Climate change in Northern Norway: toward an understanding of socio-economic vulnerability of natural resource- dependent sectors and communities. CICERO Report 2008:04, CICERO, Oslo, Norway, p 37

Yohe G, Tol R (2002) Indicators for social and economic coping capacity moving toward a working definition of adaptive capacity. Glob Environ Change 12:25–40

Young OR, King LA, Schroeder H (eds) (2008) Institutions and environmental change: principal findings, applications, and research frontiers. MIT, Cambridge, MA

Chapter 8
Climate Change, Adaptive Capacity, and Governance for Drinking Water in Canada

Rob de Loë and Ryan Plummer

8.1 Introduction

Water managers have always had to deal with daily, seasonal, and annual changes in precipitation, stream flows, lake levels, and other characteristics of the water cycle (McDonald and Kay 1988; Cech 2003). A key factor contributing to their ability to successfully adapt has been the *predictability* of climatic variability (Kabat and van Schaik 2003). To illustrate, although individual flooding events could not be accurately predicted, the frequency and expected magnitude of floods was knowable based on the observed record of past flood events. Knowing how often floods of a certain magnitude can be expected in a particular area allows – in theory at least – for the design of appropriate responses to the flood risk.

The tendency of natural systems to fluctuate within a predictable envelope of variability is known as *stationarity*. Unfortunately, as Milly et al. (2008) have noted, the assumption of stationarity in the context of climate change is no longer valid. Anthropogenic climate change has already produced measurable changes in patterns of precipitation, evaporation, and runoff. Anticipated future changes in these variables are highly likely to fall outside of the observed range of variability. Even aggressive mitigation of CO_2 and other greenhouse gas emissions will only slow the rate of climate warming (Intergovernmental Panel on Climate Change (IPCC) 2007). This means that a new "predictable envelope of variability" is

---

R. de Loë
Environment and Resource Studies, Faculty of Environment, University of Waterloo, N2L 3G1 Waterloo, ON, Canada
e-mail: rdeloe@uwaterloo.ca

R. Plummer
Department of Tourism and Environment, Brock University, 500 Glenridge Avenue, L2S 3A1 St. Catharines, ON, Canada
Stockholm Resilience Centre, Stockholm University, Stockholm, SE-106 91, Sweden
e-mail: rplummer@brocku.ca

unlikely to emerge (Bergkamp et al. 2003; Bates et al. 2008; Milly et al. 2008). Milly et al. (2008) refer to this situation as the "death of stationarity".

The implications of the death of stationarity for water management are profound. Simply put, past capacity to adapt to the observed climatic variability should not provide confidence in future adaptive capacity. Water managers may have to find ways to deal with much greater complexity and uncertainty than has previously been experienced. How do we cope in this new environment? Milly et al. (2008) suggest that the answer is to improve the sophistication of modeling. They argue that "We need to find ways to identify nonstationary probabilistic models of relevant environmental variables and to use those models to optimize water systems" (Milly et al. 2008, p. 573). Improvements to modeling capabilities certainly will be an important part of any strategy to adapt to the impacts of climate change on water resources and human societies (Kundzewicz et al. 2007). However, in an environment where water management increasingly takes place through collaborative governance involving a shifting mix of state and nonstate actors (de Loë and Kreutzwiser 2006), technological innovation cannot be the only focus for adaptation. Instead, it is also essential to strengthen the capacity of organizations, communities, and societies to adapt to the climate change (Ivey et al. 2004), and to address pressing concerns relating to governance.

Canada provides an ideal context for exploring these concerns. The water management challenges being faced in different parts of the country are extremely diverse, and capacity for addressing those challenges is highly variable. Climate change is expected to have profound impacts on Canada's water resources, with attendant threats to socio-ecological systems (Lemmen et al. 2008). And, importantly, water governance in Canada is in flux, with the roles and responsibilities of state and nonstate actors shifting and changing as new, more collaborative and distributed approaches to governance are adopted (Plummer et al. 2005; de Loë and Kreutzwiser 2006). Hence, Canada's experiences offer insights pertinent to many parts of the world.

In this chapter, we explore challenges associated with adapting to climate change in Canada in the context of drinking water. We adopt the integrative perspective promoted in the introduction to the book. Hence, in the next section, we draw on literature pertaining to climate change adaptation, complex systems, and water governance to identify key concerns that emerge through synthesizing insights from these areas. These concerns are then explored in the context of drinking water supply in two very different settings: urban water supply in small and large centers; and drinking water quality in Aboriginal communities. Exploring the complexity of climate change adaptation from the perspective outlined in the next section permits for nuanced insights into the challenges faced, and highlights the significance of governance.

8.2 Adaptation and Adaptive Capacity

In climate change research and policy making, a distinction normally is drawn between *mitigation* and *adaptation*. The former involves efforts to reduce the greenhouse gas emissions that produce climate change. The latter refers, in the language of the IPCC, to "Adjustment in natural or human systems in response to actual or expected climatic stimuli or their effects, which moderates harm or exploits beneficial opportunities" (Parry et al. 2007, p. 869). Adaptation involves countless actions by governments, individuals, firms, and nongovernment organizations. These can be grouped according to intent, timing, and scope (Smit et al. 1999; Lemmen et al. 2008). In the context of the water sector, numerous adaptation options have been identified. For example, in reference to the challenge of providing water for human uses, adaptation options on the supply side could include expanding storage, desalinizing sea water, capturing rainwater, and prospecting for groundwater. On the demand side, options could include recycling water, expanding water markets, implementing agricultural water management measures, and increasing use of water efficient fixtures (de Loë et al. 2001; Kundzewicz et al. 2007).

In many respects, climate change "adaptation options" such as these are tools that have long been part of the water management toolkit. For example, dams and reservoirs have been used for millennia to capture water when it is relatively abundant so that it can be used when it would be scarce under natural flow conditions (McDonald and Kay 1988). Addressing water shortages in urban, agricultural, and industrial settings by influencing demand is a recent approach when compared with millennia-old supply-focused approaches. Nonetheless, adaptation to water scarcity through demand management techniques has been part of contemporary water management for several decades (Vickers 2001).

From historical perspective, water managers have made considerable progress in addressing problems relating to sanitation, water supply, water quality contamination, flood plain management, and, more recently, the effects of human development on ecological systems. Perhaps the best example in support of this claim can be found in the history of drinking water supply, where improvements in treatment technologies and practices have led to tremendous improvements in the quality of life (e.g., Melosi 2000). Unfortunately, a recent report of the United Nations World Water Assessment Program (UNWWAP) (2006) reinforces the fact that significant water-related problems persist in all parts of the world. More seriously, as noted earlier, the impacts of climate change on all aspects of the water cycle will be so pronounced in most regions and in most contexts that techniques and practices used with some success to date will not guarantee successful adaptation in future (Kabat and van Schaik 2003; Kundzewicz et al. 2007). The death of stationarity is a significant contributing factor. However, the problem is much more fundamental. Simply put, climate change is expected to overwhelm the *capacity* of those involved in water management to adapt (Bergkamp et al. 2003).

In the climate change field, the concept of adaptive capacity is conventionally defined as "the ability of a system to adjust to climate change (including climate variability and extremes) to moderate potential damages, to take advantage of opportunities, or to cope with the consequences" (IPCC 2007, p. 21). Importantly, in IPCC literature relating to water, adaptive capacity is commonly treated in an insular and relatively narrow fashion. For example, the IPCC's Fourth Assessment Report is a review of the international state of the art relating to mitigation and adaptation. To their credit, the authors of the water chapter (Kundzewicz et al. 2007) identify key constraints on adaptive capacity, including the following: the nature of the water resource itself; minimum societal needs; insufficient financial resources; political or social constraints; and a set of system-related capacity-limiting factors including ineffective governance, lack of coordination among agencies, and interjurisdictional tensions. However, these critical concerns are dealt with in a cursory fashion. Instead, the adaptation challenge is framed predominantly in terms of the need to improve models and techniques for dealing with risk and uncertainty (Kundzewicz et al. 2007; Bates et al. 2008).

A broader, more integrative perspective on the challenges of adapting to the impacts of climate change in the water sector is clearly needed. We suggest that resilience, flexibility, and adaptability are preconditions for making robust decisions that can respond to changed conditions. This is a position that the authors of the IPCC's Fourth Assessment Report water chapter themselves have briefly acknowledged (Kundzewicz et al. 2007).The complex systems perspective promoted in this volume can provide insights that complement the climate change adaptation literature. The challenge is to find a way to bridge the different literatures effectively.

Fortunately, the water management literature can provide that bridge. For example, there is a long tradition in the water field of considering the factors that shape the capacity of countries, organizations, and communities to provide water services and to protect water resources (e.g., Cromwell et al. 1992; Biswas 1996; Franks 1999; de Loë et al. 2002). This literature is directly pertinent to understanding adaptive capacity relative to climate change. At the same time, authors in the water field have been concerned specifically with questions of adaptability and resilience for over a decade. For example, in the 1990s, a small number of water researchers were drawing on the concept of Adaptive Environmental Assessment and Management (AEAM) to conceptualize new ways of dealing with problems relating to water quality (Grayson et al. 1994) and flooding (Sendzimir et al. 1999).

The impact of AEAM thinking on the mainstream water literature in the 1990s was quite modest. However, concern for complexity and uncertainty, and recognition of the importance of adaptability as a precondition for effective water management, also developed on a separate path that is having a more enduring impact. For example, Geldof (1995a, b) is among the earliest proponents of a nonequilibrium approach to water management. He argued that integrated water management (IWM) displays all the characteristics of a complex adaptive system (i.e., networks of agents acting in parallel, many levels and scales of organization, characterized by perpetual novelty). Instead of the static perspective prevailing in IWM, Geldof argued that *adaptive* water management, which embraced complexity and uncertainty, was needed.

This perspective is becoming more prominent in the water literature (Pahl-Wostl and Sendzimir 2005; Pahl-Wostl 2007). The Global Water Systems Project, for instance, defines the global water system as encompassing human, physical, and biological components and their interactions (Pahl-Wostl 2007). Impacts of climate change on water have prompted concerns about the robustness of water systems to perturbations and their ability to recover (e.g., Fowler et al. 2003). Attention is increasingly directed toward social–ecological resilience, which is concerned with the amount of disturbance a system can absorb and remain in relatively the same state, the extent to which the system is able to self-organize, and the degree to which the system can build capacity for learning and adaptation (Folke et al. 2002; Folke 2003).

Research conceptualizing the water system in terms of complexity and uncertainty provides a conceptual foundation for enhancing understanding of adaptation evident in the mainstream climate change literature. Hence, synthesizing insights from the IPCC-oriented climate change and the complex systems literatures is worthwhile. Other recent efforts in this direction include a 2003 study by the IUCN-The World Conservation Union (2003). In considering the challenge of climate change adaptation, these authors emphasized the importance of building the capacity of people and institutions, maintaining and increasing social capital, and adopting adaptive management styles that involve social learning. These themes also are prominent in the complex systems literature, which stresses the importance of multi-scale networks, cross-scale interactions, and multiple knowledge systems (Folke et al. 2003; Armitage 2005; Gunderson et al. 2006; Armitage et al. 2009).

Framing the challenge of adaptation to climate change in the water sector in terms of complexity and uncertainty has additional benefits. At the outset of this chapter, we argued that contemporary water *management* occurs in an environment where collaboration among a shifting mix of state and nonstate actors has become the norm. In this environment, *governance* – the ways in which societies make decisions – has shifted away from the traditional top-down, technocratic model of past decades. In considering how climate change has affected water resources, Kabat and Van Schaik (2003) argued that climate change has changed the *water* rules. Concomitantly, a transition to distributed and collaborative governance – typically marked by a shift from single centers of power to multiple, distributed centers of power (Plummer et al. 2005) – is changing the *water management* rules. New actors bring new values and capabilities, and new ways of making decisions about water lead to shifts in the distribution of power within society (de Loë and Kreutzwiser 2006). This simply reflects the fact that water governance is and has always been a highly political activity (Swatuk 2005). From this perspective, optimizing water systems through developing probabilistic models of relevant environmental variables – the way forward for water managers confronting the death of stationarity identified by researchers such as Milly et al. (2008) – may not be as important as the need to build adaptive capacity and strengthen governance.

8.3 Case Studies

Canada is perceived to be a water rich nation – a fact reinforced by country statistics on fresh water availability. For instance, the World Resources Institute (2009) places Canada's freshwater supply as third in the world, behind Brazil and Russia.

What is less well understood is the fact that Canada is a vast land with a population concentrated in major cities, most of which are located in the southern part of the country. This means that most of the water resources for which the country is famous are not readily accessible to the majority of its population (Kreutzwiser and de Loë 2004). Hence, despite a persistent myth of abundance (Sprague 2006), Canada does face significant challenges to adapting to climate change (Lemmen et al. 2008).

In this section, we explore the challenge of adapting to the impacts of climate change on a critical system: drinking water supply. Two examples are presented that illustrate the impacts of climate change, the multiple scales at which responses are required, and the challenges of adapting in the Canadian water resource context. These are urban drinking water supply in large and small centers, and drinking water provision in Canada's Aboriginal communities. Each example introduces the specific water management challenge, describes how solutions to climate change will require increased adaptive capacity, and highlights the importance of governance. The cases speak to a series of key questions that emerge from the broad, integrative perspective outlined in the previous section:

- *What is the context of the water system under investigation?* The water system can be defined according to one or more scales (e.g., spatial, temporal, jurisdictional) and described in terms of its human, physical, and biological components as well as their interactions.
- *How is current and projected climate change going to impact the water system and what are the anticipated outcomes of those impacts?* Modeling techniques and climate change scenarios can assist in understanding the influences of climate change on the components of the water system and highlight potential vulnerabilities. However, as noted earlier, more sophisticated modeling alone will not permit successful adaptation to climate change in the water sector.
- *What is the capability of the water system's management arrangements and practices to address the current and future impacts from climate change?* Insights into the degree to which adjustments are possible and adaptation options are feasible come from examination of supply and demand strategies, structural and nonstructural approaches, elements of institutional arrangements and components of capacity. In turn, these insights speak to the extent to which the water system can absorb disturbances and remain in the relatively same state, self-organize, and build capacity for learning and adaptation.

8.3.1 Urban Water Supply

Approximately 90.6% of Canada's population receives its drinking water supply from a water supply system (Environment Canada 2005). These range in size from small communal systems serving a handful of households, to large systems such as the one that provides water to residents of the City of Toronto through a network of four water treatment plants, 18 pumping stations, 10 underground storage reservoirs and 510 km of water mains (City of Toronto 2009).

Canadian drinking water treatment and distribution systems are operated by municipalities, public utilities, and, in a few cases, private companies (Bakker and Cameron 2005). Sources of water for these systems are diverse, and include groundwater, lakes, and rivers – with surface water sources comprising 89% of the water supplied by municipalities in 2001 (Environment Canada 2005). The importance of groundwater as a source of supply is closely associated with population. In 2001, systems in smaller communities (those with populations ranging from 2,000 to 5,000 people) took 42.7% of the water they provided from aquifers. Systems in larger communities (those with populations of more than 500,000) only drew 0.4% of the water they supplied from aquifers (Environment Canada 2005). Exceptions to this rule exist, including the Regional Municipality of Waterloo in southern Ontario, current population of approximately 507,000 (Regional Municipality of Waterloo 2009), which draws approximately 80% of the water its system supplies from groundwater sources.

Globally, climate change is emerging as a serious problem for the operators of drinking water treatment and distribution systems. Concerns relate to the impacts of climate change on both water quality and quantity (Bates et al. 2008). In Canada, climate change is expected to affect the ability of drinking water treatment and distribution systems to provide adequate supplies of safe drinking water in a host of ways. For example, in the Great Lakes Basin, the following concerns pertinent to drinking water systems have been identified based on predictions of likely impacts of climate change on the hydrologic cycle in this critical region (de Loë and Berg 2006):

- An increased frequency of extreme rainfall events is expected to contribute to a greater frequency of waterborne diseases and increased transportation of contaminants from the land's surface to water bodies. For many systems this will necessitate additional efforts to protect drinking water sources from contamination, and to treat water of potentially lower quality.
- Decreases in runoff will contribute to reduced water quality as less water becomes available for dilution of sewage treatment plant effluents and runoff from agricultural and urban land. In turn, this will contribute to increased treatment costs. Decreased runoff will also increase competition for scarce water resources during periods of low flow.
- Decreases in groundwater recharge will increase competition for scarce water resources, for instance, as users formerly reliant on surface water switch to groundwater. This may have implications for surface water resources dependent on groundwater for baseflow.
- Increases in water temperature may lead to reduced source water quality because of greater biological activity (e.g., algae production), and a greater frequency of taste and odor problems in drinking water supplies. As a result, an increased risk of disease may be expected, alongside increased customer dissatisfaction.

Climate change clearly has serious implications for Canada's drinking water treatment and distribution systems. However, these are far from the only challenges faced by these systems (Box 8.1). Thus, in considering how to respond to the challenge of climate change, it is essential that the threats it poses be viewed in a larger context. Specifically, it must be recognized that the impacts of climate change will be layered on top of a host of *existing* concerns and challenges faced by the operators of drinking water treatment and distribution systems. These concerns and challenges are diverse, and include (but are by no means limited to) the following:

Box 8.1: Drinking Water Supply in the Capital Region District

The capital region district (CRD) provides water to approximately 320,000 people living in southern Vancouver Island, making it the second largest system in British Columbia. CRD Water Services wholesales water within the Greater Victoria Drinking Water System (GVDWS), and retails it to customers in the Western Communities. Additionally, it provides system-wide services relating to water conservation and water quality protection. Water for this region is provided by a series of surface water reservoirs. The GVDWS is governed by three water supply commissions, with representatives from area municipalities. Community input into drinking water system planning and operation is provided through the Regional Water Supply, Protection and Conservation Advisory Committee.

The drinking water system faces challenges common to many systems in Canada, including pressure to meet customer demands, compliance with new regulatory requirements imposed by the provincial government, and ongoing infrastructure maintenance and upgrades. Climate change also is emerging as a related concern. Demand for water has increased in the region, and drought-like conditions have been experienced. Following the 2001 drought, reservoir capacity dropped to 73%, prompting the development and implementation of a water conservation and demand management strategy. Forecasts of population and climate indicate that existing infrastructure may be insufficient relative to demands, even with aggressive water conservation. Whether or not the system can meet this challenge depends in part on its ability to address a series of potential vulnerabilities linked to climate change. Many of these are typical of most large water systems, e.g., effects of changes in water temperature on water quality, and finding and repairing leaks in the distribution system. However, some are distinctive to the GVDWS, for instance, dependence on reservoirs that are too small, impacts of forest fires in the source catchments, an inability to control land uses and activities in source catchments, the need to account for downstream fishery needs, and uncertainty regarding Aboriginal title. The last two considerations fall within federal jurisdiction, and thus are beyond the ability of the CRD to control.

Sources: (Cameron 1998; Kolisnek and de Loë 2005; Capital Regional District 2009).

8 Climate Change, Adaptive Capacity, and Governance for Drinking Water in Canada 165

- *Serious concerns about drinking water safety in small communities.* Contamination incidents in Walkerton, Ontario (in 2000) and in North Battleford, Saskatchewan (in 2001) highlighted major weaknesses in operating procedures, standards, and regulations (Christensen and Parfitt 2001). Many problems have been addressed, but recent evaluations suggest that considerable room for improvement still exists, especially in smaller communities (Christensen 2006; Hrudey 2008). For example, over 1,700 boil water orders were in effect across Canada in 2008 – many of these being long term, and most occurring in small systems (Eggertson 2008b). Systems serving Aboriginal communities, as explored in the subsequent case, have a disproportionate number of problems (Eggertson 2008a).
- *Aging infrastructure that is in urgent need of repair and replacement.* In a recent assessment, Infrastructure Canada (2004) indicates that much of Canada's water supply and distribution infrastructure is reaching the end of its life and must be replaced. Cost estimates vary widely. The National Round Table on the Environment and Economy (NRTEE) estimated in 1996 that new capital demands for water and wastewater infrastructure would, conservatively, exceed $41 billion by the year 2015 (NRTEE 1996). Infrastructure Canada's (2004) report questions the accuracy of this and similar more recent estimates, but indicates nonetheless that a massive financial outlay is needed for replacement and maintenance of water supply and wastewater treatment infrastructure.
- *Pressure to meet the demands of urban growth.* Canada is an urban country, with approximately 80% of its population living in places with 1,000 or more people (Statistics Canada 2007). Intensifying the trend to urbanization, population growth is occurring almost exclusively in already densely populated areas such as are found in the southern parts of Ontario, Quebec, and British Columbia, and in the Calgary-Red Deer-Edmonton corridor (Statistics Canada 2007). Where this growth is occurring through low density suburban development, extension of infrastructure can be extremely costly (Infrastructure Canada 2004). At the same time, growth compounded by increased demand for water from existing and new customers adds additional stress on drinking water supply and distribution systems.
- *New regulatory requirements that increase costs for system operators.* Following the outbreaks in Walkerton, Ontario, and North Battleford, Saskatchewan, the regulatory burden on operators of drinking water treatment and distribution systems increased as provincial governments sought to improve drinking water safety (Christensen 2006). While stricter regulations are widely recognized as necessary, they do lead to increased costs for inspections, monitoring, and infrastructure upgrades (Infrastructure Canada 2004). In Ontario, for example, the costs of meeting the regulatory requirements imposed on municipalities by the provincial government following the Walkerton incident were estimated to be over $800 million as of 2005 (Water Strategy Expert Panel 2005).
- *Uneven capacity.* With the size of drinking water systems in Canada ranging from communal supplies that provide water to a few households up to large

systems serving millions of people, it should not be surprising that capacity is highly variable. The United States Environmental Protection Agency (1998) characterized the capacity of drinking water systems to provide safe water in terms of technical, financial, and managerial considerations. Systems that lack trained staff and necessary equipment have weak or inadequate administrative procedures, and are not revenue self-sufficient struggle to provide clean, safe drinking water (Kreutzwiser and de Loë 2002; Brown et al. 2005). These systems tend to be smaller, and, as noted earlier, the consumers they supply with water are disproportionately exposed to risks.

From this broader perspective, two conclusions may be drawn that are directly pertinent to the themes in this chapter. First, in the context of municipal drinking water supply and treatment systems, adaptation to climate change clearly must be undertaken in concert with responses to other considerations (and vice-versa). For example, as noted in the introduction to this chapter, efforts to upgrade or replace water treatment and distribution system infrastructure must take account of the fact that future water supply availability can no longer be predicted based on the historical hydrological record. At the same time, estimates of the number of people who can be served by the system should take account of changes in demand related to anticipated shifts in temperature, precipitation, and evaporation. In relation to capacity challenges, it is important to recognize that the ability to design effective adaptive responses to climate change will be linked to broader considerations. Thus, operators of small systems who are struggling to meet new regulations relating to drinking water safety cannot be expected to have the expertise needed to develop models that permit integrating the impacts of climate change into demand forecasts. Therefore, initiatives to build capacity to adapt to climate change clearly must be framed in terms of larger capacity concerns. And, respecting the fact that the people who manage the systems already are overburdened with concerns (Kabat and van Schaik 2003), every effort must be made to *mainstream* climate change adaptation, in other words, to build it into existing planning and decision making processes (de Loë and Berg 2006).

Second, the case of drinking water treatment and distribution systems reinforces the fact that efforts to enhance the adaptive capacity of water systems must consider a broad range of technical/engineering approaches *and* socio-economic considerations relating to their resilience, including governance. To illustrate, demand management is commonly identified as a way in which water systems can adapt to climate change (e.g., de Loë et al. 2001). This approach involves technological measures (e.g., leak detection and repair, water efficient fixtures), economic instruments (e.g., pricing, financial incentives), social measures (e.g., public education and outreach), and regulatory measures (e.g., summer water use restrictions) (Vickers 2001). Demand management is a relatively uncontroversial illustration of the way in which technological and socio-economic tools and approaches can be combined. Shifting the focus to the system level raises more divisive issues, and further clarifies the importance of governance.

For instance, in response to the Walkerton tragedy in 2000, the Province of Ontario struck an expert panel with the mandate to inquire into Ontario's drinking water system as a whole (Water Strategy Expert Panel 2005). The Panel identified a series of reforms it considered necessary to address the problems it identified in the system. Several of these related directly to governance:

- Increase the scale and capacity of systems by joining them together into regional networks that permit sharing of resources, economies of scale, and a greater capacity to manage risks.
- Improve governance of systems through forming them into larger, municipally-owned corporations that own assets, and whose finances are separated from the municipal owners.
- Permit greater flexibility through permitting systems to contract out key functions to private companies.

These are issues of governance because they speak directly to who is involved in decision making and how decisions will be made. As such, they are inherently political. The fact that these recommendations have not been pursued by the provincial government reflects their – in Ontario at least – controversial nature. Strong endorsement by the Panel of private sector involvement in what is often considered a "public" realm by many in Ontario led to immediate negative criticism of its report (e.g., Nadarajah and Miller 2005; Canadian Union of Public Employees 2006). As a result, the governance issues raised by the Panel – which could have formed the basis for a broader dialog about the robustness and resilience of drinking water treatment and distribution systems in Ontario – remain largely unaddressed.

8.3.2 Water Quality and Health in Aboriginal Communities

According to the 2006 Canadian Census, 1,172,785 people in Canada (3.75% of the Canadian population) are of "Aboriginal" identity, i.e., Indians (First Nations), Métis, and Inuit (Statistics Canada 2006). There is enormous diversity within the Aboriginal population – culturally, economically, and socially. One key distinction that is important in the context of drinking water and climate change adaptation is the fact that there are 612 recognized Indian bands, and 2,675 reservations designated under the *Indian Act;* these reservations comprise a land area of 2,685 km^2 (Statistics Canada 2006). In contrast, Inuit peoples have not historically been located on reservations. Rather, they have occupied vast traditional territories primarily in northern Canada. Nunavut – Canada's newest territory – was created to reflect this fact. Because of the heterogeneity of Aboriginal people, it should not be surprising that drinking water provision in Canada's Aboriginal communities takes place in an extremely complicated institutional and geographic context.

Access and rights to resources are paramount issues for Aboriginal peoples in Canada. Their traditional institutions to control access and use of the natural environment were destroyed through European settlement, and thereafter access has been a function of contested Aboriginal rights and Aboriginal title (Booth and Skelton 2004). Walkelm (2006, p. 304) observes that "historically, (the Government of) Canada has simply denied that any indigenous territorial rights (including water) exist" and that increasingly access to, and protection of, water has come about through "reserve water rights and Aboriginal title, Aboriginal rights, and treaty rights".

Although a review of Canadian law pertaining to Aboriginal rights and title is well beyond the scope of this chapter, it is important to recognize four key realities that influence Aboriginal water governance and shape the context for adaptation to climate change in Canada's Aboriginal communities (Booth and Skelton 2004; Walkem 2006). First, the Constitutional Act of 1982 (s. 35) protects Aboriginal title and rights and recognizes the fiduciary responsibilities of the Government of Canada. Second, modern treaties (land claims) contribute certain rights and autonomy as established via cooperative arrangements and self-government. Third, water and other "resources" hold different and broader meanings for Aboriginal people than the ones defined in Canadian law. Fourth, gaining rights does not necessarily mean gaining access.

The fourth consideration is particularly significant in this case study. Through a variety of recent Supreme Court decisions, Aboriginal people have secured certain rights that previously were not acknowledged. However, these rights have not translated into access. For instance, in 1995 Indian and Northern Affairs Canada (INAC) estimated that one in four First Nation water systems posed "significant risks to human health," and noted that despite substantial investments in system upgrades between 1995 and 2001 (approximately $560 million dollars plus operation and maintenance at $100–125 million annually) the situation still deteriorated (INAC 2007). The National Aboriginal Health Organization (2002) offered an informative critique of the national level data used in the 1995 and 2001 assessments, noting that these assessments exclusively addressed First Nations and were not representative of Aboriginal peoples; defined a community water system in such a way that provided few details about individual households and/or the 235 communities without any services; and were limited because the categories addressed appear to be arbitrarily selected and lacking explicit operational measures.

In 2003, INAC conducted a national assessment of water systems servicing five or more homes in First Nation communities in Canada (INAC 2003). Disturbingly, 29% of the 740 community water systems assessed were found to pose a "potential high risk that could negatively impact water quality" and an additional 46% were found to pose "potential medium water quality risks". Consequently, INAC and Health Canada launched a First Nations Water Management Strategy and allocated an additional $600 million dollars in funding over the next 5 years.

While these measures are laudable, their effectiveness is questionable. The Report of the Commissioner of the Environment and Sustainable Development in 2005 found that despite efforts to improve drinking water, "residents of First Nations communities do not benefit from a level of protection comparable to that of people who live off reserves" (Office of the Auditor General 2005, p. 1). Responding to this charge, INAC and the National Chief of the Assembly of First Nations in 2006 announced a Plan of Action for Drinking Water in First Nations that examined the multi-barrier approach of the First Nations Water Strategy, emphasized reducing the risk rankings of water systems, and addressed the Commissioner's recommendations (INAC 2008). Perhaps most visible among these accomplishments is the formation of the Expert Panel on Safe Drinking Water for First Nations, which considered the options for a regulatory framework to ensure water quality for First Nations communities (Swain et al. 2006). The most recent progress report on the Plan of Action highlights reductions in the number of high-risk drinking water systems, removal of two thirds of the communities identified as priorities, and several other accomplishments (INAC 2008). Nonetheless, as of November 30th, 2008, 103 First Nation communities in Canada were under a drinking water advisory (Health Canada 2008), and a recent report concluded that "unsatisfactory access to safe drinking water persists for many First Nations people..." (Harden and Levalliant 2008, p. 7).

Clearly enormous challenges exist in relation to drinking water in Canada's Aboriginal communities. Unfortunately, these challenges will be magnified by climate change. As noted earlier, the impacts of climate change will be significant and pervasive in all regions of Canada (Lemmen et al. 2008). Aboriginal people are particularly vulnerable because of their exposure and sensitivity to climatic changes. This is most evident in northern Canada (defined as lands located north of 60° latitude). In the past 50 years, the Arctic region of Canada has experienced change at an unprecedented rate; projections consistently anticipate continued increases in temperature and precipitation and significant alterations in the cryosphere, which will create abrupt changes and variations in freshwater (Bates et al. 2008; Furgal and Prowse 2008). Aboriginal peoples constitute more than half of the population of northern Canada. Additionally, a majority of the more than 100 communities in northern Canada are small (<500 residents) and located along coastlines. Most are places where traditional livelihood activities are incorporated into daily routines (Furgal and Prowse 2008). This close connection with land and water, especially in isolated communities where subsistence livelihoods strategies are employed, increases sensitivity to climate change as current and projected impacts cut across physical, social, and cultural dimensions (Center for Indigenous Environmental Resources 2006a, b). Implications for health are a particularly important concern because of the potential for reduced nutritional contributions from country foods, greater frequency, and magnitude of accidents due to changes in ice, and changes in infective agents (waterborne and foodborne) that lead to real and perceived declines in water quality (Furgal and Seguin 2006).

The impacts of climate change will not only be experienced by Aboriginal people living in northern Canada. Aboriginal people in southern Canada also are vulnerable, but sometimes in different ways. For example, in a recent study the Center for Indigenous Environmental Resources (CIER) draws attention to the impacts of changes in water quality and quantity due to climate change on First Nations people living south of 60° latitude (CIER 2008). These impacts vary enormously by ecoregion (e.g., boreal forest, Carolinian/Great Lakes, taiga/tundra). For example, in the boreal forest ecoregion, the study suggests that overall drier conditions will be experienced, while in the taiga/tundra ecoregion, decreased water quality due to melting permafrost is a particular concern. Thus, in considering the challenge of adaptation to climate change by Aboriginal people in Canada, it is important to recognize that their vulnerability is a function of the different ways in which they are exposed to the impacts of climate change. At the same time, it must not be forgotten that their vulnerability is exacerbated by persistent poverty, disadvantage, and marginalization (Royal Commission On Aboriginal Peoples 1996; Reading et al. 2007). For example, as noted earlier, drinking water quality already is a serious concern in many Aboriginal communities; declining source water quality due to climate change will worsen this problem unless the capacity of community members to provide safe drinking water is greatly enhanced.

The ability of water system managers in many Aboriginal communities to address current and future impacts from climate change is in doubt. Indeed, enhancing capacity for drinking water management in Aboriginal communities has been identified for well over a decade. Most recently, the Summative Evaluation of the First Nations Water Management Strategy (INAC 2007), as well as the most recent progress report on the Plan of Action for Drinking Water in First Nation Communities (INAC 2008), drew attention to the need for a series of capacity enhancement measures. These included development of a protocol on water standards; support for the Circuit Rider Training Program, which trains water operators; provision of technical advice and oversight; and the rectification of key problems relating to the protection of sources of water, system design, operating procedures, and training. These measures have decreased the number of high-risk systems and priority communities from 224 to 116 in a 2-year period (Indian and Northern Affairs Canada 2008). However, as was noted earlier, serious problems exist in almost all Aboriginal communities – even those that have demonstrated both capacity and resolve to address their drinking water safety challenges (Box 8.2).

Box 8.2: Drinking Water and Source Water Protection on Six Nations of the Grand River

Six Nations of the Grand River is the most populous reservation in Canada. From 1972 to 2005, the on-reserve population increased from 4,907 to 11,297, with the future population projected to reach 19,244 in 20 years. There are 2,674 housing units on the reservation, with 85% of those units being classified as rural. The present communal water treatment plant was constructed in 1989 and takes water from the Grand River. It is operated by the Six Nations Public Works Department and primarily services the town of Ohsweken (approximately 450 residential and nonresidential units), also providing a source of water to the 522 houses with cisterns (holding tanks). The community has developed considerable capacity (human, technical, financial) for the community water system. A committee has been formed to address concerns about water quality and quantity and a source water protection plan has been developed. Recognizing the importance of water management beyond the boundaries of the community has prompted several

(continued)

formal and informal mechanisms for information sharing and collaboration with organizations and governments involved in source water protection and watershed management.

Six Nations of the Grand River illustrates the importance of a broad approach to source water protection and recognition that safe drinking water extends beyond water treatment plants. In their alarming report, *Boiling Point*, Harden and Levalliant (2008) observe that water wells are considered the individual responsibility of homeowners and therefore do not factor into the assessment of a community's risk. In Six Nations of the Grand River, approximately 1,735 housing units are serviced by individual wells. Testing of 312 wells in 2005 revealed 86% had some level of fecal coliform contamination and 30% were seriously contaminated with *Escherichia coli* (Neegan Burnside Engineering and Environmental Ltd 2005). Identifying this issue was an important first step for the Source Water Protection Committee in Six Nations of the Grand River. Enhancing well stewardship through awareness and education is an important part of implementing the source water protection plan.

Sources: (Ontario Clean Water Agency 2002; Six Nations of the Grand River 2009; Six Nations of the Grand River Environmental Office 2009).

From a national perspective, Aboriginal communities clearly face systemic problems at individual, organizational, and community-wide levels (Graham 2003). The current form of water governance in Canada is a contributing factor – and will strongly influence the ability of Aboriginal communities to adapt to climate change. For example, Walkem (2006) questions the assumptions and beliefs that underpin water resource management in Canada and calls for fundamental changes that take account of indigenous laws and traditions. Despite the entrenched rights of Aboriginal peoples and the potential of climate change to impact those rights, decisions about climate change have been made without the involvement of Aboriginal peoples. At the same time, climate change itself has not been clearly recognized as an important concern in strategies designed to address drinking water safety. In the absence of a commitment from the various governments to fully and appropriately engage Aboriginal people in water governance, their ability to respond effectively to climate change clearly is limited (Center for Indigenous Environmental Resources 2006a).

Explorations of resilience and adaptive capacity of Arctic communities to climate change have occurred in place-specific case studies and have emphasized short-term responses or coping mechanisms and adaptations such as traditional knowledge and relationship networks that enhance the capacity for learning and self-organizing across levels (Berkes and Jolly 2002; Chapin III et al. 2004). Although Arctic communities have demonstrated considerable adaptability and can continue to draw upon these reservoirs of resilience, researchers who have studied these communities are concerned that future changes in conditions may exceed conventional coping capacities (Ford and Smit 2004; Furgal and Prowse

2008). In a southern Canadian context, capacity to adapt to climate change also is a concern. However, there is recognition of the potential for multiple knowledge systems, including Aboriginal traditional knowledge, to enhance the ability of First Nations communities to undertake source water protection (Chiefs of Ontario 2007). Thus, where Arctic communities already are experiencing and adapting to the impacts of climate change, First Nations peoples in southern latitudes may be able to anticipate adaptation strategies that increase resilience and reduce risks (CIER 2008).

8.4 Discussion and Conclusions

A key tool in the water management toolkit has been the ability to construct a predictable envelope of variability from the observed hydrological record. Climate change is putting this tool out of reach. Not only are future climatic conditions expected to fall outside of the observed range of variability, but also it may be that a new predictable envelope of variability cannot be created. Climate scientists such as Milly et al. (2008) would address this challenge by developing new models that do not depend on stationarity. We argue that while nonstationary probabilistic models may be necessary in adapting to climate change, they will not be sufficient. Instead, we suggest that attention must be directed to building capacity to adapt to climate change, and this, in turn, demands a broad perspective on the system whose capacity is to be developed.

The broad, integrative perspective advanced in this book provides an appropriate lens for understanding the magnitude of the climate change adaptation challenge related to water. This was illustrated in the chapter through examining two very different contexts in which drinking water provision occurs: urban centers and Aboriginal communities. Three questions focused the discussion:

- What is the context of the water system under investigation?
- How is current and projected climate change going to impact the water system and what are the anticipated outcomes of those impacts?
- What is the capability of the water system management to address the current and future impacts from climate change?

Through addressing questions such as these, a more nuanced understanding of the challenges associated with adapting to climate change can be developed, and appropriate strategies can be revealed. This is illustrated in the following synthesis of insights from the two case studies considered in this chapter.

Drinking water systems in Canada are enormously heterogeneous. They draw on different kinds of sources, face varying pressures from development, and must deal with different kinds of infrastructure-related concerns. Furthermore, their vulnerability to climate change varies widely. Thus, it should not be surprising that the impacts of climate change will be experienced differently among systems. Some urban systems already are experiencing pressures from growth, and face constraints on their ability to secure additional water resources; in many parts of Canada, climate change is expected to worsen their circumstances. In contrast, Aboriginal communities in northern Canada that are served by systems drawing from relatively abundant water sources such as large lakes and rivers may feel the effects of climate change on water quality more than on water quantity.

These facts alone reinforce the importance of taking account of contextual circumstances in any effort to create broad-scale adaptation strategies. However, varying sensitivity to climate change is not the only consideration. In Canada, drinking water provision occurs in an extremely complex social and institutional milieu. Systems vary in size from those serving a few households to those serving millions of people. Not surprisingly, their technical, financial, and managerial capacity varies enormously. Capacity challenges are particularly pronounced in systems serving Aboriginal communities. However, small systems across the country typically have less financial, technical, and managerial capacity than their larger counterparts. Adding to basic capacity challenges is the fact that municipalities in Canada are subservient to provincial jurisdiction, while Aboriginal communities located on reserves are under federal jurisdiction. Thus, in both cases the systems that will experience the impacts of climate change lack the legal and policy tools needed to address a threat that, in most respects, originates outside of their boundaries.

Finally, both cases reveal ways in which adaptive capacity is shaped by larger circumstances. In Aboriginal communities, there is little reason to be confident that drinking water systems can absorb shocks originating from climate change, and independently build capacity for learning and adaptation. This reflects the severe challenges that many Aboriginal communities in Canada face, including poverty, unemployment, loss of traditional cultures, substandard education systems, and inadequate health care. The situation of drinking water systems in large urban municipalities unquestionably is better. However, complacency is not warranted given that these systems face profound challenges relating to meeting new regulatory requirements, coping with pressures from growth and development, and renewing crumbling infrastructure. In both cases, therefore, the challenges associated with building capacity to adapt to climate change and making drinking water systems more resilient are inextricable linked to the larger contexts in which those systems are embedded.

The discussion so far has emphasized challenges. Importantly, however, the two cases also provide insights into the most appropriate strategies for addressing these challenges. Simply recognizing that in the context of drinking water systems, climate change adaptation is not just a technical challenge but is a critical first step. It certainly should not be assumed that producing more sophisticated climate models will automatically increase the adaptive capacity of drinking water systems. In systems where the capacity exists to use them, these models can play a critical role in identifying likely alternative futures. However, this capacity should not be assumed in small systems that are challenged to undertake basic maintenance. More fundamentally, sustained attention must be directed to the factors that shape capacity to provide clean drinking water and to adapt to climate change, including those associated with the systems themselves (staff resources, financial stability), but also the larger social, institutional, and biophysical circumstances in which those systems exist.

In light of these facts, we argue that climate change adaptation must be addressed in *concert with* other concerns. Presenting climate change adaptation as a distinct concern that will be added to the existing load borne by the operators of drinking water systems, the communities in which these systems exist, and the various agencies involved, simply is untenable. Indeed, relative to the jurisdictional issues noted above, this position makes it easy for people involved in drinking water provision to argue that responding to climate change simply is not within their mandate. Instead, we argue that climate change must be *mainstreamed* into existing and future planning and decision making processes that fall within the core mandate of those concerned with drinking water provision. Three straightforward examples include the following:

- Operators of large drinking water systems already undertake demand forecasting. Thus, it is not unreasonable to ask that likely future scenarios of climatic conditions should be integrated with demand forecasts.
- Communities across Canada are engaged in processes designed to identify threats and vulnerabilities to source waters. Climate change can readily be integrated into water budgeting exercises, vulnerability assessments, long range planning efforts, and other activities associated with source water protection (e.g., de Loë and Berg 2006).
- All stakeholders involved in drinking water provision in Aboriginal communities have recognized the need to build community capacity to maintain and operate systems and to protect drinking water sources. Capacity building initiatives must be holistic and address a broad range of concerns, and the impacts of climate change can be included as one of those considerations.

Posing the kinds of questions considered in this chapter highlights the complexity of climate change adaptation in the context of drinking water systems. However, it also reveals a host of additional ways in which climate change can be mainstreamed and integrated with other concerns. In that respect, the broader perspective adopted here is essential to respond effectively to climate change.

Acknowledgements Rob de Loë would like to acknowledge support provided by the Canadian Water Network, the Social Sciences and Humanities Research Council, the Walter and Duncan Gordon Foundation and the University of Waterloo.

Ryan Plummer gratefully acknowledges support for his research program from a Brock University Chancellor's Chair for Research Excellence, the Canadian Water Network and the Social Sciences and Humanities Research Council of Canada.

References

Armitage D (2005) Adaptive capacity and community-based natural resource management. Environ Manage 35(6):703–715

Armitage DR, Plummer R, Berkes F, Arthur RI, Charles AT, Davidson-Hunt IJ, Diduck AP, Doubleday NC, Johnson DS, Marschke M, McConney P, Pinkerton EW, Wollenburg EK

(2009) Adaptive co-management for social–ecological complexity. Front Ecol Environ 7(2):95–102

Bakker K, Cameron D (2005) Governance, business models and restructuring water supply utilities: recent developments in Ontario, Canada. Water Policy 7:485–508

Bates BC, Kundzewicz ZW, Wu S, Palutikof JP (2008) Climate Change and Water. Intergovernmental Panel on Climate Change, Geneva

Bergkamp G, Orlando B, Burton I (2003) Change: adaptation of water resources management to climate change. IUCN, Gland

Berkes F, Jolly D (2002) Adapting to climate change: social–ecological resilience in a Canadian western Arctic community. Conserv Ecol 5(2):18

Biswas AK (1996) Capacity building for integrated water management: summary and conclusions. Int J Water Resour Dev 12(4):513–514

Booth AL, Skelton NW (2004) First Nations access and rights to resources. In: Mitchell B III (ed) Resource and environmental management in Canada. Oxford University Press, Don Mills

Brown B, Weersink A, de Loë RC (2005) Measuring financial capacity and the effects of regulatory changes on small water systems in Nova Scotia. Can Water Resour J 30(3):197–210

Cameron VZ (1998) Source water protection in British Columbia: the conundrum of public access in community watersheds, Paper in *CWRA 51st Annual Conference, Mountains to Sea: Human Interaction with the Hydrological Cycle*. Canadian Water Resources Association, Victoria, BC

Canada E (2005) 2004 Municipal water use report. Environment Canada, Ottawa

Canada I (2004) Water infrastructure: research for policy and program development. Infrastructure Canada, Ottawa

Canadian Union of Public Employees (2006) Leaky propositions: the ontario watertight report. Summary for OMECC 2006. Unpublished

Capital Regional District (2009) Welcome to water services. http://www.crd.bc.ca/water/index.htm. Unpublished

Cech TV (2003) Principles of water resources: history, development, management and policy, 2nd edn. Wiley, New York

Centre for Indigenous Environmental Resources (2006a) First Nations' Governance and Climate Change. Report 4. Centre for Indigenous Environmental Resources, Winnipeg

Centre for Indigenous Environmental Resources (2006b) How Climate Change Uniquely Impacts The Physical, Social and Cultural Aspect of First Nations. Report 2. Centre for Indigenous Environmental Resources, Winnipeg

Centre for Indigenous Environmental Resources (2008) Climate change and First Nations south of 60: impacts, adaptation and priorities final report. Centre for Indigenous Environmental Resources, Winnipeg

Chapin FS III, Peterson G, Berkes F, Callaghan TV, Angelstam P, Apps M, Beier C, Bergeron Y, Crépin A-S, Danell K, Elmqvist T, Folke C, Forbes B, Fresco N, Juday G, Niemelä J, Shvidenko A, Whiteman G (2004) Resilience and vulnerability of northern regions to social and environmental change. Ambio 33(6):344–349

Chiefs of Ontario (2007) Aboriginal traditional knowledge and source water protection: First Nations' views on taking care of water: Chiefs of Ontario

Christensen R (2006) Waterproof 2: Canada's drinking water report card. Sierra Legal Defence Fund, Toronto

Christensen R, Parfitt B (2001) Waterproof: Canada's drinking water report card. Sierra Legal Defence Fund, Toronto

City of Toronto (2009) Toronto water at a glance. Accessed on January 22, 2009 at http://www.toronto.ca/water/glance.htm. Unpublished

Cromwell JE III, Harner WL, Africa JC, Schmidt JS (1992) Small water systems at a crossroads. J Am Water Works Assoc 84:40–48

de Loë R, Di Giantomasso S, Kreutzwiser RD (2002) Local capacity for groundwater protection in Ontario. Environ Manage 29(2):217–233 de Loë R, Kreutzwiser R, Moraru L (2001) Adaptation options for the near term: climate change and the Canadian water sector. Glob Environ Change 11:231–245

de Loë RC, Berg A (2006) Mainstreaming climate change in drinking water source protection planning in ontario. Pollution Probe, Ottawa

de Loë RC, Kreutzwiser RD (2006) Challenging the status quo: the evolution of water governance in Canada. In: Bakker K (ed) Eau Canada: the future of Canadian water governance. University of British Columbia Press, Vancouver

Eggertson L (2008a) Despite federal promises, First Nations' water problems persist. Can Med Assoc J 178(8):995

Eggertson L (2008b) Investigative report: 1766 boil-water advisories now in place across Canada. Can Med Assoc J 178(10):1261–1263

Folke C (2003) Freshwater for resilience: a shift in thinking. Philos Trans R Soc B: Biol Sci 358:2027–2036

Folke C, Carpenter S, Elmqvist T, Gunderson L, Holling CS, Walter B (2002) Resilience and sustainable development: building adaptive capacity in a world of transformations. Ambio 31(5):437–440

Folke C, Colding J, Berkes F (2003) Synthesis: building resilience and adaptive capacity in social–ecological systems. In: Berkes F, Folke C, Colding J (eds) Navigating social–ecological systems: building resilience for complexity and change. Cambridge University Press, Cambridge

Ford JD, Smit B (2004) A framework for assessing the vulnerability of communities in the canadian arctic to risks associated with climate change. Arctic 57(4):389–400

Fowler HJ, Kilsby CG, O'Connell PE (2003) Modeling the impacts of climatic change and variability on the reliability, resilience, and vulnerability of a water resource system. Water Resour Res 39(8):10(1)–10(11)

Franks T (1999) Capacity building and institutional development: reflections on water. Public Adm Dev 19:51–61

Furgal C, Prowse T (2008) Northern Canada. In: Lemmen DS et al (eds) From impacts to adaptation: Canada in a changing climate 2007. Natural Resources Canada, Ottawa

Furgal C, Seguin J (2006) Climate change, health, and vulnerability in Canadian northern Aboriginal communities. Environ Health Perspect 114(12):1964–1970

Geldof GD (1995a) Adaptive water management: Integrated water management on the edge of chaos. Water Sci Technol 32(1):7–13

Geldof GD (1995b) Policy analysis and complexity: a non-equilibrium approach for integrated water management. Water Sci Technol 31(8):301–309

Graham J (2003) Safe water for First Nations: charting a course for reform, vol Policy Brief No. 14. Institute on Governance, Ottawa

Grayson RB, Doolan JM, Blake T (1994) Application of AEAM (Adaptive Environmental Assessment and Management) to water quality in the Latrobe river catchment. J Environ Manage 41:245–258

Gunderson LH, Carpenter SR, Folke C, Olsson P, Peterson G (2006) Water RATs (resilience, adaptation, and transformability) in lake and wetland social–ecological systems. Ecol Soc 11(1):16, [Online] URL: http://www.ecology and society.org/vol1/iss1/art16/

Harden A, Levalliant H (2008) Boiling point: six community profiles of the water crisis facing First Nations within Canada. Polaris Institute, Ottawa

Health Canada (2008) Drinking water advisories. Accessed 7 Jan 2008. Unpublished

Hrudey SE (2008) Safe water? Depends on where you live! Can Med Assoc J 178(8):975

Indian and Northern Affairs Canada (2003) National assessment of water and wastewater systems In First Nation communities: summary report. Indian and Northern Affairs Canada, Ottawa

Indian and Northern Affairs Canada (2008) Plan of action for drinking water in First Nations communities: progress report. Indian and Northern Affairs Canada, Ottawa

Indian and Northern Affairs Canada and Health Canada (2007) Summative evaluation of the First Nations water management strategy. Indian and Northern Affairs Canada, Ottawa

Intergovernmental Panel on Climate Change (2007) Summary for policymakers. In: Parry ML et al (eds) Climate change 2007: impacts, adaptation and vulnerability. contribution of working group II to the fourth assessment report of the intergovernmental panel on climate change. Intergovernmental Panel on Climate Change, Cambridge

Ivey JL, Smithers J, de Loë RC, Kreutzwiser RD (2004) Community capacity for adaptation to climate-induced water shortages: linking institutional complexity and local actors. Environ Manage 33(1):36–47

Kabat P, van Schaik H (2003) Climate changes the water rules: how water managers can cope with today's climate variability and tomorrow's climate change. Dialogue on Water and Climate, The Netherlands

Kolisnek P, de Loë R (2005) Assessing the vulnerability of a regional water supply system to climate-induced water shortages in Victoria, British Columbia. Paper in Reflections on Our Future: Into a New Century of Water Stewardship.Proceedings of the CWRA 58th Annual Conference, Banff, Alberta, Cambridge, ON: Canadian Water Resources Association, 14–17 June 2005

Kreutzwiser R, de Loë R (2004) Water Security: From Exports to Contamination of Local Water Supplies, Chapter in *Resource and Environmental Management in Canada: Addressing Conflict and Uncertainty*. In: Mitchell B (ed), 3rd edn. Oxford University Press, Don Mills

Kreutzwiser RD, de Loë RC (2002) Municipal capacity to manage water problems and conflicts: the Ontario experience. Can Water Resour J 27(1):63–83

Kundzewicz ZW, Mata LJ, Arnell NW, Doll P, Kabat P, Jimenez B, Miller KA, Oki T, Sen Z, Shiklomanov IA (2007) Freshwater resources and their management. In: Parry ML et al (eds) Climate Change 2007: impacts, adaptation and vulnerability. contribution of working group II to the fourth assessment report of the intergovernmental panel on climate change. Intergovernmental Panel on Climate Change, Cambridge

Lemmen DS, Warren FJ, Lacroix J, Bush E (2008) From impacts to adaptation: Canada in a changing climate 2007. Government of Canada, Ottawa

McDonald AT, Kay D (1988) Water resources: issues and strategies. Longman Scientific and Technical, Harlow, Essex

Melosi MV (2000) Pure and plentiful: the development of modern waterworks in the United States, 1801–2000. Water Policy 2(4–5):243–265

Milly PCD, Betancourt J, Falkenmark M, Lettenmaier D, Stouffer RJ (2008) Stationarity is dead: whither water management. Science 319(5863):573–574

Nadarajah R, Miller S (2005) Comments of the Canadian environmental law association regarding 'watertight: the case for change in Ontario's water and wastewater sector', Publication NO. 522. Canadian Environmental Law Association, Toronto

National Aboriginal Health Organization (2002) Drinking water safety in Aboriginal communities in Canada. Brief. Unpublished

National Round Table on the Environment and the Economy (1996) State of the debate on the environment and the economy: water and wastewater services in Canada. National Round Table on the Environment and the Economy, Ottawa

Neegan Burnside Engineering and Environmental Ltd (2005) Six Nations of the Grand River hydrogeological study. Six Nations Council, Ohsweken

Office of the Auditor General (2005) Drinking water in first nations communities. Report of the commissioner of the environment and sustainable development to the house of commons. Minister of Public Works and Government Services. Canada, Ottawa

Ontario Clean Water Agency (2002) Six Nations of the Grand River (Band No. 121). Assessment study of water and wastewater systems and associated water management practices in Ontario First Nation communities. Ontario Clean Water Agency, Toronto

Pahl-Wostl C (2007) Transitions towards adaptive management of water facing climate and global change. Water Resour Manage 21(1):49–62

Pahl-Wostl C, Sendzimir J (2005) The Relationship Between IWRM and Adaptive Water Management. NeWater working paper 3, NeWater

Parry ML, Canziani OF, Palutikof JP, van der Linend PJ, Hanson CE (2007) Impacts, adaptation and vulnerability: contribution of Working Group II to the fourth assessment report of the Intergovernmental Panel on Climate Change. Cambridge, UK

Plummer R, Spiers A, FitzGibbon J, Imhof J (2005) The expanding institutional context for water resources management: the case of the Grand River watershed. Can Water Resour J 30(3):227–244

Reading JL, Kmetic A, Gideon V (2007) First Nations wholistic policy and planning model: discussion paper for the world health organization commission on social determinants of health. Assembly of First Nations, Ottawa

Regional municipality of Waterloo (2009) population profile. Accessed at http://www.region.waterloo.on.ca/web/region.nsf/DocID/0776E1882A72B3DC85256B1B006F8ADB? Open Document. Unpublished

Royal Commission On Aboriginal Peoples (1996) Restructuring the Relationship, Chapter in Report of the Royal Commission On Aboriginal Peoples. Ottawa: Report of the Royal Commission On Aboriginal Peoples

Sendzimir J, Light S, Szymanowska K (1999) Adaptively understanding and managing for floods. Environments 27(1):115–136

Six Nations of the Grand River (2009). Six Nations community profile. Available http://www.sixnations.ca/CommunityProfile.htm. Unpublished

Six Nations of the Grand River Environmental Office (2009). Source water protection. Available http://www.sixnations.ca/SWP/. Unpublished

Smit B, Burton I, Klein RJT, Street R (1999) The science of adaptation: a framework for assessment. Mitig Adapt Strat Glob Change 4:199–213

Sprague JB (2006) Great Wet North? Canada's Myth of Water Abundance. In: Bakker K (ed) Eau Canada: the future of Canadian water. University of British Columbia Press, Vancouver

Statistics Canada (2006) Profile of Aboriginal peoples for Canada, provinces, territories, Census divisions and census subdivisions. Accessed January 7, 2008. Unpublished

Statistics Canada (2007) Portrait of the Canadian Population in 2006, 2006 Census. Population and Dwelling Counts, 2006 Census, vol. Catalogue no. 97-550-XIE. Statistics Canada, Ottawa

Swain H, Louttit S, Hrudey S (2006) Report of the expert panel on safe drinking water for First Nations. Minister of Indian affairs and northern development and federal interlocutor for métis and non-status Indians, Ottawa

Swatuk LA (2005) Political challenges to implementing IWRM in Southern Africa. Phys Chem Earth 30:872–880

United Nations World Water Assessment Programme (2006) Water: a shared responsibility. UNESCO and Berghahn Books, Barcelona

United States Environmental Protection Agency (1998) Guidance on implementing the capacity development provisions of the safe drinking water act amendments of 1996. United States Environmental Protection Agency, Office of Water, Washington, DC

Vickers A (2001) Handbook of water use and conservation: homes, landscapes, business, industries, farms. Waterplow, Amherst

Walkem A (2006) The land is dry: Indigenous peoples, water, and environmental justice. In: Bakker K (ed) Eau Canada: the future of Canadian water. University of British Columbia Press, Vancouver

Water Strategy Expert Panel (2005) Watertight: the case for change in Ontario's water and wastewater sector. Queen's Printer for Ontario, Toronto

World Resources Institute (2009) EarthTrends Environmental Information. Water Resources and Freshwater Ecosystems. Data tables. Accessed January 12, 2009. http://earthtrends.wri.org/datatables/index.php?theme=2. Unpublished

Chapter 9
Institutional Fit and Interplay in a Dryland Agricultural Social–Ecological System in Alberta, Canada

Johanna Wandel and Gregory P. Marchildon

9.1 Introduction

The Special Areas of Alberta are an administrative district of 2.1 million hectares in the dry short grass region of the Canadian Great Plains. The Great Plains have endured recurring dry periods that present a challenge for the social–ecological systems present in the area; however, the human settlement of the region has been influenced by a variety of nonclimatic external factors including the in-migration of European peoples and associated population change, land use policy, and global demand for natural resources and agricultural commodities. The indigenous bison-based system was fundamentally changed by European contact during the latter part of the nineteenth century. Since then, the primarily Euro-Canadian social–ecological system can be divided into three distinct periods of steady progress and adaptation, separated by abrupt system changes prompted by both climate and nonclimatic stimuli. This type of abrupt change in social–ecological systems is frequently associated with political upheaval and prompt institutional and societal reorganization (Folke et al. 2005). Global climate change has the potential to challenge the current period of steady progress and adaptation in the Special Areas of Alberta.

This paper examines three successive phases of agricultural and settlement systems in Alberta's Special Areas in the context of drought and other stresses. Shifts between social–ecological systems are viewed through the conceptual lens of institutional fit and interplay in seeking to explain the transition, from range-based agriculture to intensive crop farming, and finally to a predominantly sedentary,

---

J. Wandel
Department of Geography and Environmental Management, University of Waterloo, 200 University Ave. W., Waterloo, ON, Canada N2L 3G1
e-mail: jwandel@uwaterloo.ca

G.P. Marchildon
Johnson-Shoyama Graduate School of Public Policy, University of Regina, 110- 2 Research Drive, Regina, SK, Canada S4S 0A2
e-mail: greg.marchildon@uregina.ca

grazing-based agricultural system. We contend that the Special Areas have enjoyed a high capacity for adaptation to external perturbation since the last major shift in land use, avoiding a wholesale social–ecological reorganization since the 1940s. This high adaptive capacity, as evidenced by the span of the current agricultural and settlement phase, reflects a high degree of institutional fit and interplay to the conditions of the twentieth and early twenty-first centuries. However, projected climate change, coupled with global-scale nonclimatic forces (e.g., rising oil prices), may lead to decreases in fit and interplay and, consequently, adaptive capacity.

Adaptive capacity is generally used to refer to a system's ability to cope with – or adapt to – external stimuli (Smit and Wandel 2006). Recently, scholarship on the determinants of adaptive capacity has emphasized the importance of including governance, regulations, legislation, and the role of formal and informal institutions as part of the analysis (Keskitalo and Kulyasova 2009; Young 2008). Assessing the role of institutions in the adaptive capacity of a system involves the concepts of "fit" and "interplay." Fit describes how a single institution matches a resource problem, while interplay addresses how institutions are linked both horizontally and vertically, and the implications of these links for adaptive capacity.

The term fit relates how well the temporal, spatial, and functional scales of an institution match the system it manages (Folke et al. 1998, 2007; Galaz et al. 2008). Spatial fit refers to the congruence between the geographical extent of a biophysical system and an institution's management area. A "problem of fit" here can be indicative of too small or large an institutional jurisdiction, as well as an institutional inability to cope with external drivers of change (Moss 2007; Galaz et al. 2008). For example, Ablan and Garces (2005) illustrate the spatial mismatch between national exclusive economic zones and marine systems in the South China Sea. In this example, overfishing in one country affects resources available in the exclusive economic zone of another, because fish are not bound by jurisdictional delineations. In this situation, the institutional jurisdiction is too small, as national-level institutions have little control regarding the overall population of migratory fish stocks. However, within large national jurisdictions, such as the Philippines, the nationwide implementation of a single fisheries policy based on the distance from the shore does not fit depth-determined variations in the composition of fish communities (Ablan and Garces 2005). In the second case, the institutional jurisdiction is too large, since the "one size fits all" approach to management is inappropriate for a local context (Galaz et al. 2008).

Temporal fit refers to an institution's management horizons relative to biophysical system change. Temporal fit can be either too short or too long. For example, if an institution's approach to resource management is dictated by four-year election cycles, the temporal fit may be too short to adequately manage depleted resources, which can require decades to recover. Conversely, the speed of ecosystem change can be very rapid, so the temporal fit may be considered too long when institutions do not have processes for a sufficiently quick response. For example, Walters and Maguire (1996) note that a slow response in Canadian fisheries policy to early indications of declining North Atlantic cod stocks contributed to the collapse of the cod fishery in 1992.

Functional fit defines how well an institution's mandate is matched to the system of interest. For example, Bauer (2004) reviews the impacts of Chile's 1981 Water Code, which established water as a commodity. The Water Code, as an institution, has been criticized for failing to promote equity among water users; however, this can also be seen as a case of poor functional fit between an institution designed to manage water quantity allocation and a system in which insecure land tenure and lack of access to securing rights in water markets precluded equitable participation.

The concept of institutional interplay is used to describe how institutions interact with each other (Lebel 2005; Gehring and Oberthür 2007). Interplay can be used to highlight conflicting mandates or other potential sources of conflict, both horizontally and vertically. Horizontally, interplay expresses the relationship between organizations and the various rules they implement (Lebel 2005). Vertical interplay is concerned with cross-scale linkages, for example among regional, provincial, and federal institutions (Lebel 2005).

9.2 The Special Areas of Alberta, Canada

The North American Great Plains have been occupied for the past 11,000 years. Throughout the history of the region, livelihoods have been tied to harvesting and managing natural resources. Prior to European contact, the First Nations peoples primarily survived by hunting bison. Early European settlers established extensive grazing methods, followed by crop-based agriculture. Currently, this region includes a range of agricultural systems, ranging from intensive, irrigated crop agriculture to large grazing operations. This coupling of the native prairie ecosystem and the various social systems –including their associated settlement systems and governance structures – can be viewed as a social–ecological system, as outlined in Berkes and Folke (1998). Analyzed at the regional scale, this social–ecological system is known as the Special Areas administrative unit (commonly and hereafter referred to simply as the "Special Areas").

The Special Areas constitutes a large part of what climatologist Vilmow (1956) originally described as Canada's Dry Belt (Marchildon et al. 2009). The Special Areas covers 2.1 million hectares of sparsely populated land in the Canadian province of Alberta, north of the Red Deer River between 50.6° and 53.3° north and 110° and 112.5° west (see Fig. 9.1). Most of the Special Areas receives less than 350 mm of precipitation each year, making it one of the driest environments in the Canadian Great Plains (Jones 2002). Evaporation generally exceeds precipitation throughout this area, so its contemporary agricultural systems rely on storing winter moisture, groundwater, and the surface water that originates in the Eastern Slopes of the Canadian Rocky Mountains. The area primarily consists of fertile brown and dark brown chernozem soils, with frequent instances of solonetzic soils, which are a product of low precipitation coupled with high evaporation, and are characterized by subsoil that is hard when dry, swollen and compact when wet, and consequently difficult to cultivate (Wandel et al. 2009). The area is part of the dry

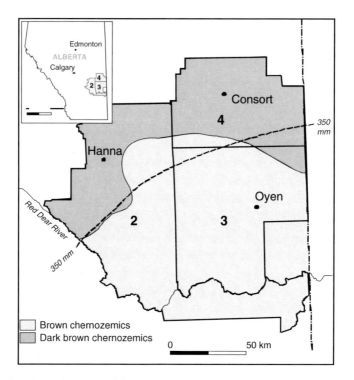

Fig. 9.1 Alberta's special areas, 1941
Source: Wandel et al. 2009

shortgrass prairie ecozone, which is marked by treeless prairie with few trees or shrubs except in particularly wet areas.

Southeastern Alberta was first settled by Europeans beginning in 1881, and was opened for homesteading in 1909. Like much of the Great Plains, the region was subjected to the "dust bowl" years of the 1920s and 1930s (Khandekar 2004), which were accompanied by high rates of out-migration and the attendant collapse of the rural settlement system. The Special Areas administrative unit originated with an institutional reorganization that transferred local government authority to the provincial government as a temporary solution to the crisis of the 1920s and 1930s (Gorman 1988; Marchildon 2007). This administration survives today in the portion of the region illustrated in Fig. 9.1.

Bison (estimated to number between 30 and 60 million) were the staple resource for the First Nations inhabitants of the North American Great Plains for approximately 11,000 years prior to European contact. By AD 100, the bow and arrow had begun to supplement the spear and throwing stick for hunting bison. Pedestrian hunters also herded bison into jumps (cliffs and other natural drops) and pounds (constructed corrals). Bison grazed on the short grass prairie, vegetation that the First Nations peoples altered only when they burned it in order to drive bison herds to organized kill locations (Daschuk 2009).

While the northern Great Plains have always been relatively dry, the area has been subject to major climatic perturbation in the past. The warming of the Neo-Atlantic Climatic Episode between AD 850 and 1200 allowed for the establishment of Aboriginal maize-based farming in the southern parts of the Great Plains. The following Little Ice Age from the thirteenth to the early nineteenth centuries prompted Aboriginal in-migration to the northern Great Plains after the collapse of maize farming increased the appeal of the regularity and reliability of bison herd movements in this region.

European contact further changed the Aboriginal social–ecological systems. The introduction of horses and guns facilitated bison hunting, and the European-based fur trade created a demand for industrial-scale production of pemmican (dried and pounded bison meat and fat) by both Aboriginal groups living on the Plains, and Métis (people of mixed European and Aboriginal descent) hunters who moved into the western Great Plains as they began to deplete their traditional bison hunting areas in the Red River Valley.

An increasing demand for bison hides and preserved bison tongues, coupled with advances in hunting technology in the form of fast-loading and accurate rifles, led to the near extinction of the bison in the late nineteenth century (Arthur 1984). The collapse of the bison hunt, and its associated economic and food insecurity prompted the settlement of Treaty 6 (1876) and Treaty 7 (1877) with the Plains Cree and other First Nations. The treaty process re-settled Aboriginal groups spread throughout the Canadian Plains to smaller plots of land known as reserves. Shortly after this re-settlement, the predominantly European settlement systems and attendant formal institutional structures were established in the present-day region of the Special Areas.

9.2.1 Phase I: Open-Range Ranching, 1880–1906

The first European social–ecological system to be established in southeastern Alberta was open-range cattle ranching. Aboriginal re-settlement onto reserves left the present-day Special Areas vacant, and ranchers began to establish large range-based operations. The establishment of large-scale ranching was facilitated by a number of economic and institutional developments. A large market for live cattle existed in Britain, and major cattle businesses began to emerge in the United States to meet this demand (Breen 1983). The Canadian federal government, eager to stimulate similar export-oriented businesses in Canada, passed an order-in-council to permit 21-year leases of land of up to 100,000 acres (40,468 ha) at the rate of one cent per acre per year. These leases were "closed", meaning that they explicitly prohibited homestead settlement and thus resulted in a very low population density. Furthermore, in order to encourage cattle ranching in the Canadian Great Plains, the Canadian federal government permitted ranchers to bring cattle duty-free from the United States for 2 years. It is estimated that approximately one-third of all the land in southwestern Alberta was soon being used for grazing in this manner (Jameson 1986).

Open-range ranching reached its zenith by the late 1890s, but institutional changes coupled with a climatic stimulus in the early twentieth century precipitated a crisis causing the collapse of open range ranching in the Special Areas (Evans 1983). There were a number of factors that caused this crisis. First, there was a major change in the market for live cattle after 1900. With improvements in chilling methods, British imports of chilled beef from Argentina began to supplant live cattle imports from Canada. Second, in 1904, the federal government abolished large leases, and shifted toward policies that heavily favored homestead settlement over ranching. Finally, the particularly cold "Killer Winter" of 1906–1907 decimated the cattle herds of the Canadian West. The portion of the Great Plains most affected by this was the short grass prairie of the Dry Belt. Evans (1983) estimates that between 60 and 65% of cattle stock in short grass ranching operations perished.

Large ranch operations were in a poor position to adapt to the Killer Winter, largely because of the shift in institutional attitude to favor smaller, crop-based homesteading operations. Smaller, more marginal ranchers were in a better position to survive stock losses and less secure land tenure arrangements (Elofson 2000). Adaptation strategies included the introduction of grain cultivation alongside now smaller cattle operations, or even a wholesale shift to grain farming (Evans 1983)

The open-range social–ecological systems described above illustrates a shift from relatively high institutional fit and interplay to an institutional reorientation, which in conjunction with a climatic stimulus, was a poor fit for open range cattle ranching. The original 21-year, 100,000 acre grazing leases were a good spatial and temporal fit for open-range ranching. Functionally, the federal government of the 1880s sought to facilitate extensive cattle ranching with the removal of import duties on cattle in the first 2 years. Domestically, however, the area was under federal jurisdiction, and thus there was a lack of interplay, with (largely nonexistent) local and regional institutions. Local-level institutions were unlikely to emerge with the relatively low population density, as the area had a total settler population of 75 in 1901 (Wandel et al. 2009). Furthermore, there were no long-range international trade negotiation and thus no replacement market when Britain shifted its trade focus to Argentina. In the late nineteenth century, the federal government and its institutions were a poor spatial fit for the local social–ecological system. Federal administration was poorly connected to local realities due to a lack of local institutions (and thus no vertical interplay), that left ranching vulnerable to the cyclical droughts of the Dry Belt. Indeed, the abolishment of the large leases and the shift toward homesteading were a poor functional fit for ranching.

9.2.2 Phase II: Monoculture Wheat-Crop Cultivation, 1908–1920s

As noted above, federal policy shifted to emphasize homesteading in the northern Canadian Great Plains in the early nineteenth century. In 1906, two years after the abolishment of the large leases, the area had a population of 800. In 1909, the present-day Special Areas became the last extensive region of the Canadian Great Plains that opened to homestead settlement under the Dominion Lands Act. Settlement was guided by the Dominion Land Survey, which mapped out six mile by six mile by six mile (9.65 by 9.65 km) townships, which were further subdivided into 36 sections of 640 acres (259 ha). Settlers were granted 160 acres (65 ha) on the condition that 25% of the land was cultivated and a permanent homestead established within 3 years. Furthermore, two sections per township were reserved for schools, with the proceeds of land sales expected to fund the school buildings. School districts were formally established by 1911, and municipalities and local improvement districts had formed by 1914. This reorganized the social ecology of the region, now focused on small, intensely cultivated family-run homesteads governed by numerous rural and urban municipalities. This shift to a high density wheat-farming system was reinforced by external climate and macro-economic stimuli during 1915 and 1916. The First World War was accompanied by inflated wheat prices by a rapid inflation in wheat prices – the world price would more than double between 1914 and 1918 – and above average rainfall produced bumper crops in 1915 and 1916 (Jones 2002; Marchildon 2007; Marchildon and Anderson 2006). In-migration continued during this time, and the 1921 Census lists an all-time high of 26,000 settlers in the Special Areas.

Climatic stimuli, however, ultimately triggered the decline of the wheat-farming phase. In 1917, farmers experienced the first in a series of successive droughts. With the exception of one wet year in 1927, droughts persisted every year in the Special Areas until 1939 (Marchildon et al. 2008). Federal response focused on immediate, short-term aid in the form of a total $300,000 of loans for seed grain, fodder, and relief for Alberta (Jones 1985). However, most drought response strategies were left to the provincial governments. This placed tremendous strain on the provincial government of Alberta, which had spent $7 million in relief loans by 1924 (Jones 1985). In 1921, the United Farmers of Alberta were elected to the provincial legislature, promising to address the catastrophic drought situation in the Dry Belt. The new government passed the Drought Relief Act in 1922 and the Debt Adjustment Act in 1923, which instituted debt moratoria and consequently contributed to the high loan default rate (Jones 1985). However, the 1922 Tax Recovery Act forced homesteaders to pay tax arrears or forfeit their land (Jones 1985). As the droughts continued, the provincial government helped fund farmers to move their families, equipment, and supplies into irrigated areas further west and north of Calgary.

The climatic stimulus of prolonged and severe drought meant that municipalities went bankrupt, roads were abandoned, and schools were closed due to the collapse in local property revenues. As noted above, the establishment of the wheat-farming social–ecological system was primarily the result of an institutional re-orientation to homesteading. The system was characterized by unprecedented human population density in the region and a reliance on monoculture cultivation, with an attendant emergence of local institutions. The institutions and policy instruments which guided the emergence of this system, especially the Dominion Lands Act and Dominion Land Survey, were designed to fit a relatively small farm size of 65 ha

(160 acres), and only had provisions to increase farm size to 130 ha (320 acres). Furthermore, the conditions of obtaining secure land tenure prescribed cultivation. Consequently, these instruments had a high degree of spatial and functional fit only if climatic and commodity price conditions were such that small scale crop farming was profitable. Due to higher-than-average rainfall before 1917 and high wheat prices before and during the First World War, small scale farming was financially attractive to farmers in the Special Areas in this period.

Although local-level institutions such as municipal governments and school districts had been facilitated by the federal government policy – although the province of Alberta was created in 1905, it did not gain control over public lands and resources until 1930 – there was little vertical interplay. Generally, local level institutions relied on local revenues, while federal drought response was limited to providing minimal relief (i.e., basic foodstuffs and seed) to farm families in only the most extreme circumstances. Provincially, there was insufficient interplay between local institutions and provincial aid strategies to prevent the collapse of local institutions, and provincial strategies exacerbated this by facilitating out-migration. Generally, the institutions of the early wheat-farming phase were designed to fit a federally designed and desired outcome for the Canadian Plains as a whole, but they did not fit the long-term climatic and macro-economic reality of the Dry Belt of southeastern Alberta.

9.2.3 Phase III: The Special Areas and Mixed Ranching, 1930s to Present

The prolonged droughts and associated dramatic failure of the wheat-based crop farming social–ecological system of southeastern Alberta prompted experimentation with strategies beyond short-term relief at the provincial government level. The first of these experiments was in the Tilley East Area immediately north of the city of Medicine Hat (see Fig. 9.2). In 1927, the provincial government passed a law, placing the area under the stewardship of a single board that would facilitate taking over land from bankrupt farmers and publicly manage access to both land and water in order to allow the more sustainable, larger ranches and ranch-farms to expand. The major impediment to the scheme was the fact that, as in the case of the two other prairie provinces of Saskatchewan and Manitoba, the federal government – not the province of Alberta – owned public (Crown) lands as their resources. Only after the 1930 federal transfer of public lands and resources to the three provinces, could the government of Alberta execute its first Special Areas scheme (Marchildon 2007).

The experiment proved successful enough that the Alberta government extended the administrative control of the existing board to the Berry Creek Area directly north of Tilley East in 1932. In 1935, under the Special Municipal Areas Act, Sounding Creek, Neutral Hills, and Sullivan Lake Special Areas were added to Tilley East and Berry Creek. Two years later, the Bow West Special Area was also added. Finally, in 1938, the provincial government amalgamated all of the

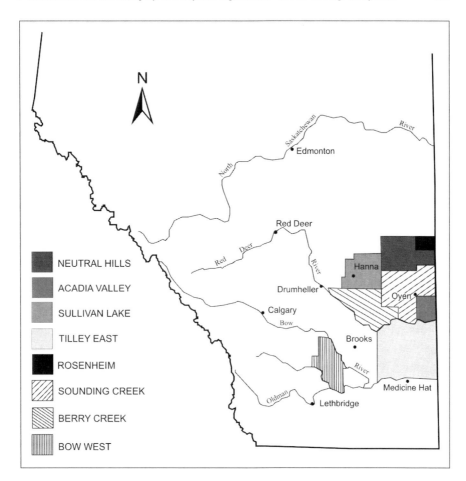

Fig. 9.2 Alberta's present-day special areas
Source: Marchildon 2007

municipalities and improvement districts into a single Special Areas administration headquartered in Hanna. With local government eliminated, the new administrative entity was mandated to manage all public resources, including land and water resources, as well as roads and schools throughout the Special Areas. Appointed by the provincial government in Edmonton, the Special Areas Board was conferred all the legal and administrative tools required to manage the area in the manner it deemed most efficient for the remaining residents (Gorman 1988; Marchildon 2007).

The main objective of the Special Areas Board and its predecessors was to reduce the population of the region while transforming small wheat farms into larger ranches and ranch-farms. The government administrators took over bankrupt farm lands and then leased these lands at inexpensive rates; in 1938, for example, grazing lands were leased for 2.5 cents per acre, while crop lands could be rented for a one-sixth share of the annual crop. The boards also created community pastures out of Crown lands, offering everyone within a particular district inexpensive access to grazing lands. By 1941, the population in the Special Areas was 15,000, roughly one-half of what it had been 10 years before, while the average farm was double the provincial average in size, and would be 3.6 times larger than the provincial average by the mid-1950s (Marchildon 2007).

Through most of the postwar era, rural society in the Special Areas, including the leases, community pastures, roads, and schools upon which ranchers and farmers are dependent, have been administered by the Special Areas Board. In effect, the Special Areas Board takes the place of local, democratically elected, government. Despite the fact that residents have not faced droughts on the scale suffered before the Second World War, and despite commissions that have offered the possibility of the restoration of local self-government, the majority of the people in the region have consistently supported this unique institutional arrangement, and it has persisted to the present day in the former Neutral Hills, Sullivan Lake, Sounding Creek, and Berry Creek Special Areas. It is noteworthy that the portions of the Special Areas that benefited from irrigation (Acadia Valley, Bow West) or military base development (Tilley East) after the 1930s are no longer part of this institutional arrangement.

Currently, the land tenure system in the Special Areas is comprised of 60% public land, 91% of which is administered through grazing and cultivation leases. Lease land within the Special Areas has been assessed for both suitability for cultivation and carrying capacity for grazing, and only a few parcels in particularly moist areas in river valleys can be cultivated under the terms of the lease. The bulk of the land is under 20-year renewable grazing leases, in which the terms of the lease specify a maximum carrying capacity, and lease land is inspected twice annually for signs of overgrazing (Slemp 2006). Consequently, the current institutions explicitly specify a particular land use, ranching, on a large portion of the land base of the region. Furthermore, the provincial government of Alberta directly subsidizes ranching through the maintenance of five community pastures on grazing reserves within the Special Areas. The community pastures are designed to help farmers build herds through access to both land and bulls for breeding (Hyland 2006). Priority is given to ranchers with fewer resources for cattle raising, thus facilitating new entrants into ranching operation. There is no similar support system for dryland farming in the region, and thus the current institutional structure implicitly promotes a conversion to an extensive land use which does not involve cultivation. These institutional strategies for promoting ranching are managed to fit within the Special Areas Board's mandate to "drought-proof" the area, in this case, to retain or return as much of the land as possible to permanent grass cover and to maintain healthy grassland systems on both leased and grazing reserve land.

The land tenure system in the Special Areas illustrates particularly good fit in spatial, temporal and functional terms. Spatially, the large land base allows for low density stocking rates on very large tracts of land including several farm operations over 100,000 hectares. Temporally, the renewable 20-year grazing leases allow for investment in good grass cover, with many ranchers re-seeding native grass species. In terms of ecological resilience to drought, ranching preserves the moisture in short grass prairie better than commercial crop farming. Furthermore, the Special Areas Board and its attendant administration is not subject to recurring election cycles, and administrators remain in their position for many years. Functionally, the Special Areas Board can emphasize drought-proofing in light of persistent dry periods and carry out its mandate without catering to election cycles.

In the present-day Special Areas, both horizontal and vertical interplay are well developed and match the current social–ecological system. Horizontally, the organization tasked with implementing the land tenure system, the Special Areas administration, closely matches its mandate. Vertically, the Special Areas administration is under the direct supervision of the Alberta government's Department of Municipal Affairs. Furthermore, the Special Areas administration maintains a small irrigation project in the southern portion of the region and has managed to enhance security of irrigation water in collaboration with the provincial government's Department of the Environment by installing pumps on the Red Deer River to recharge one reservoir. In doing this, the Special Areas administration works in collaboration with a private utility company, ATCO Power, to pump water from the river to another multi-purpose reservoir which serves as a cooling pond for a thermal generating station, local recreational area, and irrigation canal recharge for the residents in the region.

Vertical interplay is also evident in terms of the government of Alberta's collaboration with the government of Canada. Through a federal-provincial agreement known as the Master Agreement on Apportionment administered by the Prairie Provinces Water Board, a minimum of 50% of the natural flow of rivers in the Canadian plains must be passed on to the next downstream province. The southern portion of the Province of Alberta is increasingly water stressed, and portions of the South Saskatchewan River Basin (SSRB) are over-allocated for use. In Alberta, water diversions are subject to volume-based water licenses, and license priority is determined by its date of issue. Most of the Special Areas is part of the Red Deer sub-basin of the SSRB, and while the Red Deer River is not currently over-allocated, the more southern sub-basins of the St. Mary's, Oldman and Bow Rivers are over-allocated. As a consequence, the flow of the Red Deer River is frequently used to meet Alberta's apportionment to Saskatchewan as per the federal policy. In turn, this federal policy places limits on the volume of water which can be used – and licensed – by Alberta Environment, and consequently influences whether or not the large-scale diversions are approved.

Currently, the Prairie Association of Water Managers (PAWM), a Special Areas-based organization working in conjunction with Special Areas personnel, has proposed a large-scale diversion from the Red Deer River involving a 4.5 km long pipeline, 80 km of canals and existing reservoirs in northern Special Areas. The proposed mean diversion of 53,000 cubic meters allow for almost 40,000 hectares of supplementary irrigation as well as stock watering (to meet the drinking needs of livestock) during particularly dry years, and thus increase the resilience of the Special Areas to extreme dry conditions (Special Areas Board 2005). The license application was submitted in 2003; to date, PAWM has yet to receive an allocation from Alberta Environment. Furthermore, their project proposal has led to conflict with environmental non-governmental organizations and the upstream City of Red Deer. It is expected that, as Alberta's population continues to grow and water needs increase, conflict will also increase and vertical interplay will continue to weaken.

9.3 Discussion and Conclusion

The Special Areas of Alberta have undergone three dominant social–ecological system phases since the European settlement was established in the late nineteenth century. Institutional and societal reorganization is both triggered, and then reshaped, by a combination of environmental, economic and political factors creating, in effect, a new social–ecological system in each phase. The establishment of the open ranching phase began with the slaughter of the vast bison herds driven by the external demand for bison products and the containment and enclosure of Aboriginal peoples. This first phase came to a close as a result of the loss of institutional support for extensive ranching coupled with the extreme climatic event of the winter of 1906–1907. The second monoculture wheat-farming phase was triggered by government support of family farm settlement, furthered by favorable climatic conditions and rising wheat prices. This social–ecological system collapsed as a result of the climatic stimulus of prolonged drought combined with the inability of local, regional and national institutions to respond to the disaster. The third and current phase of Special Areas administration based on mixed ranching has lasted longer than any other phase of Euro-Canadian settlement on the Canadian Prairies. The administrative structure which was established during the most recent institutional and societal reorganization, although intended as a temporary measure, has persisted to the present day.

During the wheat monoculture phase, the fit between the ecosystem and local and regional institutions was poor relative to the first and third phases. Homesteaders were expected to subsist on 160 acres of cultivated land. The homesteading system relied on the associated high population density for the maintenance of local government and associated services such as schooling. Institutional reorganization radically changed the desired outcome from small scale crop-based farming to extensive, ranching based livelihoods. In this case, the current institutions shaping land use are comprised of a system of land tenure which retains at least half of the region's area in public ownership and administered under restricted use leases, a shift in agricultural norms from cash cropping to grazing, and the Special Areas Board's "drought-proofing" mandate. In effect, the Board is facilitating the conservation of moisture-stressed short grass prairie, the key economic element in the ecosystem.

Since their establishment, the Special Areas have remained relatively resilient to repeated short-term droughts. The average farm size in the Special Areas became much larger than the average farm size in the rest of Alberta due to provincial policy. By the census of 1941, the average farm in the Special Areas was a little more than two times the size of the average provincial size. By 1956, this average

Special Areas farm was almost four times the size of the average farm in the province (Marchildon 2007). This was accompanied by a further significant decrease in population from just over 15,000 in 1941 to 11,000 by the mid-1970s (Wandel et al. 2009). Since that time, the population of the Special Areas has remained relatively constant, indicating a relatively stable system.

The stability of the current socio–ecological system was illustrated during the most recent drought disturbance in the Special Areas. Southeastern Alberta was faced with a severe drought from 2001 to early 2003, to the point where some of the large irrigation districts in the southern part of the SSRB ran short of water. The Special Areas were subject to poor grass growth, crop failures among dryland farmers, and a concurrent grasshopper outbreak which decimated what vegetative growth there had been. The drought event did not, however, lead to widespread bankruptcy, in part due to the already high resource accumulation which allowed individual farmers to draw on costly adaptation strategies such as hauling stock water and importing feed for livestock. Furthermore, several dryland farmers converted their operation to mixed ranching or abandoned cropping in favor of extensive grazing operations altogether (Wandel et al. 2009). Furthermore, the current social–ecological system of the Special Areas seems resilient to more than climatic stimuli. Despite changing commodity prices, including the crash of beef prices following the confirmed report of *bovine spongiform encephalopathy* (BSE) in Alberta in 2003, and recurring dry periods, the area has continued to support a viable agricultural economy.

However, southeastern Alberta has not been faced with prolonged and severe drought on the scale of the 1920s and 1930s. Both farmers and regional institutional representatives note that they tend to plan for drought events lasting a maximum of 3 years (Wandel et al. 2009). Climate records show that there have been no severe multi-year droughts since the mid twentieth century. Furthermore, the last 60 years have exhibited one of the most favorable moisture regimes in the past 400 years (Sauchyn 2007). Climate scenarios, however, project warmer, drier summers and decreased flow levels in the South Saskatchewan River's tributaries throughout the twenty-first century (Lapp et al. 2009). Given the expected climate change coupled with the insights from the 400-year moisture record presented in Sauchyn (2007), it is likely that the region will be subject to longer, more severe droughts than it has experienced over the past 70 years. A social–ecological system and its attendant institutions designed to match drought events lasting three years or less may face unprecedented challenges in the coming decades.

Another major challenge for the current social–ecological system is Alberta's oil and gas industry. Currently, many farm and ranch operations in the area lease small parcels of land to oil and gas companies for wells and pumps. To date, the revenues from these leases has increased capital accumulation on farm operations and provided off-farm employment to those wishing to supplement their incomes. However, the oil and gas industry places high demands on Alberta's water resources, and this introduces potential conflict. Since 2005 alone, surface water use in Alberta's petroleum industry has increased by 3.67% (CAPP 2009). Furthermore, as oil and gas revenues take on increasing importance in the regional economy, the institutional fit may decrease. Schindler and Donahue (2006) note that, to date, the cumulative effects of drought and human use of Alberta's water resources have seldom been considered in institutional arrangements. The Special Areas administration and its associated land tenure system were designed for the specific purpose of managing an agricultural resource in light of the recurring drought. Increasing resource use conflicts have already become evident in the case of concerns over coal bed methane contaminating the groundwater resource (Bester 2007) and maintaining water reserves for upstream growth in the City of Red Deer (Williams 2006). As the conflict among the various institutional and noninstitutional actors increases, the currently successful institutional interplay will be challenged.

There are a number of broad public policy lessons or implications that can be drawn from this case study of the Special Areas of Alberta. The first is that even in countries with well-developed civil society institutions such as Canada, the state plays a major, if not determinative, role in shaping the social–ecological system through the policies, laws, regulation, subsidies and other instruments at its disposal to support or reshape institutions. Although civil society also engages in institution-building, governments have certain advantages in that they can pass and enforce laws, impose and collect tax revenues, use the funds so collected to subside or encourage activities, as well as actively manage land, water and other resources in which the state has some ownership or regulatory control. Moreover, because government have a mandate to act for all of its citizens or residents unlike civil society organization whose mandate is limited to its membership, the state (whether defined as local, regional or national government, or some combination of all) has an obligation to act in the public interest in the event of social–ecological crisis.

The second lesson is that state involvement and intervention can be critical in determining the nature and quality of the institutional fit. If the state intervenes to encourage a mode of socio-economic habitation that conflicts with, or destroys, the long-term ecological sustainability of land and water resources, for example, then this will produce social and ecological dislocation. Based upon a management horizon that was too short-term, the federal government recklessly encouraged farm settlement and monoculture wheat cropping in the Dry Belt of the Palliser Triangle. Despite evidence that this region was subject to cyclical bouts of prolonged drought, the federal government treated this high-risk part of the Palliser Triangle the same as other, lower-risk areas, of the Canadian Plains for the purposes of settlement policy (Marchildon 2009). It must also be recognized, however, that the crisis produced by such poor spatial and temporal fit may elicit a more desirable set of public policies from the state, thereby facilitating a better institutional fit. For example, the government of Alberta's response to the crisis of drought in the 1920s and 1930s produced an innovative response through the creation of the Special Areas and the rapid transition from monoculture agriculture to mixed farm-ranch agriculture that fitted much better with the soil, water and long-term climate features of the region.

The third lesson is that the state is not monolithic. The state is made up of various levels of government, each with its own roles, responsibilities and capacities. Therefore, it is important to determine the level of government that is most appropriate, given such roles, responsibilities and capacities as well as the size and peculiarities of the region and population that most requires state intervention in order to avoid or mitigate a social–ecological catastrophe. Of course, the appropriate or effective level of government to ensure effective spatial fit depends on numerous factors including constitutional authority, effective responsibility, fiscal capacity, governance competency, and proximity to the population most affected by the lack of the institutional fit. The government of Alberta was prevented from instituting the Special Areas until it gained constitutional authority over land and resources from the federal government in 1930. At the same time, the convergence of a fiscally limited local government with a federal government that was too distant from the population in this unique region within the Palliser Triangle meant that the provincial government was the most logical choice for intervention. At the same time, given the fact that rivers cross provincial boundaries, the federal government must play a role in avoiding beggar-thy-neighbor policies in order to protect the national interest. This has been achieved through intergovernmental collaboration through federal-provincial agreements such as the Master Agreement on Apportionment or inter-provincial institutions such as the PAWM.

References

Ablan MCA, Garces LR (2005) Exclusive economic zones and the management of fisheries in the South China Sea. In: Ebbin SA, Hoel AH, Syndes AK (eds) A sea change: the exclusive economic zone and governance institutions for living marine resources. Springer, New York, pp 136–149

Arthur G (1984) The North American Plains bison: a brief history. Prairie Forum 9(2):281–289

Bauer CJ (2004) Results of Chilean water markets: empirical research since 1990. Water Resources Research 40:1–11

Bester D (2007) Personal communication with Chair of Butte Action Committee, Innisfail, Alberta, February 2007

Breen DH (1983) The Canadian Prairie West and the ranching frontier, 1874–1924. University of Toronto Press, Toronto

Berkes F, Folke C (1998) Linking social and ecological systems for resilience and sustainability. In Berkes F, Folke C (eds), Linking Social and Ecological Systems: Management Practices and Social Mechanisms for Building Resilience. Cambridge: Cambridge University Press, pp 1–25

Canadian Association of Petroleum Producers (CAPP) (2009) Using water. http://www.capp.ca/environmentCommunity/water/Pages/UsingWater.aspx. Accessed 9 Jul 2009

Daschuk J (2009) A dry oasis: the Canadian plains in late prehistory. Prairie Forum 34(1):1–29

Elofson WM (2000) Cowboys, gentlemen and cattle thieves: ranching on the western frontier. McGill-Queen's University Press, Montreal, Kingston

Evans SM (1983) The end of the open range era in Western Canada. Prairie Forum 8(1):71–87

Folke C, Pritchard L, Berkes F, Colding J, Svedin U (1998) The problem of fit between ecosystems and institutions. IHDP Working Paper No. 2. International Human Dimensions Program on Global Environmental Change, Bonn, Germany. Online at http://www.ihdp.uni-bonn.de/html/publications/workingpaper/wp02m.htm

Folke C, Hahn T, Olsson P, Norberg J (2005) Adaptive governance of social–ecological systems. Annu Rev Environ Resour 30:441–473

Folke C, Pritchard L Jr, Berkes F, Colding J, Svedin U (2007) The problem of fit between ecosystems and institutions: ten years later. Ecol Soc 12(1):30, [online] URL: http://www.ecologyandsociety.org/vol12/iss1/art30/

Galaz V, Olsson P, Hahn T, Folke C, Svedin U (2008) The problem of fit among biophysical systems, environmental and resource regimes, and broader governance systems: insights and emerging challenges. In: Young OR, King LA, Schroeder H (eds) Institutions and environmental change. MIT, Cambridge, pp 147–186

Gehring T, Oberthür S (2007) Interplay: exploring institutional interaction. In: Young OR, King LA, Schroeder H (eds) Institutions and environmental change. MIT, Cambridge, pp 187–223

Gorman J (1988) A land reclaimed: a story of the special areas in Alberta. Special Areas Board, Hanna, AB

Hyland R (2006) Personal communication with Manager of Bullpound Pastuer, Special Areas. November 2006

Jameson SS (1986) The ranching industry of Western Canada: its initial epoch, 1873–1910. Prairie Forum 11(2):229–242

Jones DC (1985) The Canadian Prairie dryland disaster and the reshaping of 'expert' farm wisdom. J Rural Stud 1(2):135–146

Jones DC (2002) Empire of dust: settling and abandoning the prairie dry belt. University of Calgary, Calgary

Keskitalo ECH, Kulyasova AA (2009) The role of governance in community adaptation to climate change. Polar Res 28:60–70

Khandekar ML (2004) Canadian prairie drought: a climatological assessment. Alberta Environment Publication T/787, Edmonton

Lapp S, Sauchyn D, Toth B (2009) Constructing scenarios of future climate and water supply for the SSRB: use and limitations for vulnerability assessment. Prairie Forum 34(1):153–180

Lebel L (2005) Institutional dynamics and interplay: critical processes for forest governance and sustainability in the mountain regions of northern Thailand. In: Huber UM, Bugmann HKM, Reasoner MA (eds) Global change and mountain regions. Springer, Berlin, pp 531–540

Marchildon GP (2007) Institutional adaptation to drought and the special areas of Alberta, 1909–1939. Prairie Forum 32(2):251–272

Marchildon GP (2009) The Prairie farm rehabilitation administration: climate crisis and federal–provincial relations during the great depression. Can Hist Rev 90(2):275–301

Marchildon GP, Anderson C (2006) Economic, demographic and agricultural statistics for Alberta, Saskatchewan and Manitoba, 1905–2005. Institutional Adaptation to Climate Change working paper. http://www.parc.ca/mcri/iacc039.php. Accessed 16 Jul 2009

Marchildon GP, Kulshreshtha S, Wheaton E, Sauchyn D (2008) Drought and institutional adaptation in the great plains of Alberta and Saskatchewan, 1914–1939. Nat Hazards 45(3):391–411

Marchildon GP, Pittman J, Sauchyn D (2009) The dry belt and changing aridity in the Palliser Triangle, 1895–2000. Prairie Forum 34(1):31–44

Moss T (2004) Institutional drivers and constraints of floodplain restoration in Europe. International Journal of River Basin Management 5(2):121–130

Sauchyn D (2007) Climate change impacts on agriculture in the prairies. In: Wall E, Smit B, Wandel J (eds) Farming in a changing climate: agricultural adaptation in Canada. UBC, Vancouver, pp 67–80

Schindler DW, Donahue WF (2006) An impending water crisis in Canada's western prairie provinces. PNAS 103(19):7210–7216

Slemp J (2006). Personal communication with the Chair of the Special Areas Board, Hanna, Alberta, November 2006

Smit B, Wandel J (2006) Adaptation, adaptive capacity and vulnerability. Glob Environ Change 16:282–292

Special Areas Board (2005) Special areas water supply project: project summary. Special Areas Board, Hanna, Alberta

Vilmow JR (1956) The nature and origin of the Canadian dry belt. Ann Assoc Am Geogr 46(1):211–232
Walters C, Maguire JJ (1996) Lessons for stock assessment from the northern cod collapse. Rev Fish Biol Fish 6:125–137
Wandel J, Young G, Smit B (2009) Vulnerability and adaptation to climate change: the case of the 2001–2002 drought in Alberta's special areas. Prairie Forum 34(1):211–234
Williams D (2006) Personal communication with Chair of Prairie Association of Water Managers, Hanna, Alberta, November 2006
Young OR (2008) Institutions and environmental change: the scientific legacy of a decade of IDGEC research. In: Young OR, King LA, Schroeder H (eds) Institutions and environmental change. MIT, Cambridge, pp 3–45

Section II
Frontiers in Adaptive Capacity

Chapter 10
The Learning Dimension of Adaptive Capacity: Untangling the Multi-level Connections

Alan Diduck

10.1 Introduction

As discussed in Chap. 1, social–ecological systems are complex and dynamic, and characterized by cross-scale feedback, rapid change, nonlinearity, surprises, and uncertainty. The governance[1] of such systems is typified by high decision stakes, conflicting interests and knowledge claims, levels of uncertainty that are often in the realm of ignorance and indeterminacy, multiple probative scales of analysis, and seemingly intractable management situations. In such a post normal context, there is a need for governance institutions[2] with high levels of adaptive capacity. Defined generally, adaptive capacity is the capability of a social–ecological system to cope with disturbances and changes while retaining critical functions, structures and feedback mechanisms (Folke 2006; Olsson et al. 2004b). Governance institutions with high levels of adaptive capacity are flexible in problem solving, innovative in developing solutions, and responsive to feedback. What's more, they are capable of learning at and across multiple levels of social organization; i.e., adaptive institutions reflect learning at individual and various collective levels (Armitage 2005; Folke et al. 2003; Walker et al. 2002).

This chapter focuses on the processes that link learning at multiple levels. Governance that facilitates multi-level learning fosters relationships, builds trust,

---

A. Diduck
Environmental Studies Program, The University of Winnipeg, 515 Portage Avenue, R3B 2E9 Winnipeg, Canada
e-mail: a.diduck@uwinnipeg.ca

[1]My definition of governance is similar to the one adopted in Chap. 1: the processes and institutions used to address challenges and create opportunities in society (Armitage et al. 2009).

[2]I have adopted Young's (2002, p. 286) definition of institutions, which is similar in approach and breadth to the definition used in the first chapter: "institutions are the conventions, norms and formally sanctioned rules of a society. They provide expectations, stability and meaning essential to human existence and coordination. Institutions regularize life, support values and produce and protect interest".

reconciles diverse views and interests, resolves conflict, and develops shared understandings of problems and potential solutions. It encourages participants to monitor the outcomes of management initiatives, reflect upon those outcomes, and make necessary adjustments, thus enabling decision making in the face of uncertainty. Additionally, adaptive governance promotes broad-based partnerships conducive to assembling the resources (human, financial, scientific, and political) needed to generate enduring change in the face of management complexity. Further, and most importantly for the purposes of this chapter, adaptive governance involves receptive and deliberative vertical connections across various levels of social organization.

Learning has been the subject of research in resource and environmental governance for several decades. Notable early studies examined the implications of learning through public involvement in management (Glasser et al. 1975), conceptualized international development projects as learning processes (Korten 1980), and advanced the notion of learning through adaptive management (Holling 1978). In the 1990s the field continued to grow, and began to rely increasingly on conceptual or theoretical frameworks drawn from social learning (Webler et al. 1995), collaborative learning (Daniels and Walker 1996), critical pedagogy (Diduck and Sinclair 1997), and transformative learning (Alexander 1999). Since 2000, studies on learning have proliferated, branching into organizational (Fitzpatrick 2006) and policy learning (Haas 2000), and deepening earlier lines of inquiry into communicative action (Wiklund 2005), transformative learning (Diduck and Mitchell 2003), and adaptive management and governance (Armitage et al. 2007). The cross-cutting topic of social learning has attracted particular attention, yielding rich models of collaborative inquiry, concerted action and learning for sustainability (Keen et al. 2005; Leeuwis and Pyburn 2002; Pahl-Wostl et al. 2008; Schusler et al. 2003).

Despite the proliferation and diversity of recent research, numerous uncertainties and challenges remain, including lack of precision in defining basic terms and units of analysis (How is learning defined? Who or what is doing the learning?) (Armitage et al. 2008; Easterby-Smith et al. 2000), lack of attention to multi-level learning, i.e., learning at multiple levels of social organization (What are the relationships among individual and the various levels of collective learning?) (Bapuji and Crossan 2004; Pelling et al. 2008), and also lack of attention to the influence of power (How do power differentials affect learning outcomes and processes?) (Armitage et al. 2008; Ferdinand 2004).

The purpose of this chapter is to help fill these gaps by presenting a conceptual framework of connections linking learning outcomes across multiple levels of social organization. The framework was developed through an integrative (rather than a synoptic) literature review (Montuori 2005; Torraco 2005) that selectively synthesized learning constructs from various disciplines. The framework is theoretical, although aspects of it are supported by empirical evidence from resource and environmental governance experiences.

Section 10.2 offers five pertinent definitions and explanations of learning and lays out the social units of analysis used in the chapter. Section 10.3 reviews and synthesizes literature on processes that link learning at multiple levels. The term multi-level learning connections is used throughout the chapter, and is defined as those institutions that facilitate the transmission of learning outcomes from one level of social organization to another (by either up-scaling or down-scaling the outcomes). Section 10.4 problematizes multi-level connections and other learning processes that do not adequately accommodate contested values, power imbalances, economic constraints, and social objectives of sustainability. Section 10.5 concludes the chapter with a discussion of the implications for policy and practice, and for research regarding multi-level learning and adaptive capacity in social–ecological governance.

10.2 Defining Learning and the Social Units of Analysis

The chapter does not adopt a single definition of learning applicable to multiple levels of organization. I agree with Levy (1994) that the reification of individual learning to the collective level is not analytically fruitful. Action groups, organizations and networks do not literally learn in the same sense that individuals do. It is more helpful, particularly in understanding multi-level learning connections, to follow the lead of early organizational learning scholars (e.g., Argyris and Schön 1978; Hedberg 1981) in distinguishing among conceptions of learning applicable to various levels of social organization.

This chapter distinguishes among five conceptions of learning: individual, action group, organizational, network, and societal. Table 10.1 provides a summary offering definitions for each conceptualization along with descriptions of the pertinent levels of organization (or social units of analysis). I explain the table below, but before doing so, four caveats are in order. First, for the sake of precision, I have not used the term social learning, despite the importance and growth of the recent literature on the subject. This term continues, as Parson and Clark (1995) said over 10 years ago, to mask great diversity and complexity of meaning (see the recent reviews by Muro and Jeffrey (2008) and Blackmore (2007)). Recent studies have used the term to mean, or encompass, each or combinations of what I have called action group, organizational and societal learning (e.g., Bouwen and Taillieu 2004; Hayward et al. 2007; Holden 2008; Woodhill 2002). Additionally, the best established conception is likely Bandura's (1986), which is essentially a theory of individual learning. Second, for the sake of simplicity, I have restricted the framework to five conceptions of learning and five units of analysis, excluding important concepts like government and policy learning (Levy 1994), and important units like communities of practice (Wenger 1998) and informal organizations (Pelling et al. 2008). In doing so, my intent was not to diminish the importance of the excluded items, but rather to offer a relatively simple framework to help scrutinize the linkages among the units. On a related note, the third caution is that the discussion below and Table 10.1 are somewhat hierarchical and only hint at the full range of messy, or complicated, interactions among the various levels of social organization. A fuller portrayal of the untidy mesh of relationships, acting across all levels

Table 10.1 Five conceptions of learning, and related social units of analysis

Individual learning: the process through which a person's knowledge, skills, beliefs, or behaviors are changed as a result of experience	*Individual*: much of the discussion in this chapter applies most directly to adults rather than children
Action group learning: the processes by which individual learning outcomes become part of a web of distributed and mutual outcomes in a collection of individuals	*Action group*: a cohesive but relatively informal association of individuals focused on specific objectives and tasks, often with a short lifespan
Organizational learning: the processes by which individual or action group learning outcomes are stored in and brought forth from organizational memory, such as routines, practices, procedures, and cultures in the organization	*Organization*: like an action group but often with a longer lifespan and more complex mandate, and usually framed by formal membership and institutions
Network learning: the processes through which organizational learning outcomes become part of a web of distributed and mutual outcomes in a collection of organizations (and thus effect change in network-level properties)	*Organizational network*: a collection of organizations lacking a trans-organizational structure but sharing political, social, economic or cultural interests
Societal learning: the democratic processes by which core societal institutions are modified in response to social and environmental change	*Society*: the community of people living in a particular region or country having shared customs, organizations, and laws

simultaneously, was beyond the scope of the chapter. Finally, although the distinctions among the units of analysis are conceptually sound and analytically useful, in practice some might be fuzzy and prove difficult to apply in any given case of multi-level learning, e.g., the degree of formality of structure differentiating an action group from an organization.

10.2.1 Individual Learning

In an attempt to take a complementary or integrated approach, I have defined individual learning broadly: the process through which a person's knowledge, skills, beliefs, or behaviors are changed as a result of experience (Levy 1994; Merriam et al. 2007; Parson and Clark 1995). The related social unit of analysis is the individual, and as noted in Table 10.1 the discussion applies most directly to adults rather than children. I make this distinction because the analytic focus in this section and throughout the chapter is on adult learning processes, such as critical reflection, deliberation and dialog, and socio-political empowerment (Merriam et al. 2007).

Although the emphasis here is on changes taking place at the level of the individual, individual learning is a highly social process, often facilitated in a collective setting, dependant on concerted inquiry and action, or deeply embedded in specific socio-cultural practices (Bandura 1986; Mezirow 2000; Wenger 1998). Borrowing from Salomon and Perkins (1998), it is helpful to view the influence of social variables as being on a continuum from, say, highly solitary learning (reliant on social artifacts like cultural beliefs or historical relics and documents) to highly situated communities of practice (in which it is difficult to separate individual from collective learning outcomes).

At the social end of the spectrum, it is difficult to define and explain learning separate and apart from explaining multi-level learning connections, the subject of Sect. 10.3. In such instances, the central processes of learning themselves are the institutions that up-scale or down-scale learning outcomes. However, closer to the solitary end of the spectrum, it can be more helpful to split the definitions and explanations of learning processes from an examination of multi-level connections, which of course is what has been done in this chapter. In that vein, the following adds greater depth to the definition of individual learning presented earlier by offering insights from transformative theory, a leading theory of adult learning.

Transformative theory is in the midrange of the influence of social determinants continuum, but it is likely best placed nearer to the individual than the social end. The theory provides a detailed explanation of cognitive changes taking place at the level of the individual, but at the same time it emphasizes that communicative processes (modeled on Habermas' ideal speech situation) and social engagement are highly influential in shaping individual cognitive changes (Mezirow 1990, 1991, 2000). Additionally, the theory, which has a small but growing research base in the field of environmental governance (e.g., Alexander 1999; Diduck and Mitchell 2003; Sinclair and Diduck 2001), is helpful for the purposes of this chapter because it presents a construct, namely emancipatory learning, that assists with the politicization of learning discourse.

Transformative theory describes three forms of learning: instrumental, communicative, and emancipatory.[3] Instrumental learning provides competence in coping with the external world through technical control of natural forces. Communicative learning helps people negotiate their own meanings, intentions and values, rather than merely accepting those of others. Disagreements in the realm of communicative learning are typically resolved by resort to force, authority, or discourse. Resolving disagreements through discourse is emancipatory if it frees the learner

---

[3]Organizational theory describes three similar forms of learning (Argyris 1977, 1990; Argyris and Schön 1978; Easterby-Smith et al. 2000; Flood and Romm 1996). Single-loop learning involves improving efficacy, or getting better at fulfilling existing purposes in the context of a given set of fundamental governing variables. Double-loop learning involves evaluation of and changes to both instrumental means and ends and fundamental governing variables. Triple-loop learning asks if power structures act too much in support of selected and privileged definitions of rightness. Additionally, emancipatory learning was influenced by Paulo Freire's "pedagogy of the oppressed" (Freire 1970, 1973). This educational approach is intended specifically to counter power asymmetries and hegemonic influences. It looks to empower the disenfranchised, challenge socio-political and economic presuppositions, foster emancipatory learning, and mobilize concerted action for structural change.

from oppressive social relations (e.g., personal empowerment enabling a critique of inequitable resource sharing). Emancipatory learning outcomes are most likely to occur under the ideal conditions of learning: accurate and complete information; freedom from coercion; openness to alternative perspectives; ability to reflect critically upon presuppositions; equal opportunity to participate; and, abilities to assess arguments as objectively as possible and to accept a rational consensus as valid (Mezirow 1990, 1991, 2000).

The theory further suggests that learning occurs through changes (or transformations) in a person's frame of reference. A frame of reference consists of meaning perspectives (broad epistemic, psychological and socio-cultural predispositions) and meaning schemes (specific beliefs, feelings, attitudes and value judgments). Transformation occurs through critical reflection on the underlying assumptions of the various elements of the meaning perspective or meaning scheme. Critical reflection involves processes such as assessment of role assumptions and social expectations, recognition that one's problem may be shared by others in the community, exploring new patterns of behavior, provisional efforts to try new roles and gain feedback, and social engagement with a new perspective (Mezirow 1990, 1991, 2000). One can see that social variables are part of this process, but the extent to which they might influence the learning outcome in any particular instance is contingent upon individual and contextual circumstances.

10.2.2 Action Group Learning

Building on the conceptualization of individual learning presented above, action group learning is defined as the processes by which individual learning outcomes become part of a web of distributed and mutual outcomes in a collection of individuals. Borrowing from Friedmann (1987), the related social unit of analysis is the action group, defined here as a cohesive but relatively informal association of individuals focused on specific objectives and tasks, and which often has a short lifespan.

What individuals learn is vital to what action groups learn, and consistent with the discussion in the preceding section, the reverse is true: what action groups learn is vital to what individuals learn. Section 10.3 provides details of multi-level connections spanning individual and action group learning, but what follows is a brief overview of a selected model of action group learning, to add further depth to the basic definition provided above. Röling's (2002) elements of cognition model is summarized because its cognitive focus complements the emphasis in the previous section. Additionally, like Friedmann's (1987), Röling's model emphasizes praxis, or the connections between theory and practice. Further, it encompasses the essential idea of leading models of social learning (or what I have called action group learning) found in the resource and environmental governance literature. This idea is that through facilitated social interaction and concerted action, differing frames of reference are likely to be adapted, possibly leading to mutual or complementary frames, further concerted action, and the development of a common social reality (Keen et al. 2005; Rist et al. 2007; Schusler et al. 2003; Steyaert and Jiggins 2007; Webler et al. 1995).

For Röling (2002), social learning consists of facilitating a transition in a collection of individuals from a state of multiple cognition to one of distributed or collective cognition. Cognition is broader than the mental process of acquiring knowledge and understanding; it encompasses perceptions and theories of one's environment (or context), and values, emotions and goals respecting the environment/context. Multiple cognition emphasizes the presence, in any given situation, of different cognitive agents with multiple perspectives. Distributed cognition emphasizes different but complementary attributes that permit concerted action. Collective cognition stresses shared attributes plus collective action. Coherence among the components of cognition is a fundamental driver of cognitive development. Another driver is correspondence between a person's cognitive components and his or her environment. In the model, the transition from multiple to collective cognitive states and the development of cognitive coherence and correspondence are facilitated by intentionally designed platforms for learning (Röling and Maarleveld 1999). Such platforms require various problem-solving and decision-making processes (see Sect. 10.3) to identify and assess interdependencies in the group, resolve conflicts stemming from the interdependencies, build enduring trust relations, and enable concerted action toward a common goal.

The ensuing section builds on the foregoing introductions to individual and action group learning by situating them in the context of organizational learning. However, before turning to that task I want to quickly mention the issue raised by Levy (1994) in his discussion of what he called the accuracy criterion of learning. I do so now because the issue often arises in definitions of social (or what I have called action group) learning. The issue is that definitions of learning are sometimes restricted to processes that result in outcomes that are empirically accurate or otherwise normatively desirable, according to a given analytical framework or set of normative criteria. An example is the efficiency conception of learning described by Tetlock (1991) in which learning involves acquiring the ability to match ends and means more effectively – either by employing more appropriate means or by pursuing goals that are more realistic. Another example is Woodhill's (2002) definition of social learning, which was restricted to outcomes that optimize the collective wellbeing of current and future generations.[4] Yet another can be seen in how Keen et al. (2005) defined social learning with reference to outcomes that improve the management of human and environmental interrelations.

There are advantages to defining learning using restrictive criteria, such as being able to more clearly distinguish learning from other forms of cognitive, organizational and social change. However, a fundamental problem also arises, particularly in complicated governance situations (which is true of most pressing resource and

---

[4] I adopt Woodhill's (2002) definition of social learning as my definition of societal learning minus the restrictive sustainability criterion – see Sect. 3.2.5.

environmental problems), namely the lack of agreed upon standards to measure whether learning outcomes are congruent with the pertinent criterion, be it accuracy, efficiency or sustainability. This lack of standards creates a real danger of analysts using the term learning in so self-serving a manner that it becomes synonymous with the accrual of outcomes that the analyst deems correct. Additionally, narrow definitions of learning can decrease the chances of encountering useful empirical findings regarding the nature of learning outcomes (Levy 1994; Tetlock 1991). For these reasons, as the reader would have already noted, I have not taken a restrictive approach in this chapter; I have adopted a process-centered orientation and have not defined learning in reference to whether a particular type or direction of outcome has resulted.

10.2.3 Organizational Learning

I define organizational learning as the processes by which individual or action group learning outcomes are stored in, and brought forth from, organizational memory, including the practices, procedures, plans, conventions, strategies and cultures of the organization (Argyris and Schön 1978; Hedberg 1981; Levitt and March 1988). The unit of analysis is the organization, defined as being like an action group but with a longer lifespan, more complex mandate, and formal membership and institutions. As implied in the definition of learning, organizations are frequently composed of subunits, including action groups as I have defined that term (Crossan et al. 1999; Easterby-Smith et al. 2000; Pelling et al. 2008). As above, it is hard to distinguish the multi-level connections in organizational learning from fundamental definitions and explanations (e.g., Crossan et al.'s (1999) integrated model of multi-level learning discussed in Sect. 10.3). Still, to provide a slightly fuller introduction, what follows is a brief elaboration of the aspect of organizational learning that sets such learning apart from individual and action group learning, namely organizational memory.

Organizational memory is an under-researched topic empirically, but its conceptual and theoretical foundation is relatively rich. The essential idea is that, as Hedberg (1981, p. 6) put it, although members and leaders of organizations come and go, "organizations' memories preserve certain behaviors, mental maps, norms and values over time". Argyris and Schön (1978) viewed organizational memory as consisting of public maps (e.g., organization charts, diagrams of workflow, management plans, informal institutions) and private images (i.e., mental models of self in relation to others and in relation to the organization). Levitt and March (1988) emphasized routines (e.g., forms, rules, procedures, conventions, strategies, and technologies) that frame and operationalize an organization, plus the organizational cultures and paradigms through which the organization's members interpret the routines.

Olivera's (2000) framework is pertinent for our purposes here because of its emphasis on multi-unit organizations (and by implication multi-level learning variables). Olivera defined organizational memory systems as knowledge retention devices that collect, store and provide access to an organization's experience. He developed a framework for mapping and understanding the complex knowledge environment of multi-unit organizations. The primary criteria in the framework were content (i.e., knowledge categories), structure (both the location and index of knowledge) and processes for collecting, maintaining and accessing knowledge. A nice feature of the framework is that the third criterion focuses attention on an important issue raised in the discussion on power differentials in Sect. 10.4: the equity of the processes used to construct and access organizational memory.

10.2.4 Network Learning

My conception of network learning is analogous to the conception of action group learning described earlier. Scaling up the analysis from the level of the single organization, I define network learning as the processes through which organizational learning outcomes become part of a web of distributed or mutual outcomes in a collection of organizations (and thus effect change in network-level properties). The unit of analysis is the organizational network, defined as a collection of organizations lacking a trans-organizational structure but sharing political, social, economic or cultural interests. I selected this unit of analysis not simply because it furnishes a convenient and rational entity located between the organizational and societal units of analysis. I chose it because networks play an important and growing role in both business and public management, including government learning since organizational networks are part of the complex of entities and people involved in such learning (Knight and Pye 2005; Levy 1994).

This section relies on Knight and Pye's (2004, 2005) model of network learning. That model places less emphasis on cognitive processes than most of the discussion in this chapter, but it is still applicable and has value for understanding network learning over a period of time, and this is important because more informed longitudinal perspectives are essential to adaptive governance of complex social–ecological systems. The model explained how learning occurred by a network of organizations involved in the provision of prosthetic limbs in England as the member organizations interacted and gradually solved problems that arose during five discrete learning episodes. The researchers developed descriptive and conceptual models consisting of context, process and outcome variables. The context variables, e.g., purpose, actors, history, and operations, were highly situated, and their impact on the way learning episodes developed varied from episode to episode. The process variables reflected three major themes: developing mutual meaning schemes, developing commitment for new organizational means or ends, and developing methods for achieving organizational ends. Three major themes also reflected the content variables: changes in network practices (both behavioral and cognitive), changes in structures (explicit and systemic economic priorities) and changes to aspects of the network's culture having consequences for the identity of the network. The content variables are indicators of network learning; in the absence of changes to network-level properties, such as shared practices and structures, there is no network learning. The process elements are particularly insightful regarding multi-level learning connections.

10.2.5 Societal Learning

As noted, I have refrained from using the term social learning, and I did this in part to distinguish between group-level conceptualizations of the term and macro conceptualizations that attempt to describe and explain large scale societal change processes. To encapsulate the latter conceptualization, I have used the term societal learning, defined as the democratic processes through which core societal institutions are modified in response to social and environmental change. The unit of analysis is society, defined simply as the community of people living in a particular region or country having shared customs, organizations, and laws.

Discussions of large-scale learning of the sort contemplated here often begin with reference to the politics and epistemology of John Dewey, who argued that public policy decisions should be viewed as a series of experiments. He argued that, guided by the principles of scientific inquiry and bounded by democratic debate, experimental politics will yield progressive social improvement (Dewey and Sidorsky 1977; Friedmann 1987). Habermas' (1979, 1984) universal pragmatics and communicative action, concerned with social change processes driven by citizen engagement in deliberative processes in public spaces, provide yet another important theoretical foundation. Recently, Waddell (2005) framed societal learning as the development of new relationships, strategies, and organizational structures that span the public, private and civic sectors and which foster innovation to address chronic problems and develop new opportunities. The approach taken in this chapter is consistent with the broad parameters of each of these frameworks. I based the approach on Woodhill's (2002) conception of social learning, although a definitional difference is that I did not restrict learning as he did to those outcomes that optimize the collective wellbeing of current and future generations.

As noted earlier, institutions are defined broadly in this chapter, and are inclusive of social norms and values and political and economic structures. Institutions and their interplay create incentives for individuals, action groups, organizations and networks to act in particular ways. Some institutions, either formal (e.g., policy and law) or informal (e.g., customs and codes of conduct) are deeply embedded and relatively impervious to change, and thus lock societies into a particular development trajectory. Reforming such core institutions can rely on traditional authority, existing institutional imperatives, political expediency, vested interests, economic or political power, or technocratic and instrumental thinking. Alternatively, reform can be driven by societal learning processes: open dialog, democratic constraint of inequality, investment in education and social capital, the establishment of mediating forums, open policy processes, questioning of basic assumptions, and greater democratization of politics and the technocratic sphere (North 1990; Woodhill 2002; Young 2002).

10.3 Multi-level Learning Connections

As mentioned earlier, it is difficult to separate the discussion of multi-level connections from the discussion above, which attempted to define and elaborate five conceptions of learning and related social units of analysis. More often than not, multi-level linkages are integral to basic definitions and explanations of learning. Nevertheless, I split the discussion to underline the importance of the multi-level aspects of the various conceptualizations.

Pelling et al. (2008) have described a promising approach for examining multi-level connections. In their study of a local dairy farmers association and two supporting public sector organizations in Great Britain, they developed a conceptual framework consisting of pathways for learning and adaptation which are established by the interplay between formal and informal institutions. Crossan et al. (1999) have also offered a valuable framework, involving the articulation of four fundamental learning processes and the explanation of how these processes create feed-forward and feed-back loops spanning the individual, action group and organizational levels of analysis. Figure 10.1 is adapted from their model, modified to encompass network and societal units of analysis and to envision the linkages as multi-level learning connections. Accompanying the figure is Table 10.2, which summarizes the five conceptions of learning connections depicted in the graphic.

10.3.1 Social-Cognitive Filters

Social-cognitive filters, shortened to filters in Fig. 10.1, encapsulate a range of social and psychological influences on individual learning (recall Salomon and Perkins' (1998) influence of social variables continuum). The term is adapted from Andrews and Delahaye (2000) who investigated individual-level factors influencing the flow of knowledge in organizations. They developed the concept of the psychosocial filter (individuals' perceptions of approachability, credibility and trustworthiness) as a mediator of how knowledge is imported and shared in organizations. I view socio-cognitive filters broadly; they encompass Andrews and Delahaye's (2000) psychosocial factors as well as a host of processes and mechanisms that influence both informal and nonformal adult learning.[5]

---

[5]Nonformal adult learning results from deliberate education for adults occurring outside of educational institutions, such as facilitated activities found in community groups and organizations. Nonformal learning is different from informal learning, which refers to the experiences of everyday living from which individuals learn something (Merriam et al. 2007).

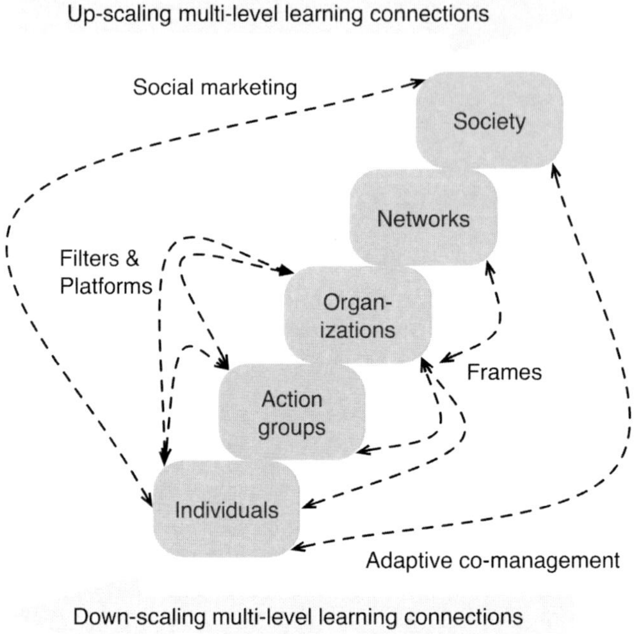

Fig. 10.1 A conceptual framework of multi-level learning connections, showing basic social units of analysis and types of up- and down-scaling linkages. Adaptive co-management arrangements and community-based social marketing touch every level of organization, but for the sake of simplicity are only shown as connecting the individual and societal levels

Table 10.2 Five conceptions of multi-level learning connections

Social-cognitive filters	Psychological and social mediators of individual and action group/organizational learning; e.g., communication, peer engagement and social action
Facilitated platforms	Deliberate interventions in which interdependent stakeholders are brought together to interact in a forum for collective decision making towards concerted action
Organizational frames	Cultural, strategic and structural arrangements that enable individual and action group learning within the organization plus the development and use of organizational memory
Adaptive co-management arrangements	Flexible, community-based systems of management tailored to specific places and situations; supported by and working with various groups and organizations at different scales
Community-based social marketing	Principles, strategies and practices for influencing human behavior to achieve public goals; emphasizes the involvement of the people whose behavior is targeted

The bidirectional arrows representing the filters in the figure reflect that the filters often act as both up- and down-scaling multi-level connections. That is, they show the mutuality of many instances of individual learning that occur in social situations. Not only do social variables affect individual learning, individual learning in turn affects learning processes and outcomes at higher levels of social organization. The figure shows the filters as linking: (a) the individual with the action group level (e.g., the development of mutual from multiple cognition); (b) the individual with the organizational level (e.g., development and use of organizational memory by one of the organization's leaders); and, (c) the action group with the organizational level (e.g., when action groups are part of or otherwise engaged with organizations).

Examples of filters that operate largely by down scaling to the individual from the action group or organization during informal learning include communication, peer engagement and socially oriented action, as contemplated by transformative theory (e.g., Diduck and Mitchell's (2003) study of individual learning outcomes brought about by involvement in the environmental assessment of a large-scale hog processing plant). Filters that are bidirectional across the individual, action group and organizational levels include deliberation, collective inquiry and concerted action, which are key in more communally centered, context focused theories (e.g., Holden's (2008) study of group learning in the Sustainable Seattle initiative and Elkjaer's (2004) application of Dewey's pragmatism to organizational learning). Such filters are influential during informal learning episodes, but are more likely to be manifest in facilitated nonformal education. I view facilitation, with its suite of techniques and mechanisms (e.g., dialog, negotiation, visioning, mediation), as a particularly important subset of social-cognitive filters. The following section discusses facilitation in more detail, focusing on a specific form developed in the context of resource and environmental governance.

10.3.2 Facilitated Platforms

Researchers at Wageningen University in Holland developed the concept of facilitated platforms, which are the vehicles by which individual learning outcomes become part of a web of distributed and mutual outcomes (Leeuwis and Pyburn 2002; Maarleveld and Dangbégnon 1999; Röling and Maarleveld 1999). As described in Sect. 10.2.2, platforms are planned interventions in which a set of interdependent stakeholders in some resource are brought together to interact in a forum for conflict resolution, negotiation, and collective decision making towards concerted action. In effect, platforms are contrived situations of nonformal education for individual and action group learning.

Platforms are a subset of social-cognitive filters, and as described above, link individual and action group, individual and organization, and action group with organization. The platform arrow in the figure is best thought of as a rather messy set of iterative facilitation processes. An important example is helping stakeholders recognize their interdependence. Another is helping participants resolve conflict (through negotiation, mediation and other methods), given that interdependencies can highlight differences and lead to disjoint and counterproductive action. A third, linked to conflict resolution, is helping to design shared goals and the means of accomplishing those goals, which sets the stage for concerted action and collective inquiry. Maarleveld and Dangbégnon's (1999) investigation of fisheries in Benin and water resources in Holland provides practical instances of these platform processes. Important for the purposes of generalization, similar multi-level connections are seen in studies that do not explicitly adopt the notion of facilitated platforms (e.g., Webler et al.'s (1995) Swiss case study of the environmental assessment of a waste disposal facility).

10.3.3 Organizational Frames

The preceding sections alluded to multi-level connections that affect organizational learning, such as psychosocial factors influencing the uptake and transmission of knowledge, social-cognitive filters affecting the development and use of organizational memory, and learning by an action group embedded in an organization. Organizational frames, called frames in Fig. 10.1, encompass such variables in a broad way, but their emphasis is slightly different. They center on organization-level arrangements that enable individual and action group learning within the organization plus the development and use of organizational memory. The section draws from a recent review of the literature on facilitators of organizational learning. Bapuji and Crossan (2004) classified facilitators as being cultural, strategic, structural and environmental. Relying on this work, I have emphasized facilitators with obvious implications for multi-level connections.

Like social-cognitive filters, organizational frames influence both informal and nonformal learning. Further, as with filters, they often act as both up- and down-scaling multi-level connections. Figure 10.1 shows frames connecting organizations with individuals, action groups and networks. It is helpful to think of these frame arrows as overlapping sets of complex organizational arrangements. Important framing arrangements across the organizational and individual levels include a culture of openness, a learning orientation, participative decision making, transformational leadership, cognitive diversity, positive organizational support, and goal and supervision autonomy. At least four of these arrangements were factors in Hayward et al's (2007) case study of the connections between individual and organizational learning in the environmental assessment of a major flood control infrastructure in southern Canada.

Key framing arrangements linking the organizational and action group levels are cross-functional communication, stability of team membership, and manufactured crises to foster innovation. While not an explicit study of the connections between action group and organizational learning, Fitzpatrick's (2006) investigation of organizational learning in the assessment of diamond mines in the Canadian north confirmed the importance of two aspects of cross-functional communication, namely information sharing and information interpretation, as internal organizational arrangements that enable learning.

Finally, arrangements that connect organizations and networks include an emphasis on both internal and external learning, access to knowledge resources such as talent, collaboration partners and research institutions, and knowledge sharing with national and global innovation systems (Knight and Pye 2004, 2005).

10.3.4 Adaptive Co-management Arrangements

Adaptive co-management is a complex, incipient concept, but its defining features are flexible, community-based systems of management tailored to specific places and situations, and supported by and working with various action groups and organizations at different scales (Olsson et al. 2004a). Working with this definition and looking through a lens emphasizing multi-level learning, I view adaptive co-management arrangements as up- and down-scaling multi-level learning connections reaching from the societal to the individual level, and touching all levels of organization in between. Figure 10.1 shows only the direct connections between the individual and societal levels for the sake of graphical simplicity, and to emphasize adaptive co-management's potential for influencing societal learning.

As above, the bidirectional arrows represent the mutuality of learning across the various levels of organization. Further, as with social-cognitive filters and organizational frames, adaptive co-management arrangements are highly influential for both informal and nonformal learning situations. A conceptual strength of adaptive co-management for the purposes of this chapter is that it provides a logical means for expanding on processes of societal learning. It does so by focusing attention on relationships among collaboration, management, governance and institutional reform. These relationships are evident in the conditions for successful adaptive co-management identified by Armitage et al. (2009), which include having:

- Identifiable stakeholders with shared interests
- Access to a varied assortment of governance options
- A long-term commitment to shared governance
- Resources for enhancing stakeholder capacity
- Individuals and collectives who champion the process
- Openness of participants drawing upon multiple knowledge systems and sources
- A national and regional policy environment supportive of collaborative governance

Given Woodhill's (2002) definition of societal learning adopted in the chapter, with its emphasis on institutional reform through democratic processes, the conditions for successful adaptive co-management help illuminate the path for societal learning. Empirical evidence of aspects of this proposition is found in Olsson et al.'s (2008) case study of governance and institutional reforms affecting the Great Barrier Reef, which included analysis of interplay among actors at the individual, action group, organizational, and governmental (a form of network, as defined here) levels.

10.3.5 Community-Based Social Marketing

Community-based social marketing involves principles, strategies and practices for influencing human behavior in order to achieve public goals and for doing so in a manner that emphasizes the participation of those whose behavior is targeted (McKenzie-Mohr and Smith 1999). These principles, strategies and practices represent another bundle of multi-level connections spanning the societal and individual levels of analysis. Similar to adaptive co-management arrangements, these connections reach all levels of organization discussed in the chapter, but Fig. 10.1 depicts just the direct links between the individual and societal levels for the sake of simplicity and to highlight social marketing's potential in down-scaling societal learning to the level of individual behavioral change. As above, social marketing is influential in both informal and nonformal learning contexts. The bidirectional arrow used in the figure represents reciprocity in learning outcomes because, despite social marketing's important potential in down scaling societal learning, community-based approaches include multiple up-scaling feedback loops.

Based on a rich literature of peer reviewed studies and informal reports on how to foster behavior for resource conservation and sustainable development, McKenzie-Mohr and Smith (1999) described four basic steps in community-based social marketing. The first is identifying barriers and benefits to the selected socially desirable activity, typically done through research using multiple methods (including situated, place-based studies). The second step is designing a strategy with a mix of techniques for aligning existing behaviors with the socially desirable activity. Techniques that have proven to be effective include providing regular prompts, securing pledges or commitments, creating incentives, altering community norms, and using vivid communications tools with engaging messages and images. The third step is piloting the strategy in a segment of the community, and the fourth is evaluating it after implementing the strategy on a broader scale.

10.4 Power Differentials

The previous section presented five conceptions of learning applicable to five levels of social organization, and then introduced five sets of multi-level learning connections (i.e., institutions that facilitate the transmission of learning outcomes from one level of organization to another). Such a learning orientation has advantages for understanding uncertainty, complexity and change, and for building adaptive capacity, but learning discourse can too often mask important political and power-related variables. This result is unfortunate because politics and power[6]

---

[6]I have adopted an expansive definition of power: an expression of human agency in the context of enduring structural preconditions that has coercive, constraining, and systemic consent-producing dimensions (Raik et al. 2008).

are central factors in governance, especially the decentralized forms seen with increasing frequency in the resource and environmental field (and which are essential for building adaptive capacity) (Raik et al. 2008). Moreover, power issues, particularly finding ways of leveling differentials, are the key to achieving important normative endpoints of resource and environmental governance, such as certain economic and social objectives of sustainability (e.g., poverty alleviation, equity, empowerment and social justice).

Two recent studies have highlighted important power-related issues affecting individual and action group learning. Muro and Jeffrey's (2008) review and critique revealed fundamental points requiring further attention in the theory and practice of social learning in resource and environmental governance, including:

- Consensus and mutual cognition can hide how less powerful members of learning platforms changed their views to match those of the others.
- In some instances there is "an irreducible plurality" of viewpoints, and better solutions are found because of this (through conflict and competition) rather than because of consensus, compromise and mutual understandings.
- In cases of highly contested issues, strategies other than collaboration, learning and mutuality (such as penalties or incentives) might be more appropriate for initiating new practices and social interests.

Similarly, Armitage et al.'s (2008) comparative case study of experiences in Canada, Vietnam and Cambodia identified basic concerns needing attention in community-based resource management, including how to:

- Lower barriers and create incentives to encourage participation in learning platforms, given that the ability and willingness to experiment and learn are not likely to be distributed evenly in heterogeneous communities.
- Design effective and safe platforms for deliberation and conflict resolution that enable different segments of heterogeneous communities an opportunity to transform traditionally disadvantageous political relations.
- Establish protections for marginal stakeholder groups (especially in rural, resource-dependent regions) who become involved in learning processes.

To the extent that action group learning is linked to an organization, the preceding issues need to be accommodated in descriptions, explanations and prescriptions regarding organizational learning. In addition, although power is an under-researched topic in organizational learning, the literature on this subject reveals a further complicated set of political and power-related variables (Argyris 1990; Blackler and McDonald 2000; Coopey and Burgoyne 2000; Ferdinand 2004). The following factors are important in triggering, blocking and shaping learning, and should be considerations in most comprehensive accounts of organizational learning:

- Technical, social and economic structures
- Resource dependencies
- Shifting coalitional patterns

- Bargaining and exchanges
- Unilateral control of problems
- Inequitable access to organizational routines
- Fragmentation of interests and values
- Bureaucratic inertia, rigidity and co-optation

Scaling up the analysis, the foregoing is pertinent to network and societal learning. For example, many of the issues related to power asymmetries raised in the context of action group learning are directly relevant to the design, implementation and evaluation of social marketing strategies, and create a powerful imperative for the participative, community-based approach to social marketing. In addition, the literature on international and foreign policy learning underlines a fundamental issue that needs to be considered in societal learning and in some forms of network learning, specifically government learning (Brown 2006; Haas 2000; Levy 1994; Tetlock 1991). The issue is deceptively simple, but is of primary importance. Since government and societal learning involves policy and institutional reform, there is a risk of confounding all policy and institutional change with learning outcomes. Not all such changes are reasonably attributable to learning, when political and power-based explanations are more trustworthy and meaningful.

10.5 Conclusion

This chapter has presented an integrative review, selectively synthesizing constructs to help conceptualize the processes that link learning at multiple levels of social organization. The review was interdisciplinary, covering works from adult education and learning, organization and management studies, political studies and foreign policy analysis, environmental and resource management, and planning. A more comprehensive review would undoubtedly have revealed alternative, reasonable conceptions of learning, units of analysis, and processes that link the various units.

However, the conceptual framework presented here is useful for its implications for adaptive capacity in social–ecological governance. First, it develops the notion of multi-level learning, which is an essential feature of adaptive governance systems. It summarizes learning processes at five important levels of social organization, and provides details respecting some of the connections that link learning outcomes across those levels. Understanding such processes and connections can enable flexible, innovative and responsive governance initiatives. Further, it can facilitate initiatives founded on transformative and emancipatory (or double- and triple-loop) learning intentions or experiences. Second, it builds on an important lesson from Young's (2002) work that a key to success in multi-level governance regimes is to ensure that cross-scale interactions produce complementary rather than conflicting results. By providing details of learning-related interactions, the framework helps crystallize ways to produce complementarity, e.g., setting compatible learning objectives, establishing congruous learning environments, and using consistent methods. Third, at the conceptual level it supports Pelling et al.'s (2008) conclusion that relational spaces in and among organizations are essential for creating adaptive capacity. In that study, the authors found that such spaces yielded six discrete pathways, the potential or actual existence of which they interpreted as indicators of adaptive capacity. Deliberative and communicative processes (which produce relational spaces) are a unifying theme among the five sets of multi-level connections reviewed here.

The chapter is also helpful for revealing important research needs at the interface of multi-level learning, adaptive capacity and social–ecological governance, including the requirement for conceptual and theoretical development. Pelling et al. (2008) and Crossan et al. (1999) offer valuable frameworks and excellent entry points, but more work needs to be done in accounting for network and societal levels of analysis, assessing promising linking institutions such as community-based social marketing and adaptive co-management, and addressing power asymmetries in learning dynamics. On this last point, a promising avenue lies in giving more attention to learning's flip side, education, and particularly critical, nonformal education. (For a broader discussion of the implications of education for learning, sustainability and resilience, see Lundholm and Plummer's (2010) synopsis in which they lay out, a resilience agenda in environmental education.) Opening the discourse in this manner would provide access to well developed theory, methods and practice on how to enhance capacity by fostering personal and socio-political empowerment. A leading approach to critical education, one that has already made inroads in resource and environmental analysis (e.g., Diduck 1999; Diduck and Sinclair 1997; Fitzpatrick and Sinclair 2003), is Paulo Freire's "pedagogy of the oppressed" (Freire 1970, 1973). This approach is intended specifically to counter power asymmetries and hegemonic influences. It looks to empower the disenfranchised, challenge socio-political and economic presuppositions, foster emancipatory learning, and mobilize concerted action for structural change. Adopting this pedagogy or a similar framework would help counter power imbalances, and thereby enrich multi-level learning and adaptive capacity in resource and environmental governance. It would also improve the prospects of achieving economic and social objectives of sustainability, such as poverty alleviation, equity, empowerment and social justice.

In addition to theoretical and conceptual development, the chapter uncovers the need for place-based empirical studies of existing institutions. Too little of the conceptual framework is grounded in empirical evidence from resource and environmental governance experiences. In line with Lundholm and Plummer's (2010) take on education and learning for resilience and sustainability, research on multi-level learning and adaptive capacity can be reasonably guided in the short run by basic who, what, how and why questions:

- Who are the learners, e.g., people, action groups, organizations, and networks?
- What motivated the learning experiences?
- What was learned, e.g., knowledge, skills, beliefs, behaviors, and routines?

- Did the learning outcomes lead to governance effectiveness, flexibility or innovation?
- How did the learning occur, i.e., what were the social, political, economic and organizational variables that enabled and inhibited the learning?
  - What was the influence of multi-level factors?
- What was the influence of power asymmetries?
  - Was there a role for critical nonformal education in offsetting power differentials?

In the long run, the research of course will branch out, ideally seeking greater depth of inquiry, breadth of application, and a convergence of theory and practice respecting multi-level learning and adaptive capacity in resource and environmental governance.

Acknowledgments I want to thank Bruce Mitchell of the University of Waterloo, Derek Armitage of Wilfrid Laurier University and John Sinclair of the University of Manitoba. Our work together in 2008, supported by the Social Sciences and Humanities Research Council of Canada, was highly influential in crystallizing aspects of the framework presented in the chapter. I also want to thank Mark Pelling of Kings College London, Ryan Plummer of Brock University and an anonymous reviewer for their helpful suggestions on an earlier version of the chapter.

References

Alexander D (1999) Planning as learning: sustainability and the education of citizen activists. Environments 27(2):79–87
Andrews KM, Delahaye BL (2000) Influences of knowledge process in organizational learning: the psychosocial filter. J Manage Stud 37(6):797–810
Argyris C (1977) Double loop learning in organizations. Harv Bus Rev 55:115–125
Argyris C (1990) Overcoming organizational defenses: facilitating organizational learning. Allyn and Bacon, Wellesley
Argyris C, Schön DA (1978) Organizational learning: a theory of action perspective. Addison-Wesley, Reading
Armitage D (2005) Adaptive capacity and community-based natural resource management. Environ Manage 35(6):703–715
Armitage D, Berkes F, Doubleday N (eds) (2007) Adaptive co-management: collaboration, learning, and multi-level governance. UBC, Vancouver
Armitage D, Marschke M, Plummer R (2008) Adaptive co-management and the paradox of learning. Glob Environ Change 18(1):86–98
Armitage D, Plummer R, Berkes F, Arthur R, Davidson-Hunt I, Diduck AP, Doubleday N, Johnson D, Marschke M, McConney P, Pinkerton E, Wollenberg L (2009) Adaptive co-management for social–ecological complexity. Front Ecol Environ 7(2):95–102
Bandura A (1986) Social foundations of thought and action: a social cognitive theory. Prentice-Hall, Englewood Cliffs
Bapuji H, Crossan M (2004) From questions to answers: reviewing organizational learning research. Manage Learn 35(4):397–417
Blackler F, McDonald S (2000) Power, mastery and organizational learning. J Manage Stud 37(6):833–851

Blackmore C (2007) What kinds of knowledge, knowing and learning are required for addressing resource dilemmas?: a theoretical overview. Environ Sci Policy 10(6):512–525

Bouwen R, Taillieu T (2004) Multi-party collaboration as social learning for interdependence: developing relational knowing for sustainable natural resource management. J Community Appl Soc Psychol 14:137–153

Brown ML (2006) What have we learned about organizational learning? In: Brown ML, Lenney M, Zarkin M (eds) Organizational learning in the global context. Ashgate, Hampshire, pp 255–261

Coopey J, Burgoyne J (2000) Politics and organizational learning. J Manage Stud 37(6):869–885

Crossan MM, Lane H, White RE (1999) An organizational learning framework: from intuition to institution. Acad Manage Rev 24(3):522–537

Daniels SE, Walker GB (1996) Collaborative learning: improving public deliberation in ecosystem-based management. Environ Impact Assess Rev 16(2):71–102

Dewey J, Sidorsky D (1977) John Dewey: the essential writings. Harper & Row, New York

Diduck AP (1999) Critical education in resource and environmental management: learning and empowerment for a sustainable future. J Environ Manage 57(2):85–97

Diduck AP, Mitchell B (2003) Learning, public involvement and environmental assessment: a Canadian case study. J Environ Assess Policy Manage 5(3):339–364

Diduck AP, Sinclair AJ (1997) The concept of critical environmental assessment (EA) education. Can Geogr 41(4):294–307

Easterby-Smith M, Crossan M, Nicolini D (2000) Organizational learning: debates past, present and future. J Manage Stud 36(6):783–796

Elkjaer B (2004) Organizational learning: the 'third way'. Manage Learn 35(4):419–434

Ferdinand J (2004) Power, politics and state intervention in organizational learning. Manage Learn 35(4):435–450

Fitzpatrick P (2006) In it together: organizational learning through participation in environmental assessment. Environ Assess Policy Manage 8(2):157–182

Fitzpatrick P, Sinclair AJ (2003) Learning through public involvement environmental assessment hearings. J Environ Manage 67(2):161–174

Flood RL, Romm NRA (1996) Diversity management: triple loop learning. Wiley, Chichester

Folke C (2006) Resilience: the emergence of a perspective for social-ecological systems analysis. Global Environ Change 16(3):253–267

Folke C, Colding J, Berkes F (2003) Synthesis: building resilience and adaptive capacity in socio-ecological systems. In: Berkes F, Folke C, Colding J (eds) Navigating social–ecological systems: building resilience for complexity and change. Cambridge University Press, Cambridge, pp 352–387

Freire P (1970) Pedagogy of the oppressed. Seabury, New York

Freire P (1973) Education for critical consciousness. Seabury, New York

Friedmann J (1987) Planning in the public domain: from knowledge to action. Princeton University Press, Princeton

Glasser R, Manty D, Nehman G (1975) Public participation in water resource planning. International Water Resources Association and UNESCO Seminar, Paris and Strasbourg

Haas PM (2000) International institutions and social learning in the management of global environmental risks. Policy Stud J 28(3):558–575

Habermas J (1979) Communication and the evolution of society. Beacon, Boston

Habermas J (1984) The theory of communicative action. Beacon, Boston

Hayward G, Diduck AP, Mitchell B (2007) Social learning outcomes in the Red River Floodway environmental assessment. Environ Pract 9(4):239–250

Hedberg B (1981) How organizations learn and unlearn. In: Nystrom PC, Starbuck WH (eds) Handbook of organization design, vol 1, Adapting organizations to their environments. Oxford University Press, New York, pp 3–27

Holden M (2008) Social learning in planning: Seattle's sustainable development codebooks. Prog Plann 69(1):1–40

Holling CS (ed) (1978) Adaptive environmental assessment and management. Wiley, Chichester

Keen M, Brown VA, Dyball R (2005) Social learning: a new approach to environmental management. In: Keen M, Brown VA, Dyball R (eds) Social learning in environmental management: towards a sustainable future. Earthscan, London, pp 3–21

Knight L, Pye A (2004) Exploring the relationships between network change and network learning. Manage Learn 35(4):473–490

Knight L, Pye A (2005) Network learning: an empirically derived model of learning by groups of organizations. Hum Relat 58(3):369–392

Korten D (1980) Community organization and rural development: a learning process approach. Public Adm Rev 40(5):480–511

Leeuwis C, Pyburn R (eds) (2002) Wheel-barrows full of frogs: Social learning in rural resource management. Koninklijke Van Gorcum, Assen

Levitt B, March JG (1988) Organizational learning. Annu Rev Sociol 14:319–340

Levy JS (1994) Learning and foreign policy: sweeping a conceptual minefield. Int Organ 48(2):279–312

Lundholm C, Plummer R (2010) Resilience and learning: a conspectus for environmental education. Environmental Education Research (in press)

Maarleveld M, Dangbégnon C (1999) Managing natural resources: a social learning perspective. Agric Human Values 16:267–280

McKenzie-Mohr D, Smith W (1999) Fostering sustainable behavior: an introduction to community-based social marketing. New Society, Gabriola Island

Merriam SB, Caffarella RS, Baumgartner LM (2007) Learning in adulthood: a comprehensive guide. Wiley, San Francisco

Mezirow J (ed) (1990) Fostering critical reflection in adulthood: a guide to transformative and emancipatory learning. Jossey-Bass, San Francisco

Mezirow J (1991) Transformative dimensions of adult learning. Jossey-Bass, San Francisco

Mezirow J (ed) (2000) Learning as transformation. Jossey-Bass, San Francisco

Montuori A (2005) Literature review as creative inquiry: reframing scholarship as a creative process. J Transformative Educ 3(4):374–393

Muro M, Jeffrey P (2008) A critical review of the theory and application of social learning in participatory natural resource management processes. J Environ Plann Manage 51(3):325–344

North DC (1990) Institutions, institutional change and economic performance. Cambridge University Press, Cambridge

Olivera F (2000) Memory systems in organizations: an empirical investigation of mechanisms for knowledge collection, storage and access. J Manage Stud 37(6):811–832

Olsson P, Folke C, Berkes F (2004a) Adaptive comanagement for building resilience in social-ecological systems. Environ Manage 34(1):75–90

Olsson P, Folke C, Hahn T (2004b) Social–ecological transformation for ecosystem management: the development of adaptive co-management of a wetland landscape in southern Sweden. Ecol Soc 9(4). http://www.ecologyandsociety.org/vol9/iss4/art2/

Olsson P, Folke C, Hughes TP (2008) Navigating the transition to ecosystem-based management of the Great Barrier Reef, Australia. Proc Natl Acad Sci USA 105(28):9489–9494

Pahl-Wostl C, Tàbara D, Bouwen R, Craps M, Dewulf A, Mostert E, Ridder D, Taillieu T (2008) The importance of social learning and culture for sustainable water management. Ecol Econ 64(4):484–495

Parson EA, Clark WC (1995) Sustainable development as social learning: theoretical perspectives and practical challenges for the design of a research program. In: Gunderson LH, Holling CS, Light SS (eds) Barriers and bridges to the renewal of ecosystems and institutions. Columbia University Press, New York, pp 428–460

Pelling M, High C, Dearing J, Smith D (2008) Shadow spaces for social learning: a relational understanding of adaptive capacity to climate change within organisations. Environ Plann A 40:867–884

Raik DB, Wilson AL, Decker DJ (2008) Power in natural resources management: an application of theory. Soc Nat Resour 21(8):729–739

Rist S, Chidambaranathan M, Escobar C, Wiesmann U, Zimmermann A (2007) Moving from sustainable management to sustainable governance of natural resources: the role of social learning processes in rural India, Bolivia and Mali. J Rural Stud 23(1):23–37

Röling NG (2002) Beyond the aggregation of individual preferences. In: Leeuwis C, Pyburn R (eds) Wheel-barrows full of frogs: social learning in rural resource management. Koninklijke Van Gorcum, Assen, pp 25–47

Röling N, Maarleveld M (1999) Facing strategic narratives: an argument for interactive effectiveness. Agric Hum Values 16(3):295–308

Salomon G, Perkins DN (1998) Individual and social aspects of learning. Rev Res Educ 23(1):1–24

Schusler TM, Decker DJ, Pfeffer MJ (2003) Social learning for collaborative natural resource management. Soc Nat Resour 16(4):309–326

Sinclair AJ, Diduck AP (2001) Public involvement in environmental assessment in Canada: a transformative learning perspective. Environ Impact Assess Rev 21(2):113–136

Steyaert P, Jiggins J (2007) Governance of complex environmental situations through social learning: A synthesis of SLIM's lessons for research, policy and practice. Environ Sci Policy 10(6):575–586

Tetlock PE (1991) Learning in U.S. and Soviet foreign policy: in search of an elusive concept. In: Breslauer GW, Tetlock PE (eds) Learning in US and Soviet foreign policy. Westview, Boulder, pp 20–61

Torraco RJ (2005) Writing integrative literature reviews: guidelines and examples. Hum Resour Dev Rev 4(3):356–367

Waddell SJ (2005) Societal learning and change: how governments, business and civil society are creating solutions to complex multi-stakeholder problems. Greenleaf, Pensacola

Walker B, Carpenter S, Anderies J, Abel N, Cumming G, Janssen M, Lebel L, Norberg J, Peterson GD, Pritchard R (2002) Resilience management in social-ecological systems: a working hypothesis for a participatory approach. Conserv Ecol 6(1). http://www.ecologyandsociety.org/vol6/iss1/art14/

Webler T, Kastenholz H, Renn O (1995) Public participation in impact assessment: a social learning perspective. Environ Impact Assess Rev 15(5):443–463

Wenger E (1998) Communities of practice: learning, meaning, and identity. Cambridge University Press, Cambridge

Wiklund H (2005) In search of arenas for democratic deliberation: a Habermasian review of environmental assessment. Impact Assess Proj Appraisal 23(4):281–292

Woodhill J (2002) Sustainability, social learning and the democratic imperative. In: Leeuwis C, Pyburn R (eds) Wheel-barrows full of frogs: social learning in rural resource management. Koninklijke Van Gorcum, Assen, pp 317–331

Young OR (2002) Institutional interplay: the environmental consequences of cross-scale interactions. In: Ostrom E, Dietz T, Dolsak N, Stern PC, Stonich S, Weber EU (eds) The drama of the commons. National Academy, Washington, DC, pp 263–291

Chapter 11
Adaptive Capacity as a Dynamic Institutional Process: Conceptual Perspectives and Their Application

Ralph Matthews and Robin Sydneysmith

11.1 Introduction

Whereas most scientific approaches are inherently reductionist, the primary stance of environmental analysis is synthetic. Its roots are in ecology, a late-modern scientific development that emphasizes the importance of understanding environmental observations within a *systems perspective* of integrated organisms. Modern environmental knowledge extends that perspective to incorporate social phenomena into a comprehensive, cross-scale analysis. Gunderson and Holling (2002) have labeled this new systematic and synthetic approach "panarchy" and, of particular relevance given the focus on institutional analysis in this paper, argue that such a perspective:

> ... must be capable of organizing our understanding of economic, ecological, and institutional systems. And it must explain situations where all three types of systems interact (Holling et al. 2002, p. 5).

Indeed, many approaches to ecological knowledge regard environments as systems of natural and social processes that have *resilience* and *adaptive* qualities that permit them to withstand *exposures* that would otherwise leave them *vulnerable* and at *risk*. From this perspective, *adaptation* and *adaptive capacity*, whether seen from a biological or social perspective, are also seen as embodying system assumptions. For example, Smit and Wandel state:

> Adaptation in the context of human dimensions of global change usually refers to a process, action, or outcome in a system (household, community, group, sector, region, country) in

R. Matthews (✉)
The University of British Columbia, 6303 N.W. Marine Drive, V6T 1Z1 Vancouver, BC, Canada
e-mail: ralph.matthews@ubc.ca

R. Sydneysmith
Department of Sociology, The University of British Columbia, 6303 North West Marine Drive, V6T 1Z1 Vancouver, BC, Canada
e-mail: robin.sydneysmith@ubc.ca

order for the system better to cope with, manage, or adjust to some changing conditions, stress, hazard, risk or opportunity. (2006, p. 282)

As will be shown later in this paper, approaches dealing with the *adaptive capacity* of complex socio-ecological systems tend to look for such qualities within the system itself, while hazards are seen largely as external to the fundamental integrity of ecosystems, thereby making them vulnerable and threatening their resilience (cf. Hall and Taylor 1996, p. 7).

While many concepts used in environmental studies are derived from ecology, this is not true of the concept of *institutions*. Institutional analysis has its providence squarely in social science. Used primarily by historians, political scientists, and sociologists, the concept is most frequently used to refer to the habituated and customary dimensions of social life. From such a perspective, institutions are to society what habits are to individuals, namely the largely patterned and taken-for-granted processes whereby things are done within a societal and organizational context. Institutions constitute something equivalent to social glue, a "means for holding society together, for giving it a sense of purpose and for enabling it to adapt" (O'Riordan and Jordan 1999, p. 81). From such a perspective "institutions have to involve rules, regulations and legitimating devices" (1999, p. 82).

This institutional perspective fits nicely with the ecological concepts previously mentioned, as it also inherently embodies a systems perspective. However, perhaps because institutions are the only distinctly social element in the environmental toolbox, they are also frequently accorded the potential to be the regulatory dimension of socio-ecological processes and the agent of potential change. When this happens, institutions are seen as the basis for overcoming the (largely social) forces that are seen as threatening ecological well-being. This is probably best expressed in the Brundtland report when it states, "This real world of interlocked economic and ecological systems will not change; the policies and institutions concerned must" (World Commission on Environment and Development 1987, p. 9).

When institutions are conceptualized in this way (i.e., as change), their meaning and role is seen differently from being just the normative social glue and a body of cultural constraints. With this formulation, institutions become the basis for adaptation to the ecological changes that are occurring. Just as resilience becomes the antithesis or response to ecological vulnerability, so institutions are seen as the "mechanisms" for providing adaptive capacity within socio-ecological systems (cf. Ostrom et al. 2002; Brunner et al. 2005). That is, in environmental analysis, institutions and adaptive capacity are ineffably linked. Any analysis of that relationship requires one to focus on how institutions operate so as to bring about mitigation and/or adaption to environmental changes.

The focus on institutions as a fundamental mechanism of adaptive capacity requires a conceptual framework for examining how such dynamic institutional processes occur. Thus, an underlying tenet of this paper is that a focus on institutions simply as normative constraints, while consistent with systems assumptions about equilibrium, cannot address the ways in which actors and groups operate in institutional contexts. What is required is an institutional perspective that links culture, organizations, and the actions of individual actors.

This forms the basis for our advocacy of an approach to institutional processes, developed mostly in political sciences and sociology, known as *New Institutional Analysis* (NIA). In the following pages we explore this perspective, focusing on its ability to provide a framework for assessing dynamic behavioral processes within organizational contexts. NIA focuses particularly on how actors behave within organizational settings. While not ignoring the cultural dimensions of institutions, it focuses on whether the institutional culture of such settings constrain actors from dealing effectively with new circumstances, or whether such organizational cultures can actually facilitate adaptive capacity. We see this approach as offering the often missing dynamic social component in much ecological analysis. As part of our presentation of this perspective, we will also briefly link NIA with recent work on the institutional dimensions of global environmental change (IDGEC), including relevant work by Ostrom (2005) and by the International Human Dimensions Program (IHDP) (Young et al. 2008).

Furthermore, there are relatively few attempts to empirically utilize this NIA approach, and those that do provide little in the way of a systematic operationalization of it. Therefore, we conclude this paper with a (necessarily brief) presentation of the framework for our study of governance responses to climate change in the sub-Arctic Canadian city of Whitehorse, Yukon. In this ongoing study, we are taking preliminary steps to operationalize aspects of the NIA approach as the basis of our research on the dynamic aspects of adaptive capacity of governance institutions. Although we cannot here present much in the way of the findings of our Whitehorse study, we can demonstrate how we apply the NIA perspective in our analysis.

11.2 Adaptive Capacity in Context

Many conceptual analyses treat adaptive capacity largely as a cultural and tautological "black box" in which adaptation is seen as a function of the adaptive capacity of socio-ecological systems, with little explanation of how this takes place, It is, as Yohe and Tol (2002, p. 25) put it, treated as "an organizing concept." Even an impeccable source like the Fourth Assessment Report (AR4) of the Intergovernmental Panel on Climate Change (IPCC) defines adaptive capacity as "the ability of a system to adjust to *climate change*...to moderate potential damages, to take advantages of opportunities, or to cope with the consequence" (2007, p. 869; emphasis in the original), without articulating how this capacity is put into action. In fact, throughout the Report, adaptive capacity is treated categorically. There is simply "more or less of it," as a result of pre-existing conditions. Similarly, Adger (2006, p. 270) declares, "adaptive capacity is the ability of a system to evolve in order to accommodate environmental hazards or policy changes and to expand the range of variability with which it can cope." As Smit and Wandel (2006, p. 285) state (albeit in another context), an approach such as this "does not attempt to identify the processes, determinants or drivers" of adaptive capacity.

Table 11.1 Proposed determinants of adaptive capacity

	Smit et al. (2001, quoted in Swanson et al. 2007)	Yohe and Tol (2002, p. 26)	Brooks et al. (2005, p. 168)	IPCC (2001, as quoted in Albernini et al. 2006, p. 124)
Economic resources		Available resources and their distribution across the population	Resources	Available technological options
Technology		Structure of critical institutions and the allocation of decision-making authority	Financial capital	Resources
Information and skills		Stock of human capital	Social capital (e.g., strong institutions, transparent decision-making systems, formal and informal networks that promote collective action)	The structure of critical institutions and decision making authorities
Infrastructure		System's access to risk spreading	Human resources (e.g., labor, skills, knowledge and expertise)	The stock of human capital
Institutions		Way in which decision makers maintain and distribute information	Natural resources (e.g., land, water, raw materials, biodiversity)	The stock of social capital including the definition of property rights
Equity		Public's attribution of the source of stress		System's access to risk spreading processes
		Significance of exposure in the local situation		Information management and the credibility of information supplied by decision makers
				Public perceptions of risks and exposure

However, while there are few efforts that seek to identify the *processes* of adaptive capacity, there are numerous works that seek to categorize its *analytic dimensions*. Thus, there is a literature on what are declared to be the *determinants* that influence whether or not a place can be considered to have adaptive capacity. Most such lists overlap (see Table 11.1) and frequently the identification of such dimensions leads directly to the development of a matrix in which they are related to one another and to a range of other variables.

Prime among these "other" variables are *vulnerability* and *exposure*. Hence, Swanson et al. (2007, p. 13) state:

> The *vulnerability* of a socio-economic and environmental system to climate change is conceptualized as a function of a system's *exposure* to climate change effects, and its *adaptive capa*city to deal with those effects.

Smit and Pilifosova (2003, p. 13) capture that relationship in an equation in which vulnerability is stated to be a function of exposure, sensitivity, and adaptive capacity. Furthermore, the three concepts of *vulnerability*, *exposure* and *adaptive capacity* are also related to *resilience*. Resilience and vulnerability are defined as opposites. Following what is clearly an ecological comparison, resilience is defined as "the magnitude of a disturbance that can be absorbed before a system changes to a radically different stage, as well as the capacity to self-organize and the capacity for adaptation to emerging circumstances" (Adger 2006, p. 268), Conversely, vulnerability "is usually portrayed in negative terms as the susceptibility to be harmed" (Adger 2006, p. 269).

Notably, both vulnerability and exposure are seen as conditions of the adaptive capacity of a community or region so that "the key parameters of vulnerability are the stress to which a system is exposed, its sensitivity and its adaptive capacity" (Adger 2006, p. 269). Whether or not they can be identified as antecedents or consequences is not made clear.[1]

Another body of literature focuses on the appropriate geographic scope for any adaptive capacity analysis. Some contend that adaptive capacity is primarily a consequence of *local conditions.* Thus, for Smit and Wandel adaptive capacity is "context specific, and varies from country to country, from community to community, among social groups and individuals, and over time" (2006, p. 287). Likewise, Yohe and Tol (2002, p. 28) emphatically declare, "We argue that adaptive capacity is a local characteristic," although there is little attempt to be specific about just what constitutes "local." Smit's and Wandel's statement quoted above includes everything from groups to countries as potentially local units. What is implied is not so much that adaptive capacity is only influenced by local events, but rather that it can only be understood in the context of whether "local" areas (be they communities, regions, or nations) have the appropriate economic, social, cultural, and political resources to respond in ways that enable them to reduce their level of exposure to

---

[1] We are aware that it is something of a misnomer to refer to vulnerability as if it was in some sense a singular variable. Indeed, vulnerability, like adaptive capacity, is usually conceptualized as a very complex array of intervening dimensions and processes. However, to dwell further on it here would deflect us from our primary focus on adaptive capacity.

risk, recover from the impacts of stressful events, or, in some cases, to take advantage of opportunities that may emerge (Vincent 2007). The fundamental issue here seems not to be so much whether a phenomenon originates locally, but rather whether the local area (however defined) has the capacity to deal with the challenges that it must face, either through its own internal resources or through its ability to access external sources of support, information, and actual physical assistance.

This means that, almost invariably, any analysis of adaptive capacity involves a *geographic scale* element, in that it is necessary to examine both the exposures and the capacity of the local region, and also the resources that this community or region can draw upon from a larger area. An isolated community with limited resources is far less likely to withstand environmental exposure than a similarly affected community that can call upon a wide range of internal and external ecological and social resources to assist it. The potential adaptive capacity of each differs dramatically.

Furthermore, the lists of determinants affecting the adaptive capacity of localities that were outlined above, all emphasize that it is primarily (if not exclusively) a social process. Thus, the primary resources necessary to enhance adaptive capacity are such fundamentally social processes and products as social capital, human capital, financial capital, decision-making, and trust. Of particular note, Pelling and High (2005) make a strong case that it is the social capital of communities that is critical to the development of effective adaptive capacity. By this, they infer that the ability of a local community to be adaptive depends very much on its ability to use its established social networks to access various human, social, and economic resources. Only by knowing the extent of resources (economic, social, political, cultural, and knowledge) that the local unit and those responsible for its governance have at their disposal, are we likely to be able to assess the adaptive capacity of any locality.[2]

However, space is not the only scale variable related to adaptive capacity; time *is also a critical scalable dimension*. Risk is usually defined in terms of exposure over time. Some exposures are sudden and overwhelming, making the temporal dimension irrelevant. Others involve long-term processes. Climate change is both. Climates may change only gradually, but such changes are often manifested in sudden catastrophic events such as floods or droughts, fires, or pestilence. The adaptive capacity of a region then is a function of its ability to withstand both long-term and sudden threats.

To summarize, the perspective on adaptive capacity that emerges from this body of literature regards it as the outcome of a wide range of other social variables, local and distant, immediate and/or temporally remote. These serve to determine the ability of a local area to respond to climate change challenges. However, for the most part, these attempts to identify variables and their impact as well as to locate them in geographic and temporal contexts, contain few attempts to actually depict or understand the

---

[2]cf. Matthews work on social capital (Enns et al. 2008; Matthews 2003; Matthews and Côté 2005; Page et al. 2007; Matthews et al. 2009) strongly supports this position.

processes through which these factors work to produce the adaptive capacity of any locality. As we have noted already, neither the categorization of the attributes of adaptive capacity nor the identification of influences such as spatial resources and time, provide much of a basis for understanding how adaptive capacity takes place.

Yet, not all approaches to adaptive capacity have ignored the process whereby it is achieved. For example, Brooks (2003) contended that the assessment of adaptive capacity requires understanding of both how it is constituted and how it is translated into adaptation, "...in other words, we must understand the adaptation process (2003, p. 11). In making this distinction, he broke out of the tautological tendency to define adaptive capacity as the ability to adapt. Yet, despite recognizing the importance of process, Brooks' analysis focuses little on it. Rather, his concern is primarily the one already identified as the role of geographic scale factors in effecting adaptive capacity. Brooks argues (2003, p. 11) that the factors determining adaptation processes depend especially on the scale of the "systems that are adapting," (i.e., households, communities vs. nation states). Brook's attention largely is directed to the intersection of local versus distant factors in the adaptive capacity process. He argues that it is a trap to focus merely on local, endogenous factors to the exclusion of broader political and economic forces. Such "exogenous" factors are sometimes characterized as "political will" and, though often poorly defined, can have a powerful influence on how, or even whether the adaptation takes place (Brooks 2003). However, the processes whereby adaptive capacity occurs still remain largely unexplored.

In contrast, Smit and Wandel (2006) advocate an approach to understanding adaptive capacity that is process oriented. They contend that the appropriate way to undertake such analysis is to begin at the community level and reason "from the bottom-up." In doing so, they eschew the notion of measuring vulnerability or attempting to establish indicators or measures of adaptive capacity. Rather, they advocate strongly for a dynamic understanding of adaptive capacity that takes into account the local social processes involved. As this fits closely with the approach we take here and throughout the remainder of this paper, we quote them at some length:

> This body of work ... tends not to presume the specific variables that represent exposures, sensitivities, or aspects of adaptive capacity, but seeks to identify these empirically from the community. It focuses on conditions that are important to the community..... It employs the experience and knowledge of community members to characterize pertinent conditions, community sensitivities, adaptive strategies, and decision-making process related to adaptive capacity or resilience. It identifies and documents the decision-making processes into which adaptations to climate change can be integrated. (Smit and Wandel 2006, p. 285)

Finally, some consideration needs to be given to the methodological conundrums involved in identifying or measuring the adaptive capacity of any social unit. Smit and Wandel capture this best with their depiction of a "nested hierarchy model of vulnerability", involving exposure-sensitivity on the one side and adaptive capacity on the other, as essentially the outcome of the intervening complex array of processes (2006, p. 286). To complicate this further, it is not clear whether adaptive capacity can, in any way, be regarded as a linear process. Indeed, there is evidence that it is frequently discontinuous and, depending on the rate of exposures, may operate at variable speeds (cf. Holling et al. 2002). This seems to be particularly true of the adaptive capacity of social units, though it may also be true of ecological ones. Given this, no set of mechanisms or determinants, no matter how thorough the resulting matrix, is likely able to do more than generally assess what is essentially a social process of considerable complexity. For that, we need to focus on the ways in which the more dynamic relationships of adaptive capacity are developed and expressed. To do so, we will turn to the role of institutions.

11.3 Adaptive Capacity and Institutional Structures

Earlier, in our discussion of the concepts used in understanding environmental change, we identified "institutions" as the one widely used concept derived from social science. As just demonstrated, it is frequently linked to adaptive capacity. In a comprehensive recent analysis of "The Role of Local Institutions in Adaptation to Climate Change", Agarwal summarizes this position:

> Adaptation to climate change is inevitably local and ...institutions influence adaptation to climate change in three critical ways: (a) they structure impacts and vulnerability, (b) they mediate between individual and collective responses to climate impacts and thereby shape outcomes of adaptation, (c) they act as the means of the delivery of external resources to facilitate adaptation and thus govern access to such resources (2008, p. 1)

That is, institutions structure the way in which risks impact people and communities, and they channel (or enable) various processes of response. Whether or not risks constitute vulnerabilities is largely determined by the capacity of existing local institutional arrangements to: (1) provide access to various local and more distant physical, social, and economic resources, and, (2) enable their effective application. It follows that the adaptive capacity of any community to changing environmental conditions is inevitably a social process that is largely guided by institutional relationships. The response to climate change is a social process no matter whether that response is identified as mitigation, adaptation, or some hybrid of the two. Institutional arrangements and institutional processes constitute the foundation of all social types of responses to environmental threats.

But, if that is what institutions do, it is necessary to address how institutions actually work i.e. how they do this. If adaptive capacity is largely determined by institutional arrangement, then attention needs to be given to the ways in which such institutional arrangements operate. North (1990) was among the first to recognize the important role played by institutional arrangements in shaping human action and social change, though his focus was on economic and not ecological change. He saw institutions largely in terms of the constraints that they put on entrepreneurial actors as they sought to create new economic organizational practices (1990, pp. 84–88). North argued that institutions generally created transaction costs for such persons as they attempted to bring about change, and he emphasized the slow and incremental character of institutional change relative to changing economic demands (1990, p. 89). North regarded institutions less as the glue holding society together, than as the impediments holding entrepreneurial actors back (1990, pp. 103–104).

Some subsequent analysts in the ecological/climate change area have largely adopted North's stance. For example, Fiori (2002, p. 1026) provides an extensive discussion of formal and informal institutional constraints, arguing that formal constraints are more amenable to change than informal ones that remain essentially hidden if not unconscious. Still others examine the design of institutional structures (cf. Roland, 2004). For example, Pelling and High (2005) focus on the ways in which these designs operate to facilitate or impede change. They argue that the economic and social development of lower and middle income countries is hampered by the influence of "slow moving" cultural institutions that do not adapt or keep pace with economic institutions that are declared to be "fast moving" and adaptive. In a similar vein, Portes distinguishes between "slow moving institutions like culture" or "fast moving institutions like legal rules and organizational blueprints" (Portes 2006, p. 235). Unfortunately, such explanations appear to have a bias toward the cultural fabric of non-western societies found in earlier approaches to modernization (cf. Inkles and Smith 1974).

Of particular interest are those papers that explicitly address the relationship between institutional forms and organizational practices, if only because of the confusion that exists in the use of the two terms. Organizations are the social entities that are created to accomplish tasks. Institutions are the cultural norms, values, and accepted practices that govern how behaviors in and between these organizations takes place. Fiori (2002, p. 1028), again based on North's work, declares that institutions determine the opportunities in a society, whereas organizations are created to take advantages of those opportunities. Likewise, he discusses the constraining role of institutions on organizations, but argues that organizational change can, in time, bring about institutional readjustments. Such work, focusing on how institutions operate in relation to organizational structures and human action, invites further discussion of how institutions influence both the actions of individuals and the adaptive capacity of organizations and communities. We suggest that New Institutional Analysis (NIA), through its focus on institutions as theatres for individual action and decision-making, provides a basis for doing just that.

11.4 Adaptive Capacity and Institutional Dynamics

Hall and Taylor (1996, p. 7) state that the fundamental question for any institutional analysis is whether institutions can affect human behavior. Somewhat akin to the discussion between the normative/cultural and process/action approach we have been developing above, they distinguish between a "cultural approach" and a "calculus approach" to this question. Like our prior discussion, the cultural approach is depicted as having a focus on path dependence and cultural norms (1996, p. 7). As Hall and Taylor note, such a perspective leaves little room for explanations of events as being directed by individual choice (1996, p. 8). In contrast, the calculus approach focuses on the way in which individuals act, within the cultural framework provided by institutions, so as to achieve benefits for themselves and others. In particular, according to Hall and Taylor, this approach focuses on how individuals, operating within organizations, engage in strategic interaction in the determination of outcomes (1996, p. 12). Actors are expected to act strategically and instrumentally within a set of preferences influenced by their institutional location and its normative expectations, their social locations, and their own values and goals. That is, while actors operate strategically and with calculus, they do so in ways that are influenced by their institutional and personal preferences.[3]

The Hall and Taylor approach to institutions emphasizes the role of individual choices and strategizing in determining the resilience and adaptive capacity of communities and regions. As such, it moves analysis away from a focus on adjustments and forces within systems. Now the concern is with the way in which individuals, operating within the context of institutional frames, significantly affect adaptive capacity through their behavior. From a focus on the architecture of institutions, the balance has shifted to a concern with organizations as institutional arenas that shape behavior, albeit in culturally influenced ways.

A focus on how actors make "choice within constraints" (Brinton and Nee 2001, p. xv; Nee 2005) constitutes a new approach to institutional analysis that has come to be identified generally in the sociological literature as "new institutionalism" or "new institutional analysis". Nee (2001, p. 1) sums up the basic stance of NIA as follows:

> ... the new institutionalism seeks to explain institutions rather than simply to assume their existence. In this endeavour, new institutionalists in the social sciences generally presume purposive action on the part of individuals, albeit under conditions of incomplete information, inaccurate mental models, and costly transactions.

Hall and Taylor (1996, p. 15) describe this as a sociological perspective:

> The new institutionalists in sociology ... have a distinctive understanding of the relationship between institutions and individual action.... that is to say, they emphasize the way in which institutions influence behaviour by providing the cognitive scripts, categories and models that are indispensible for action.

Furthermore, as Hall and Taylor (1996, p. 6) point out, this is fundamentally a question of legitimacy, authority, and power. In their words:

> Central to this approach, of course, is the question of what confers "legitimacy" or "social appropriateness" on some institutional arrangements and not others. Ultimately, this is an issue of cultural authority. (1996)

---

[3]In case this statement leads to confusion of our intention, we emphasize that we are not arguing that individuals always act to maximize benefits. Much research in sociological, social psychology demonstrates that this is not the case. Rather, people seek to achieve a satisfactory level of benefit by engaging in what is called "satisficing." Likewise, we are not contending that actors always engage in objectively "rational" action. Rather than engage in action that is "rational to them," given their own values and goals for themselves and others.

This point has profound implications for the assessment of how institutions play a role in adaptive capacity, as it raises the issue of how authority and legitimacy is enacted within institutional processes. It opens up the issues of control and power within institutional contexts. If institutions are now seen as cultural forms constraining human action and in which individuals construct what they perceive to be appropriate courses of behavior, then whose values dominate in the determination of what are legitimate values? Similarly, whose values and goals guide the perception of what are the appropriate courses of behavior to take in varying circumstances?

With this, we have embedded the discussion of institutions and adaptive capacity into the broader issues related to social structure of particular concern to sociologists. Institutional values do not exist in a vacuum. They exist because they are the product of certain social relations that give them legitimacy, and they support the interests of those with the resources to make them dominant. Thus, to talk about the relationship between institutions and adaptive capacity without reference to these broader issues of control, power, and resources is to leave out significant aspects of adaptive capacity. Twenty-seven years ago, DiMaggio and Powell (1983) argued that the power of elites influences institutional genesis, reproduction, and transformation in at least two ways: in the initial shaping of institutional "premises", and at key moments or turning points where elites are favorably positioned to make choices and decisions that may have persistent temporal influence over the course of organizational goals and policy. More recently, Hotimsky et al. (2006) maintain that the nature and distribution of power in society effects the functional roles of institutions, such that it is not enough to ask for what purpose institutions exist, but also in whose interest they exist, persist, or change.

Power takes many forms. It can be obtained through coercion, or it can be achieved through knowledge. It can be realized through force, or it can be granted through legitimate political processes that are institutionalized in a society. Yet, there can be no doubt that power relations are important to the adaptive capacity of any community – including the capacity to deal with the risks produced by climate change. For example, the extent to which a community is vulnerable to environmental risk is, to a large degree, a product of the extent to which those in control have the knowledge and legitimacy to respond appropriately. We see this particularly clearly at the national and international level where, in some countries, those with the most power to act to ameliorate the negative effects of global warming develop "solutions" more inclined to protect particular economic interests than to develop policies aimed at reducing greenhouse gas emissions. However, it is also true at the community and regional level. Portes (2006, p. 243) best captures this set of issues. Institutions, he argues, need to be understood with respect to clear definitions of the role of *culture*, including norms and associated social roles, and *social structure* as these arise from the distribution of power and social class (Portes 2006, p. 239).

It is our position that, when studying the adaptive capacity of communities or larger locations, it is not enough to define institutions as simply "blueprints" or "rules of the game." One must also examine what groups or individuals have the power, legitimacy, and authority to act within, or outside of, these institutional blueprints in ways that influence adaptive capacity. For us, the empirical analysis of adaptive capacity with respect to environmental change begins with a consideration of those organizational structures in any locality that have the responsibility, the power, and the legitimacy to respond to environmental challenges within it. In any state where the rule of law is paramount and where civil government operates at the local level, this is almost invariably the civic leadership and governance structures responsible for maintaining order and delivering services within that local area. These civic political organizations constitute the frontline in responding appropriately to climate change challenges. That is, to the extent that adaptive capacity requires local solutions, then it is local civic government that has the immediate responsibility of guiding and managing societal responses to social-ecological challenges and changes. Municipal councils are the social units most directly vested with levels of legitimacy and power to act on behalf of their locality. This is particularly true of larger and more complex centres where the responsibilities of local government are explicitly articulated, but is no less true of small local communities where the legitimated authority may be the result of custom and tradition.

In making these claims, we have been careful to respect the distinction between institutions and organizations, and the relationship between them. Institutions remain the normative architecture for action that both legitimizes governance organizations and provides the blueprint of how they appropriately operate. For example, when bureaucrats talk about governmental organization as being constructed into silos, they are essentially talking about the underlying institutional norms that led to this form of organizational structure. However, behavior within these organizational forms is also structured in institutionalized ways. In part, this is simply the result of the way in which these organizations have been normatively structured over time. Institutions shape organizations, and organizations shape behavior. However, behavior within these organizations is also very much a product of what is deemed normatively appropriate. Some actors operate totally within the normative expectation of their role responsibilities. Others go beyond the expectations of their positions and act in unique, and sometimes creative, ways to achieve complex organizational goals.

That is, actors not only occupy roles, they also construct them. In the context of the concern here, namely the roles played by civic officials and administrations, it is this ability to reconstruct roles in ways not institutionally prescribed that may make the difference between a community with resilience and adaptive capacity, and one that remains vulnerable and threatened by environmental changes. Hence, by examining processes of governance and governing at the local community level, we hope to reveal some of the complex ways in which governance institutions effect or contribute to adaptive capacity. We are particularly interested in the institutionalized ways in which the power to make decisions is unevenly distributed throughout organizations and how this distribution influences the ability to take action, the paths in which action occurs, and the likelihood of appropriate and successful outcomes being achieved. Our approach, based on the perspective of NIA, places its emphasis on how regulatory structures actually work in terms of: (a) the accepted and expected "habits" (i.e., the habitus) for operating within them;

(b) what interests are served by them; and, (c) the capacity of these organizational structures to respond to the risks related to climate change. In the context of this approach, the adaptive capacity of a community is influenced by the way in which its fundamental governance organizations and the behavior of those within them are constrained by the institutional requirements of the organization. In some cases, the institutionalized ways of dealing with situations can contribute to the resilience and adaptive capacity of communities. In other cases, where new approaches are required but the capacity of undertaking them is limited by institutionally based processes, the community remains vulnerable in the face of changing conditions (Berkes 2003). We focus, therefore, on the way in which organizations embody institutionalized cognitive maps and normative expectations that either facilitate or block adaptive responses to climate change. By examining these processes in and through the actions of individuals in civic governance positions, we are better able to understand whether, how, and how successfully, a community can face climate change vulnerabilities now and in the future.

Though we have outlined an approach that has its roots in sociological analysis, its basic stance is compatible with several recent works that also seek to provide an action orientation to the institutional analysis of climate change. Of particular relevance is Ostrom's (2005) *Institutional Analysis and Development* (IAD) framework, and the work on Institutional Dimensions of Global Environmental Change (IDGEC) that has been developed by the International Human Dimensions Program (IHDP) (Young et al. 2008).

Ostrom's (2005) analysis focuses on what she calls "action arenas." These are said to include two essential elements (called holons), an action situation, and the participant in that situation (2005, p. 14). Her work links both the cultural and the action approaches identified above through a focus on how rules affect decision outcomes (2005, p. 29). In such situations, actors are seen to choose an appropriate course of action. Such "choices" are governed by normative expectation but also involve "decision points" (2005, pp. 44–45). A strong focus of Ostrom's work is on the "strategy" used by actors within institutional contexts. As we have similarly noted, such strategies are influenced by trust and by the social capital relations developed through interaction in institutional settings (2005, pp. 70–78).

The IDGEC project summarizes a decade of analysis and research by an international and interdisciplinary team of social scientists. It has deliberately sought to incorporate an NIA perspective. As they state:

> Our research has sought from the outset to take advantage of the intellectual capital of the *new institutionalism* in formulating our research agenda.... The project shares with the new institutionalism a strong interest in what are known as collective action problems, or situations in which seemingly rational choices on the part of individual members of a group lead to societal results that are understandable from the perspective of all the members of the group (Young et al. 2008, p. 6).

The main independent variable in the IDGEC framework is institution, and the focus is on how institutions are "embedded in more comprehensive process of learning and response" (Young et al. 2008, pp. 53–55). Their analysis provides something akin to a template of what aspects to examine when considering the role of institutions in adaptive capacity. In particular, they highlight six dimensions of research and analysis on which to assess institutional capacity. These are: (1) design; (2) performance; (3) causality; (4) fit; (5) interplay; and, (6) scale. While all are relevant, the last three of these are particularly germane to our own research on governance in an Arctic gateway city. *Fit* relates to the extent to which governance institutions are congruent with ecological needs. *Interplay* is the extent and manner in which different levels of governance interact. *Scale* refers to the geographic range of governance institutions from the local to the global (Young et al. 2008, pp. 26–35). The approach favors what it refers to as a "diagnostic method", by which is meant a focus on the nature of specific institutional arrangements as they guide the behavior of individual actors (Young et al. 2008, p. 120). Central to that method is an analysis of "the four Ps": (1) problems; (2) politics; (3) players; and (4) practices (Young et al. 2008, pp. 121–134). We cannot, here, present the detailed criteria provided for each of these dimensions, other than to note that they deal with actors (players), behaviours (practices), normative contexts (problems), and power relationships (politics), all of which we have identified as crucial to a dynamic institutional analysis of adaptive capacity.

As the preceding analysis demonstrates, there is now a growing body of conceptual analysis in the field of environmental analysis that identifies the adaptive capacity of institutions as a dynamic process involving strategic decisions by actors. There is also an awareness of this as a governance process involving the interplay of different jurisdictions and multiple dimensions of scale, performance, and cause. In environmental analysis, there is even something akin to a template for carrying out such analysis, focusing on the readiness for, and responses to, global warming. Yet, despite a growing call to assess the responses to climate change in such terms, there is little empirical research being carried out utilizing this "new institutional" approach. Hence, in the next section, we will provide a brief introduction to our own efforts to implement an NIA approach, in our analysis and assessment of the responses of the City of Whitehorse to the challenges of climate change in Canada's north. Our study demonstrates how the framework we have advocated here may be used in actual policy research linking issues of ecological change and social analysis.

11.5 Operationalizing *New Institutionalism* in an "Arctic Gateway City"

The City of Whitehorse, located just north of the sixtieth parallel, is the capital of Yukon Territory. With almost 26,000 residents, the City is home to approximately three-quarters of Yukon's population. The city limits encompass some 416 km^2, giving it the largest per capita land base of any city in Canada. In popular mythology, Canada's north is peopled by mostly indigenous persons living in small communities. The reality is that over two-thirds of northern Canadians live in larger administrative centres (e.g., Whitehorse, Yellowknife, Iqaluit) some hundreds of kilometres south of the Arctic Circle. These are administrative and economic development centres for the whole Arctic and, as such, constitute what we choose to call "Arctic Gateway Cities." The City of Whitehorse not only links the Territory to the amenities and opportunities that emanate from the south, but is also a gateway through which southern resources, goods, and services are channeled. For the City of Whitehorse, these are both exciting and demanding times. Whitehorse is in a period of economic and population growth, spurred primarily by the economic benefits of mineral and oil and gas exploration and sustained growth in the tourism sector. On the other hand, the western sub-Arctic region immediately north and west of Whitehorse is experiencing significant impacts from global warming and resulting ecological changes. Glaciers are melting and permafrost disappearing. Located on a low-lying plateau at a bend in the Yukon River, Whitehorse's downtown is potentially subject to flooding as a consequence. There is also the risk of increased forest fires. The climate itself is also changing. While winters are somewhat warmer than previously, in summer there is increased overcast and rain, reducing the quality of outdoor life for people who spend long winter months in sub-Arctic darkness looking forward to long, clear summer days and sunny evenings.

The City of Whitehorse's governance and management, including its Mayor, Council, and senior administration, are very much aware of the impact that global warming is having elsewhere and of the need for the City to prepare for change and to engage in sustainable practices. Accordingly, the City, with considerable community input, has developed a Sustainability Plan and has established the position of Sustainability Manager to oversee its implementation. With their concern for environmental issues, the Mayor and Council were willing to grant our request to meet with them to consider being part of a study of sustainability and adaptive capacity to climate change, and subsequently agreed to partner on the project. Over the past several months we have interviewed elected officials (Mayor and Council) and senior and middle-rank officials within the City administration. As these interviews progressed, it became clear that many of the City policies and actions were influenced by policies and actions of other levels of government and, as a result, we have also interviewed many officials within the Yukon Territorial Government (YTG) whose work activities bring them in close contact with City officials.

Throughout this paper, we have argued for an approach to adaptive capacity that focuses on institutions as arenas in which actors work. We have suggested that adaptive capacity is created not just through normative regulations, but also through the capacity of actors to operate in strategic ways within the normative contexts of organizations. In particular, we have recommended a focus on those who, at a community level, have the authority of legitimacy and the power to act, either within established normative procedures or with flexibility to respond in new ways to unique situations. We have also emphasized the importance of what Ostrom has labeled "decision points" as key windows through which to identify whether organizations are adaptive to new situations that confront them.

Our Whitehorse and YTG interviews reflect these principles. We have focused on those with various levels of power and authority to make decisions, and probed deeply into decision-making processes and practices as they are played out in the context of both routine and unusual circumstances. It is through the multiple decisions made within government agencies and organizations that processes, regulations, and priorities are applied. It is through decision-making strategies that problems are resolved and plans developed – all critical elements of governing and responding to change. At the same time, examination of the way in which decisions, are made, who makes them, and the various actions that precede and follow decisions, reveals certain cultural and social elements of institutional process and organizational structure.

We focus on two types of decision: (1) routine decisions made around regular or repeated administrative actions that in many respects define the day to day, or season to season, operations of the community; and, (2) decisions made under unusual or unique circumstances – things outside the "ordinary" or things that come about through gradual or sudden shifts either to social or ecological conditions. In our analysis, we explore multiple levels of govern*ment* in order to gague and understand how govern*ance* processes vary between and within institutions. This involves a thorough understanding of existing social and political structures and processes (Pelling and High 2005). In this respect, we focus on the institutional architecture of the City administration, and we look at the organizational relationships, flows of authority, and the allocation of resources (e.g., budgets) that occur within it. We examine the locations and applications of power and the cultural dimensions and roles that characterize relationships and interactions. To ensure that we cover these dimensions adequately, we use an interview schedule that covers the following areas:

- Workplace Roles, Relations, and Culture
- Decision Processes
- Capacity and Change – Economic, Environmental, and Climate Changes
- Sustainability – Culture and Goals, Measures and Indicators

Throughout our interviews, we seek constantly to understand how behaviour is constructed and strategies developed within the context of institutionalized relations to deal with the usual and the unusual, the normal and the unique. We remain vigilant in our efforts to determine whether the City administration retains the flexibility both to respond in well-practiced ways when these seem appropriate, and to seek new patterns of response when new approaches are required. In particular, we focus on the capacity of actors to cross institutional boundaries, both within the City administration and in contact with the YTG, as well as with civic organizations. Our reports and papers developed from these data will utilize NIA and other institutional perspectives such as those by Ostrom and the IGDHC. Through them, we will explain and evaluate the extent to which those responsible for the management and planning of Whitehorse demonstrate adaptive responses to the environmental challenges that they face.

11.6 Conclusion

This book examines what its editors describe as the quintessential contemporary question, namely the relationship between sustainability and adaptive capacity. In this chapter, we contribute to this goal by providing an analysis of the relationship between adaptive capacity and institutional processes. In particular, we argue that adaptive capacity is related to the capacity of institutional processes to adapt to unique challenges. This, in turn, is related to the ability of individuals within institutional contexts to pursue strategies that respond effectively to new situations and unique events. We also suggest that decision-making processes within various levels of governance institutions are critical to adaptive capacity, particularly at the community level. We are in agreement with Adger's observation that, "adaptive capacity is only potential until there are governance institutions that make it realizable" (2003, p. 33).

In reaching these conclusions, we have carried out four tasks. *First*, we have analysed some of the existing perspectives on adaptive capacity, highlighting that it is both an ecological and a social process. This is an important starting point given the somewhat disparate, and occasionally contradictory, literature on adaptive capacity as a social process. In our analysis, we have emphasized the extent to which much of the adaptive capacity literature rests on systems assumptions common in ecological reasoning, and has generally adopted a similar systems stance when it comes to the social sphere. Instead, we have espoused a more dynamic perspective. *Second*, we have presented a similar overview of the work dealing with the relationship between adaptive capacity and institutions. Here, our predominant concern is the extent to which institutions have been conceived largely as normative brakes on change, rather than as frameworks for strategic interaction within organizations. In contrast, we have advocated for a focus on the dynamic processes of organizational and institutional change and the processes of decision-making and strategic interaction that occurs as a result of environmental challenges. *Third*, we have outlined the NIA perspective deriving from history, sociology, and political science. We propose that it provides the basis of the dynamic institutional perspective that we see as important. Further, we have linked NIA to other recent work on institutions by Ostrom and by the IDGEC that provide analytic frameworks based in empirical research that also contributes much to this way of looking at institutions as dynamic contexts for strategic action. *Fourth*, we have provided a very brief (though we hope useful) introduction to one of our empirical research projects that is seeking to operationalize this approach to institutional analysis in a specific community context. Our aim is to show how the institutional approach we have been advocating can be operationalized as a research tool for effective empirical analysis of adaptive capacity as a social process.

This paper has involved a complexity of conceptual terms and abstract analyses. While adaptive capacity is the unifying theme under discussion, we (and we suspect other contributors to this volume) discuss at length how we define adaptive capacity, and how it relates to other key terms in the global change literature such as vulnerability, resilience, adaptation, risk, social learning, and, most particularly, institutions. The definitions are myriad and as varied as the disciplines from whence they emerge. We suspect that the civic leaders and officials, whose responses are the subject of our empirical analysis in Whitehorse, would be bemused (if not aghast) at what is presented here. We fear that they might view this as merely the intellectual pastime of academics and respond with a combination of disbelief and ridicule at our diversion into such arcane pursuits. How, they might ask, can this possibly help them chart an appropriate strategy to deal with any of the very real and multifaceted challenges and demands involved in running a mid-sized city in an isolated and harsh northern setting? It may, at this stage, be of little assurance to them that we also are constantly asking ourselves, "How will this help Whitehorse?" Our generally positive response rests on our belief that both the causes of climate change and the responses to it are inherently social processes. We are attempting to hold up a mirror for the civic leaders of Whitehorse to better enable them to see how these social processes operate in their city and identify how they might pursue effective responses to environmental and related changes.

We have argued that it is important to focus on local governance processes, as these are the keys to effective adaptive capacity. This is not to suggest that local communities must "go it alone" in responding to environmental changes. Brooks (2003, p. 12) is rightly critical of a strategy that allows us to "avoid challenging the powerful political and economic vested interests that determine the nature of the geopolitical and economic contexts within which adaptation must be carried out." We agree. On the other hand, as has been repeatedly stated in the literature, adaptive capacity occurs locally and a focus on the relationships of control, power, and governance at the local level is also critical to designing appropriate responses. Our work, then, is the appropriate first step in understanding the broader nexus of governance processes and power relationships that may ultimately determine local adaptive capacity. Adaptive capacity is ultimately the ability to respond effectively to the uncertainties of short-term hazards and long-term risks. Understanding the processes leading to effective governance responses to both is a critical aspect of achieving that capacity.

References

Adger WN (2003) Social capital, collective action and adaptation to climate change. Econ Geog 79(4):387–404
Adger WN (2006) Vulnerability. Global Environ Change 16:268–281
Agrawal A (2008) The role of local institutions in adaptation to climate change. IFRI Working Paper W081-3. Prepared for the social dimensions of climate change, Social Development Department. Washington, DC, International Forestry Resources and Institutions Program, IFRI Working Paper W081-3, School of Natural Resources and Environment. Ann Arbor, MI: University of Michigan
Albernini A, Chiabai A, Muehlenbachs L (2006) Using expert judgement to assess adaptive capacity to climate change: Evidence from a conjoint choice survey. Global Environ Change 16:123–144

Berkes F (2003) Navigating social-ecological systems: building resilience for complexity and change. Cambridge University Press, Cambridge, UK

Brinton MC, Nee V (2001) The new institutionalism in sociology. Stanford University Press, Stanford, CA

Brooks N (2003) Vulnerability, risk and adaptation: a conceptual framework. Tyndall Centre for Climate Change Research. Working Paper 38

Brooks N and Adger WN, Kelly PM (2005) The determinants of vulnerability and adaptive capacity at the national level and the implications for adaptation. Global Environ Change 15:151–163

Brunner R, Steelman T, Coe-Juell L, Cromley C, Edwards C, Tucker D (2005) Adaptive governance: integrating science, policy and decision making. Columbia University Press, New York, NY

Dimaggio PJ, Powell WW (1983) The Iron Cage revisited: institutional isomorphism and collective rationality in organizational fields. Am Socio Rev 48(2):147–160

Enns S, Malinick T, Matthews R (2008) It's not only who you know, its also where they are: using the position generator to investigate the structure of access to socially embedded resources. In: Lin N, Erickson BH (eds) Social capital: advances in research. Oxford University Press, New York, NY, pp 255–281

Fiori S (2002) Alternative visions of change in Douglas North's new institutionalism. J Econ Issues 36(4):1025–1043

Gunderson LH, Holling CS (2002) Panarchy: understanding transformations in human and natural systems. Island Press, Washington, DC

Hall P, Taylor RCR (1996) Political science and the 3 new institutionalisms. MPIFG Discussion Paper 96/6. Köln, Germany: Max Planck-Institut für Gesellschaftsforschung

Holling CS, Gunderson LH, Ludwig D (2002) In quest of a theory of adaptive change. In: Gunderson LH, Holling CS (eds) Panarchy: understanding transformation in human and natural systems. Island Press, Washington, DC, pp 1–22

Hotimsky S, Cobb R, Bond A (2006) Contracts or scripts? A critical review of the application of institutional theories to the study of environmental change. Ecol Soc 11(1):40. On-line journal available at http://www.ecologyandsociety.org/vol11/iss1/art41/

Inkles A, Smith DH (1974) Becoming modern: individual change in six developing countries. Harvard University Press, Cambridge, MA

IPCC Fourth Assessment Report (AR4) (2007) Climate change 2007: impacts, adaptation and vulnerability. In: Parry ML, Canziani OF, Palutikof JP, van der Linden PJ, Hanson CE (eds) Contribution of Working Group II to the Fourth Assessment Report of the Intergovernmental Panel on Climate Change. Cambridge University Press, Cambridge, UK

Intergovernmental Panel on Climate Change (IPCC) (2001) Impacts, Adaptation and Vulnerability. The Contribution of Working Group II to the Third Scientific Assessment of the Intergovernmental Panel on Climate Change. Cambridge University Press, Cambridge

Matthews R (2003) Using a social capital perspective to understand social and economic development in coastal British Columbia. Horizons: Policy Research Initiative-Government of Canada 6(3):25–29

Matthews R, Côté R (2005) Understanding aboriginal policing in a social capital context. In: Social capital in action – thematic policy studies. Government of Canada, Policy Research Initiative, Ottawa, CA, pp 134–152

Matthews R, Pendakur R, Young N (2009) Social capital, labour markets, and job-finding in rural and urban regions: comparing paths to employment in prosperous cities and stressed rural communities in Canada. Socio Rev 57(2):206–230

Nee V (2001) Sources of the new institutionalism. In: Brinton MC, Nee V (eds) The new institutionalism in sociology. Stanford University Press, Stanford, CA, pp 1–16

Nee V (2005) A new institutional approach to economic sociology. In: Smelser N, Swedberg R (eds). The Handbook of Economic Sociology 2nd edn. Princeton University Press, pp 49–74

North DC (1990) Institutions, institutional change, and economic performance. Cambridge University Press, Cambridge, UK

O'Riordan T, Jordan A (1999) Institutions, climate change and cultural theory: towards a common analytical framework. Glob Environ Change 9:81–93

Ostrom E (2005) Understanding institutional diversity. Princeton University Press, Princeton, NJ

Ostrom E, Dietz T, Dolšak N, Stern PC, Stonich S, Weber EU (eds) (2002) The drama of the commons. Division of Behavioral and Social Sciences and Education, National Research Council, National Academy Press, Washington, DC

Page J, Enns S, Malinick T, Matthews R (2007) Should I stay or should I go? Investigating resilience in B.C'.s coastal communities. In: Tepperman L, Dickinson H (eds) Reading sociology: Canadian perspectives. Oxford University Press, Toronto, ON

Pelling M, High C (2005) Understanding adaptation: what can social capital offer assessments of adaptive capacity? Global Environ Change 15:308–319

Portes A (2006) Institutions and development: a conceptual reanalysis. Popul Dev Rev 32(2):233–262

Roland G (2004) Understanding institutional change: fast moving and slow moving institutions. Stud Comp Int Dev 38:109–131

Smit B, Pilifosova O (2003) From adaptation to adaptive capacity and vulnerability reduction. In: Smith JB, Klein RJT, Huq S (eds) Climate change, adaptive capacity and development. Imperial College Press, London, UK, pp 9–28

Smit B, Wandel J (2006) Adaptation, adaptive capacity and vulnerability. Global Environ Change 16:282–292

Swanson D, Hiley J, Venema HD (2007) Indicators of adaptive capacity to climate change for agriculture in the Prairie region of Canada. International Institute for Sustainable Development (IISD), Draft Working Paper for Adaptation as Resilience Building. ON: Prairie Farm Rehabilitation Administration, Agriculture and Agri-Food Canada

Vincent K (2007) Uncertainty in adaptive capacity and the importance of scale. Global Environ Change 17:12–24

World Commission on Environment and Development (Brundtland Report) (1987) Our common future. Oxford University Press, Oxford, UK

Yohe G, Tol RSJ (2002) Indicators for social and economic coping capacity – moving toward a working definition of adaptive capacity. Global Environ Change 12(1):25–40

Young OR, King LA, Schroeder H (2008) Institutions and environmental change: principle findings, applications and research frontiers. MIT, Cambridge, MA

Chapter 12
Sociobiology and Adaptive Capacity: Evolving Adaptive Strategies to Build Environmental Governance

David A. Fennell and Ryan Plummer

12.1 Introduction

The relationship between evolutionary biology and adaptation is well established in the natural sciences. Adaptation fundamentally refers to both the process of improving ones fitness to an environment as well as the products of these actions (Shanahan 2004). Biological evolution is the process of gradual change of the gene pool over a period of time. All species on earth are descended from ancestral species through this process of modification, which is brought about through natural selection (Wright 2008). Natural selection is a process that takes several generations. Those individuals who are better adapted, or more "fit," outbreed those less well adapted, allowing a population to undergo slight changes in their appearance and habits over a period of time. Those individuals who maintain the characteristics that prevent them from optimizing their survival rate in any given set of environmental conditions will be eliminated, while those that change by evolving more favorable traits would survive. As observed by Wallace et al. (1981), the ratio of various alleles (a particular form of a gene at a particular location on a chromosome) in a population's gene pool can change over a period of time. As this ratio changes, the process of evolution takes place either randomly or via natural selection. In the latter case, those genes that are favored are those that help the organism not only to survive but also to pass these positive changes on to the next generation. Positive changes that promote success in a given environment are

---

D.A. Fennell
Department of Tourism and Environment, Brock University, 500 Glenridge Avenue, L2S 3A1
St. Catharines, ON, Canada
e-mail: dfennell@brocku.ca

R. Plummer
Department of Tourism and Environment, Brock University, 500 Glenridge Avenue, L2S 3A1,
St. Catharines, ON, Canada
Stockholm Resilience Centre, Stockholm University, Stockholm, SE-106 91, Sweden
e-mail: rplummer@brocku.ca

referred to as adaptive traits or simply adaptation, and translate into a greater likelihood that an individual will survive and reproduce over others who are missing the particular trait in question (Withgott and Brennan 2008).

Chapter 1 in this volume describes how the general idea of adaptation has been transferred to human systems and used to consider the roles of cultural practices and response of cultures to change (see also O'Brien and Holland 1992). It also sets forth a myriad of interpretations and applications of adaptation by social scientists, which have drawn upon this concept and applied it as adaptive capacity in environmental and resource studies. Evolutionary biology has shaped the treatment of adaptation in the social sciences. Contemporary accounts tracing the lineage of adaptive capacity recognize the origin of the concept in adaptation and the natural sciences, specifically evolutionary biology (Chap. 1, Gallopín 2006; Smit and Wandel 2006).

The relationship between evolutionary biology and adaptive capacity has garnered relatively little attention in the social sciences and the connection to novel forms of environmental governance is a required area of research. This chapter pursues insights about this relationship and the connection to environmental governance. It begins by considering the frames of reference used to understand adaptive capacity. In taking an integrative perspective, which brings together the ecological and socio-institutional contexts, we argue that evolutionary biology is foundational to adaptive capacity, and in turn, environmental governance. To explore this relationship we summarize the science of sociobiology and more specifically the theory of reciprocal altruism in explaining why some humans pursue adaptive strategies that foster collaborative processes, systematic learning and navigating resource uncertainty. Sociobiological theories like inclusive fitness (Hamilton 1964) and reciprocal altruism (Trivers 1971) were instrumental in providing a much needed theoretical basis for understanding the altruistic and self-interested tendencies of both individuals and groups. Concluding comments reflect upon the implications of this conceptual investigation for how adaptive capacity is understood and explained.

12.2 Frames of Reference to Understand Adaptive Capacity

As pointed out in Chap. 1, adaptive capacity is defined in several ways. Yohe and Tol (2002) for example underscore the value of adaptive capacity as an organizing concept and draw attention to its determinants, such as the range of technological options, availability of resources, structure of critical institutions, stocks of capital (social, human), avenues for risk spreading processes, and decision-making processes and public perception. Adger (2003) emphasizes the expansion of the range of variability in coping with change, while Olsson et al. (2004) choose to underscore the importance of coping with disturbances while at the same time maintaining critical functions, structures and feedbacks. In more detail, Folke et al. (2003, pp. 354–355) specify, four main components of adaptive capacity to include: learning to live with uncertainty and change, nurturing diversity for reorganization and renewal, combining multiple and different types of knowledge for learning, and creating opportunities for self-organization.

In light of the many interrelated dimensions of adaptive capacity, Armitage (2005) suggests that it is helpful to "unpack" the concept. He asserts that adaptive capacity can be interpreted from ecological and social–institutional perspectives. Figure 12.1 illustrates these perspectives and provides a starting point to our discussion of the frames of reference to understand adaptive capacity and our inquiry into the relationship of evolutionary biology and environmental governance.

From an ecological perspective adaptive capacity is concerned with understanding ecosystem change and informs the concept of resilience. Resilience in the conventional sense (sensu engineered resilience) emphasizes stability and the return to a single equilibrium following a disturbance; Holling (1973) conversely introduced resilience (sensu ecological resilience) to refer to the amount of disturbance an ecosystem can absorb before altering state, which highlights the presence of multiple equilibria (Gunderson 2000, 2003; Holling and Gunderson 2002). These definitions of resilience assume stationary stability domains; however, evidence now exists that the ecological processes responsible for these domains themselves are dynamic and variable (see Peterson et al. 1998; Gunderson 2000). Adaptive capacity in this context thus refers to the capability of an ecosystem to adapt to slowly changing variables which create stability domains, which in turn alter stability and resilience (Gunderson 2000, 2003).

In drawing upon three decades of evidence from ecosystem research, Holling et al. (2002) observe that it is the interactions between fast and slow processes which give rise to the organization and operation of ecosystems. While the smallest

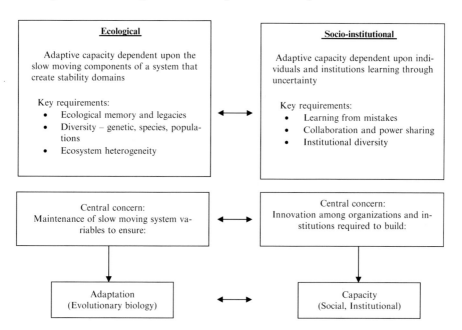

Fig. 12.1 Frames of reference to understand adaptive capacity (Armitage 2005, p. 707)

and fastest variables (e.g., insects, disease organism) tend to be dominated by biological processes, the largest and slowest variables (e.g., tree, human population) tend to alter ecological structure and functions and provide context for the other variables (Ruitenbeek and Cartier 2001; Holling et al. 2002). In reflecting upon these ideas, Holling et al. (2002 p. 73) observe that "in many ways the hierarchy and its nested adaptive cycles could well represent biological evolution". Levin (1999) similarly identifies that self-organization on ecological time scales is not different from self-organization over evolutionary time. He draws upon the title of Hutchinson's (1965) seminal work as a metaphor to describe this feedback process in terms of the ecological theater and the evolutionary play. Williams et al. (2008) recognize that: all organisms have some intrinsic capacity to adapt to change through ecological or evolutionary adaptations; ecological plasticity (plasticity refers to the ability to be shaped or to respond to changing environmental conditions) is more likely important to minimize short-term impacts on individuals, but may have an additive generic component and become fixed over a period of time and, identification of ecological correlates of evolutionary potential will require long-term studies that separate plastic from genetic response components. As Armitage (2005) points out, the central concern of adaptive capacity from the ecological perspective is the maintenance of slow moving system variables which provide key requirements like ecological memory, diversity and ecosystem heterogeneity to ensure adaptation.

From a socio-institutional perspective, adaptive capacity is concerned with the capabilities of social actors and their institutions to flexibly and innovatively respond to change. Adger (2003), writing on the social aspects of adaptive capacity in relation to climate change risks, suggests that while the capacity of individuals to adapt is a function of access to resources, the capability of a society to adapt depends on the ability to act collectively. He argues that decision making across levels (from the individual to the society) is not independent, and rather that decisions about adaptation are embedded in social processes and reflect societal goals which privilege particular interests and creates winners and losers. In this context adaptive capacity is closely connected to Putnam's (1995) work on social capital, where collaboration and cooperation are realized through networks, norms and trust (Adger 2003; Armitage 2005). Although the construct of social capital itself is subject to scholarly debate in terms of definition, the nature of endogeneity, approaches to measurement and the possibility of making generalizations; it garners attention by social scientists as a category representing various factors that come together in assorted ways to influence collective actions (Rudd 2000; Ostrom and Ahn 2003; Plummer and FitzGibbon 2007). Adaptive capacity in the socio-institutional context thus depends on "the attributes of individuals, organizations, and institutions that might foster learning in the context of change and uncertainty, such as a willingness to learn from mistakes, engage in collaborative decision-making arrangements, and encourage institutional diversity" (Armitage 2005 p. 707; see also Folke et al. 2003). Armitage (2005) asserts that the ability of humans to influence the slow (e.g., world views, values, ethics) and fast (e.g., local knowledge, operational rules) variables that shape these attributes and to consciously anticipate the associated outcomes of change leads to the central concern for innovation, despite the value of social stability. This distinction is important because adaptive capacity in the socio-institutional context can be both reactive and proactive (Smithers and Smit 1997; Gallopín 2006).

The socio-institutional context of adaptive capacity is receiving increasing attention because it suggests that social actors can exercise foresight, learn and shape change. Berkes et al. (2003) emphasize the importance of the release and reorganization phases or "backloop" of the adaptive cycle to understand the dynamics of how crises and memory build adaptive capacity for sustainability. Fabricius et al. (2007) recognize that the ability of social actors to learn and anticipate is an important characteristic of adaptive capacity of social–ecological systems which directly influences the governance of natural resources. As adaptive capacity relates to conditions for collective action, institutions and broader social processes, it has been related to governance of social–ecological systems or complex adaptive ecosystems (Folke et al. 2002, 2005; Lebel et al. 2006; Folke 2007; Nelson et al. 2007; Armitage et al. 2009).

In the ecological context adaptive capacity is concerned with ecological plasticity and evolution in relation to adaptation. In the social–institutional context adaptive capacity focuses on the ability of humans to build capacity through innovation and learning. The key to the relationship between evolutionary biology and adaptive capacity that builds environmental governance is connecting the ecological and socio-institutional contexts of adaptive capacity. Although not explicit about this connection in his conceptualization, Armitage (2005) signals this possibility by placing bidirectional arrows between the ecological and socio-institutional contexts, as shown in Fig. 12.1. There is ample evidence to support the belief that gains could be achieved by attempts to link the ecological and social sides within one framework. Gallopín (1991) thereby uses the term socio-ecological system to define a system at any scale that includes human and biophysical subsystems in mutual interaction. Berkes and Folke (1998); Berkes et al. (2003) similarly consider social and ecological systems to be linked and boundaries between them to be artificial and arbitrary; they use the term social–ecological systems and social–ecological linkages to emphasize an integrated "humans-in-nature" perspective. Adaptive capacity at this nexus encompasses the capacity of the system to adapt or deal with change and shape it as well as the capability to improve its condition and influence change toward sustainability (Berkes et al. 2003; Gallopín 2006).

The coupling of ecological and social systems permits a more robust explanation of adaptive capacity as a form of human nature or behavior. However, unlike the "humans-in-nature" perspective, we emphasize an integrated "nature-in-humans" perspective. Human beings are a product of adaptation and evolution and those selective forces have enabled us to evolve into social beings. Figure 12.2 therefore positions evolutionary biology, through the theoretical perspectives of sociobiology, as central to both ecological and socio-institutional sides of adaptive capacity. What is proposed, then, is that evolutionary biology sits as a master or metatheoretical basis which frames adaptive capacity. This insight has significant implications to

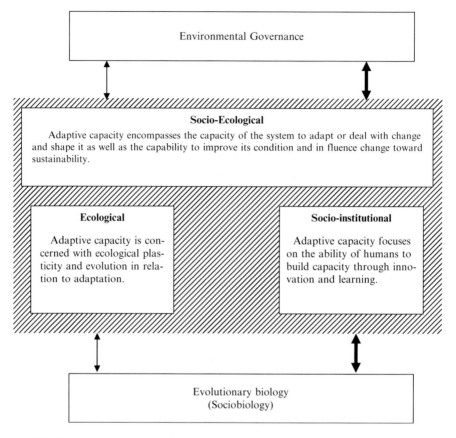

Fig. 12.2 The evolutionary foundation of adaptive capacity in environmental governance

build environmental governance that is adaptive and collaborative. The bold arrows in Fig. 12.2. serve to note that the right side of the figure, i.e., the socio-institutional side, is given more attention in Sect. 12.3.1. Figure 12.2 also illustrates that both the ecological and socio-institutional contexts (the social–ecological systems context) are essential for effective environmental governance and that, in turn, environmental governance also shapes the social–ecological system.

12.3 Evolutionary Biology and Adaptive Capacity

The relationship between evolutionary biology and adaptive capacity which builds environmental governance is set forth in Sect. 12.2 and Fig. 12.2. In this section of the chapter we further investigate this relationship. To do so, we start by summarizing sociobiology and the theory of reciprocal altruism. We then examine how it explains why some communities are more resilient as a function of kinship relations and why some humans pursue adaptive strategies that foster collaborative processes, systematic learning and navigation of resource uncertainty.

12.3.1 *A Synopsis of Sociobiology and Reciprocal Altruism*

Sociobiology, the science developed on the heels of work done by Hamilton (1964) on inclusive fitness and Trivers (1971) on reciprocal altruism, has been defined as, "the systematic study of the biological basis of all social behavior" (Wilson 2000 p. 4; see also Dawkins 1999). The sociobiologist argues that "selection and competitive advantage are as important in understanding what organisms do (particularly to each other) as they are important in understanding what organisms are" (Ruse 1988 p. 64). In this regard, Barash (1982) writes that, "It is always nice to be nice to someone else, but it is hard to see this as adaptive. If helping another individual involves some cost to the helper, then the helper is hurting itself; that is, such behavior would be selected against compared to other, selfish behaviors – unless there are compensating benefits that render helping ultimately advantageous to the performer" (p. 67). This suggests that it does not make sense, biologically, to help others in a competitive environment, because this does little to secure survival and reproductive success. However, there are countless examples, which, at face value, point to altruistic behavior in the animal world to counteract this perspective. Bats will remember other bats who have regurgitated food for them in lean times (Wilkinson 1984), chimps remember other members of the troop who have provided grooming services (De Waal 1989), and baboons will recall those who have helped them escape conflict with others (Dunbar 1980). Until the early 1960s, however, there was no theoretical basis by which to explain the nature of these seemingly altruistic behaviors.

This all changed through the seminal work of Hamilton (1964), who provided theoretical guidance through his theory of inclusive fitness. Inclusive fitness explains in reproductive terms why individuals who are related genetically choose to aid one another over others who are not biologically related (altruism is a function of relatedness). We extend altruism to those who share our genes, like parents, brothers, sisters, cousins, for the purpose of helping to increase their fitness as well as to ensure that their genes (and ours, through them) get passed on to successive generations. The closer one is related to us genetically the more we are liable to help them – we share at least 50% of our genes with siblings and parents, 25% with half siblings, and so on. So, in deciding to save one of two drowning individuals, one a relative and the other a nonrelative, we would choose to save the relative for the purpose of preserving and passing on our own genetic information. The basis of the theory of inclusive fitness is formulated as follows: altruism will be selected and will spread in a population whenever $k > 1/r$, where k is the ratio of recipient benefit to altruist cost and r is the coefficient of genetic relatedness between altruist and recipient. Here, "blood relatives cooperate to bestow altruistic favors on one another in a way that increases the average genetic fitness of the members of the network as a whole, even when this behavior reduces the individual fitnesses of certain members of the group" (Wilson 2000 p. 118). Hamilton, according to Barash, recognized that parenting is a special case of genes looking out for themselves (in the metaphorical sense) in the bodies of a special type of individual: offspring. By helping those related to us, we help to preserve our own genetic base in the population.

While Hamilton (1964) explained why kin cooperate, his colleague Trivers (1971) felt that altruistic behavior could thrive in situations involving nonrelatives through a process that he termed reciprocal altruism (RA). We cooperate with others (altruism) because there is the belief that we will realize return benefits (reciprocity) from these individuals somewhere down the road. The hope on the part of the initial altruist is that these return benefits are larger than initial costs, and that the time period between return benefits is not too great in order that cooperation might be sustained over a period of time. Early hominids living under such conditions were likely able to live longer lives because of this emergent cooperation, as RA would have been selected for such populations as a necessary tactic in achieving longevity and social harmony.

In order to demonstrate the significance of RA as a human universal, Trivers developed an elaborate model to show the underlying psychological (emotional) system that acts as the causal agent for human behavior. The model is premised on the recognition that the altruism/cheating (cheating defined as the failure to reciprocate) system is an unstable one, which needs to be regulated by one's own altruistic and cheating tendencies and responses to those of others (Trivers 1971). In this system selection favors: the tendency to like others and to act altruistically toward these others as the immediate emotional reward motivating altruistic behavior; vulnerability to those who take advantage of an altruist's willingness to bear costs in order that others may benefit; the emergence of gratitude to regulate human response to altruistic acts; friendship, moralistic aggression, guilt, sympathy, and gratitude in regulating this system; and a range of behaviors, e.g., establishment of rules and norms, designed to successfully navigate the subtleties of group interaction, given the social nature of humans (among other factors). The simplicity and underlying foundational value of this theory in explaining human nature has prompted Pinker (2002) to observe that:

> Trivers derived the first theory in social psychology that deserves to be called elegant. He showed that a deceptively simple principle – follows the genes – can explain the logic of each of the major kinds of human relationships... It offered a scientific explanation for the tragedy of the human condition (p. 241).

Current studies underscore the fact that reciprocal altruism is a highly egotistical behavior because altruistic acts come packaged with the expectation of return favors down the road (Mayr 1988), or by seeking approval, respect or admiration benefits by fellow citizens (Alexander 1987). In this regard, Griffin (1997) argues that there is a troublesome step required in moving agents, who are genetically programmed to be selfish, to having concern for others. This is because we are programmed with a primitive form of egoism that is based on the care of a small number of what Griffin calls prudential values, including our own survival, advancement over others, and gratification. We care for others, therefore, only to the degree that the well-being of these others affects us, and we must balance these costs according to the immediate and long-term nature of the benefits that may follow. But this self-interest is tempered by what Frank (1988) referred to as commitment. Agents view cheating as irrational because it erodes long-term stable relationships. It is the emotions that keep us on track in preventing us from making solely rational decisions in our own best interest.

It comes as no surprise therefore to discover that reciprocal altruism is so important in understanding human interaction because of its seminal role in the foundation of human morality (Mayr 1988). The ability to be moral, in a biological sense, is attributed to *Homo sapiens* only because of the intellectual capacities of humans to anticipate the consequences of their actions in being self-aware, make value judgments or apply the categories good and bad to events, and to choose between different courses of action in knowing that a particular act could have been suppressed (Ehrlich 2000; Ayala 1987; Kagan 1998). This capacity to be ethical is widely accepted as a function of biological evolution, but the products of the ethical capacity result from culture and society. So while it is clear that evolution has provided us with the capacity to be ethical, this system of ethics has been tested through millennia of wisdom and experience of human beings living together in communities (Keiffer 1979).

This perspective corroborates new research on gene-culture coevolution. Even though we have something in the order of 30,000 genes, the plasticity built into this "hard-wiring" comes packaged with combinatorial software that can generate countless thoughts and behaviors based on the interplay of genes and the environments in which they are situated (see Ridley 2003). The brain has thus evolved a complex reward circuit that guides us through the labyrinth of stimuli in the environment seen to be beneficial or threatening to gene survival. As such, the adaptive processes of learning and reasoning were essential in mediating between, for example, pain and pleasure (Johnston 2003, as cited in Fennell 2009). Learning allowed the individual to absorb and store new information that proved successful in interactions with changing environments, while reasoning allowed the individual to anticipate how interactions with certain aspects of the environment relate to sensations of pain and pleasure (Fennell 2009).

Consequently, it is the genes that allow us to learn and keep on learning. And even though behaviors may vary across cultures, the hardware between groups is essentially the same: we learn because we have the capacity to learn what is built in, and this ability appears to be heavily influenced by environmental stimuli (see Pinker 2002). In this regard, Wilson (1993) cautions that we often make two critical mistakes: the first is that culture, or the nurture side of the equation accounts for everything, while the second mistake is that it accounts for nothing. In this way it is not nature or nurture, but rather nature via nurture, as explained by Ridley (2003, p. 6):

It is genes that allow the human mind to learn, to remember, to imitate, to imprint, to absorb culture, and to express instincts. Genes are not puppet masters or blueprints. Nor are they just the carriers of heredity. They are active during life;

they switch each other on and off; they respond to the environment. They may direct the construction of the body and brain in the womb, but then they set about dismantling and rebuilding what they have made almost at once – in response to experience. They are both cause and consequence of our actions. Somehow the adherents of the "nurture" side of the argument have scared themselves silly at the power and inevitability of genes and missed the greatest lesson of all: the genes are on their side.

Recent work has corroborated this perspective. For example, Richerson and Boyd (2005) argue that humans are an anomaly in the natural world because of their creation of large, complex societies, invasion of just about every habitat on earth, the use of tools and subsistence techniques, and their propensity to cooperate. Ecological dominance and our emergent social structures are a function of the interactive nature of culture and biology in forming the basis of our human nature. (See also the work of Bowles (2006) who suggests how culturally transmitted practices like food sharing beyond the immediate family, monogamy, and information sharing have been influential in creating genetic predispositions to behave altruistically.)

12.3.2 *The Biological Basis of Adaptive Strategies*

Individuals, communities, organizations and societies respond to change. In synthesizing the contributions to the special feature on strengthening adaptive capacity in *Ecology and Society*, Fabricius et al. (2007) observe that "communities adapt because they face enormous challenges due to policies, conflicts, demographic factors, ecological change, and changes in their livelihood options, but the appropriateness of their responses varies." They differentiate between two types of strategies. The first is coping strategies, which focus on short-term survival through reactive or ad hoc adaptations (e.g., land use changes, shifts in assets). The second is adaptive strategies, which are proactive and aim at adaptation in the long-term (e.g., strengthening social networks, collaboration). Adaptive strategies "...are associated with social learning and institutional change based on shared experiences, often over long periods and transferred over several generations" (Fabricius et al. 2007; see also Berkes and Folke 1998; Folke et al. 2003). The ability of communities to adapt is thus described along gradients of adaptive capacity and governance capacity and influenced by their ability to exercise adaptive strategies (i.e., have the capability to adapt and to sustain and internalize adaptation over long temporal periods). Smit and Wandel (2006) argue that the adaptive capacity of a local area can be a function of several conditions, including managerial abilities, access to several types of resources, infrastructure, institutions, politics, and kinship networks. These authors also note that adaptive capacity is context-specific and thus variable from community to community and from country to country. The key questions to be analyzed here are *how* communities are more resilient as a function of kinship relations and *why* some humans pursue adaptive strategies that foster collaborative processes, systematic learning and navigation of resource uncertainty. These are deeper questions about adaptive capacity and its connection to environmental governance that have yet to be formally addressed, and have much to do with human nature.

In addressing the aforementioned questions it is important to consider that sociobiology is concerned with cooperation among kin while reciprocal altruism is a much about great deal about enlightened self-interest as it is about cooperation. Smit and Wandel (2006), for example, use strong kinship networks as an example of a local determinant of adaptive capacity in a subsistence-based society because such relationships absorb or buffer psychosocial stress, permit greater access to economic resources/livelihoods, enhance managerial ability and provide supplementary sources of labor. Adger (2003) draws attention to the concept of social capital as a way to understand societal functioning in his discussion of the social aspects of adaptive capacity by drawing upon sharing of knowledge, risk, information and reciprocity claims during a crisis. In differentiating between the different forms of social capital he observes that ties (bonding social capital) typically occur among kin or around a locality and are stronger than networking (or bridging) social capital, which tends to be weaker and based on trust and reciprocity. In times of crisis it is the kin who are likely to cooperate because it ultimately enhances their inclusive fitness (Hamilton 1964).

Adaptive strategies that involve collective action mediate this line of demarcation between self-interest and the satisfaction of the greater good. Adger (2003) argues that collective action is at the heart of decisions to adapt and that social capital is a key requirement in this regard. He continues by asserting that with regards to institutions of governance, the level of community is the most important mechanism for collective management, where, "it is the different combinations of bonding and networking social capital that allow communities to confront poverty and vulnerability, resolve disputes and take advantage of new opportunities" (Adger 2003 p. 38). We enter into symbiotic relationships with others in order to secure benefits later on in times of need. We do this by helping our friends because we recognize the reciprocal nature of this arrangement for our own benefit down the road (Levin 1999).

Whether or not these adopted strategies are selfish in nature or cooperative is the point of issue here. But we should be aware that the sociobiology literature is thick with studies that have examined the cooperative tendencies of prey species under conditions of predation. What many conclude is that the "safety in numbers" theory on animal cooperation is nothing more than individuals hiding within the group as a manner of self-preservation. According to Hamilton (1971), individual frogs will move into smaller gaps amongst a cluster of frogs if the one they inhabit happens to be larger than another gap in their efforts to avoid being eaten. In this we should heed the words of Alexander (1979) and Fox (1997) who argue that genetic self-interest is aided by the collective, where individuals freely pursue cooperative ventures with others in order to satisfy their own reproductive, family, and work objectives. For example, human groups will cooperate in times of need through the construction of seawalls in coral atolls in order to cushion settlements from the effects of nature (Withgott and Brennan 2008), and temporary human alliances are forged over short periods of time because of war and other destabilizing events (Fennell 2006). As such, changing or threatening circumstances have a way of building capacity for cooperation. A reliance on others in times of need is an adaptive strategy in coping with these changes.

This challenge in moving towards group cooperation has been broached by Fazey et al. (2007) who observe that psychological barriers inhibit changing ways of thinking and behaving and stem partially from our natural evolutionary tendencies for self interest as well as from learning via social interactions. While adaptation research often regards individuals as rational economic actors, the determinants of undertaking adaptive action are often underlying socio-psychological factors (Pelling et al. 2008). In this regard, Grothmann and Patt (2005) have emphasized the need to address psychological factors (identified risk perception and perceived adaptive capacity) as an important "bottleneck" to responding to climate changes that has received little attention. Notwithstanding the importance of this sociopsychological work, the yardstick to measure the theoretical basis for understanding group interaction continues to be the one of Trivers (1971), who argues that: we have the capacity to navigate the subtleties of group interaction, we learn from others, especially in dealing with people who are out to cheat others, or us, and we are also adept at forming multiparty exchange systems that help us to level the playing field, so to speak. We need to understand the rules of engagement in group situations because of the sensitivity we have towards costs and benefits. In this regard, Plummer and Fennell (2007) suggest that it is no surprise that comanagement systems, as one form of governance systems, are slow to emerge because multiparty arrangements require careful negotiation in mediating between individual and collective interests. The trick, therefore as noted by Plummer and Fennell (2007) is to instill a sense of value such that these individual costs (altruism by the individual) are worthy contributions to some greater whole that they themselves will benefit from, along with all of their counterparts in the system.

Uncertainty and surprise in coevolving social–ecological systems are the rule as opposed to exception (Folke et al. 2003; Gunderson 2003). While the tendency to lock-in situations has a functional purpose at cellular, individual, group, and societal levels, it may ultimately lead to pathological patterns and excessive rigidity (Scheffer and Westley 2007). In discussing the collapse of some ancient civilizations in the face of resource crises, Scheffer and Westley (2007) observe that the tendency to continue to practice habits that lead to success or coping in past times, has contributed to societal demise in times that demanded change and flexibility (see also Janssen et al. 2003; Diamond 2005; Fazey et al. in press). Adaptation can be considered a threat by the community (or society or management agency by comparison), because it necessitates drawing upon different knowledge systems, collaborating and sharing power, all of which may go across the grain of individual self-interest (Folke et al. 2003, 2005; Gunderson 2003). However, as Scheffer and Westley (2007) conclude, "...our greatest advantage is that we can analyze such patterns and learn from them". In addition to the critical factors for building adaptive capacity outlined by Folke et al. (2003) earlier, Berkes et al. (2003, p. 21) identify that "social memory is the arena in which captured experience with change and successful adaptations, embedded in a deeper level of values, is actualized through community debate and decision-making processes into appropriate strategies for dealing with ongoing changes." More robust adaptive strategies are likely in communities that can draw upon social memory to deal with uncertainty and change. Likewise, Armitage (2005) asserts that the adaptive capacity of communities depends on learning through crisis and uncertainty, maintaining memory, drawing upon different knowledge systems, collaborating and sharing power, and maintaining diverse and redundant institutions.

The connection between the individual and the collective in relation to learning, flexibility and governance is thus critical. Armitage (2007 p. 76) observes that "multilevel governance is likely required to encourage sustainable responses to these situations (where individual incentives and actions conflict with collective ones), but only if efforts are directed at building adaptive capacity at all levels in response to a commonly defined problem. There is emerging evidence (see Fazey et al. 2007; Pelling et al. 2008; Diduck, this volume) of how adaptive capacity and learning by individuals is connected to learning in multilevel environmental governance. Creating and incentivitizing pathways and relational spaces that foster learning are showing promise for enhancing adaptive capacity (Pelling et al. 2008; Diduck, this volume). Lebel et al. (2006) recognize that interventions aimed at altering resilience immediately confront issues of governance, and more specifically that pursuing social justice for vulnerable groups enhances adaptive capacity for both those groups as well as society.

We would be remiss in failing to mention important work by biologists and social scientists that has so much to do with emerging research on gene-culture evolution in the context, at least here, of how communities evolve and learn. Dawkins' (1999) concept of the meme helps to explain how cultural evolution rides parallel with genetic evolution. Memes are units of cultural transmission or imitation, and include ideas, artifacts and values that, when added to the "meme pool", help to shape the future of culture in a region. This line of thinking supports the idea that culture can be inherited and can select for corresponding genetic change in supporting gene-culture coevolution. Adaptive capacity would thus rely on the ability to change through new values, innovation, imagination, and so on, in making communities or regions more resilient to change over time.

Adaptive capacity has further been discussed by geneticist such as Wright (1932) in relation to the roles of mutation, inbreeding, crossbreeding and selection in evolution. Wright (1932) diagramed fields of gene combinations in association with two-dimensional contour lines in representing the adaptiveness of a species (landscapes of adaptation or fitness landscapes). Under constant conditions each gene will reach a period of equilibrium, represented as the occupation of a field of variation around a single peak. Evolution takes place as the species moves from lower to higher level peaks in the landscape, but they must do so by passing through valleys of maladaptive intermediate stages that occur through genetic drift, migration of genes to other populations, and mutation. Accordingly, Wright (1932) felt that adaptation was not strictly a function of natural selection, but rather a process that was influenced by these other forces on balance. Wright's (1932) shifting balance theory of evolution has been important in pointing out that adaptive capacity is very much a function of how humans respond to the environment and vice versa, and that more robust systems give rise to more opportunity for innovative change.

Most recently, Fazey et al. (in press) have incorporated Wright's (1932) perspectives on fitness landscapes in their work on adaptive strategies that needed to reduce vulnerability to future environmental changes. These authors illustrate that the capacity to cope with environmental change is a function of how combinations of adaptive strategies actually catalyze the ability of the system to cope. Higher peaks (following Wright) represent the best combination of strategies, but changing conditions necessitate the development of newer strategies, which may in turn lead to a higher peak. Response diversity corresponds to the speed and degree in which new strategies can be activated during periods of change. In this way, "as the social–ecological system changes, the combination of strategies that provides the most effective coping capacity can also change. Adaptation is therefore dynamic, with the adaptations themselves affecting the conditions that define the most effective combinations of strategies, to which further adaptation may be required" (p. 6).

12.4 Conclusion

This chapter began with a query about the relationship between evolutionary biology, adaptive capacity, and environmental governance. The ecological and social frames that are used to understand adaptive capacity as discrete entities are insufficient. Drawing upon the initial connections made between the frames by Armitage (2005) as well as the social–ecological perspective more broadly (Gallopín 1991; Berkes et al. 2003), we proposed an integrative nature-in-humans perspective of adaptive capacity and illustrated how evolutionary biology is fundamental to adaptive capacity (both ecological and social–institutional). In exploring this relationship, we briefly sketched the tenants of sociobiology and theory of reciprocal altruism and used them to explain adaptive strategies (cooperation, flexibility and learning) that foster environmental governance.

Reframing adaptive capacity through an integrative nature-in-humans perspective builds upon the conventional ways it has been understood and interpreted in both the natural and social sciences. In this chapter, we have used sociobiology to develop a very different view of adaptive capacity in social–ecological systems, one which reflects mounting evidence that human behavior is the result of nature and nurture working in concert. This integrative and coevolutionary perspective of adaptive capacity has several implications.

Reframing adaptive capacity in this way is essential to refining its relevance to the context of social–ecological systems. As Walker et al. (2006) observe, social–ecological systems are a "different thing altogether", rather than embedding humans in ecological systems or ecosystems in human systems. The perspective set forth in this chapter thus emphasizes the need to reposition the places of evolutionary biology and environmental governance (as an outcome of the socio-institutional understanding of adaptive capacity) in relation to the previous framing of the adaptive capacity. It also emphasizes the reciprocal relationship among these variables.

Adaptive capacity, in the socio-institutional sense, has received considerable attention because it relates to particular features (e.g., collaboration, flexibilityand learning) that influence governance of social–ecological systems (e.g., Fabricius et al. 2007; Folke 2007; Armitage et al. 2009). As demonstrated in Sect. 12.3.2, sociobiology and the theory of reciprocal altruism offer illuminating explanations of why some humans pursue particular adaptive strategies and how some communities are more resilient to change. Sociobiology offers explanations to these "why" and "how" questions at the foundational level of human nature, which have implications for those attempting to foster the capacity to adapt. While capacity building interventions are usually aimed at altering behavior, sociobiology suggests that some individuals, communities, societies may have particular traits that make them more apt to change in particular ways and/or have already undertaken long-term strategies. Undertaking studies of sufficient time to separate ecological plasticity from adaptive responses remain a considerable challenge; however, behavioral changes themselves are anticipated to be more important in the short-term (generational) and may, if favored by selection, become fixed (evolutionary responses) over time (Williams et al. 2008).

Biology has endowed us with the potential to adapt and the unique ability to shape that adaptation. Sociobiology points out that the capability for adaptation in humans is a coevolutionary process driven by genetic hard-wiring and shaped by complicated human devised constructs such as culture, cooperation, ethics and learning that define both our selfish and altruistic tendencies as individuals and groups. As Ostrom (1990) identified, it is often our self-interest that drives us to establish institutions for collective action. Devising means seen to be beneficial or threatening to gene survival as well as identifying key stimuli in the environment will build our adaptive capacity by appealing to the self-interest or egoism. The propensity to engage in social collectives and change behaviors will hinge on the degree to which novel governance arrangements are able to overcome immediate self-interest, catalyze individual benefits and foster the development of long-term adaptive strategies. At the same time, as Fazey et al. (in press) point out, nurturing particular human capacities (i.e., individual capacity and psychology for changing behaviors, group interactions, and societal institutions and worldviews) will enhance the realization of future adaptation. Governance strategies, such as adaptive comanagement, are receiving considerable attention as mechanisms which may foster human capacities through shared problem solving, learning processes, and interactive exchanges about values, power-sharing and ethics (Fennell et al. 2008; Armitage et al. 2009). Despite the possibilities of such strategies, the dynamic relationship between adaptation and environmental conditions necessitates ongoing attention (Fazey et al. in press).

What is perhaps most important to realize in consideration of the foregoing is that the science behind the recent changes in how we view adaptation and other aspects of human nature is very much at an incipient stage. The coupling of biology and culture as interactive agents in our ability to adapt as a species represents a significant leap of faith, and suggests that further strides will be made through interdisciplinary approaches involving the natural and social sciences.

Acknowledgements The authors would like to thank Derek Armitage, Ioan Fazey and an anonymous reviewer for their constructive feedback on an earlier version of this chapter. Ryan Plummer gratefully acknowledges support for his research program from a Brock University Chancellor's Chair for Research Excellence, the Canadian Water Network and the Social Sciences and Humanities Research Council of Canada.

References

Adger WN (2003) Social aspects of adaptive capacity. In: Smith JB, Klein RJT, Huq S (eds) Climate change, adaptive capacity and development. Imperial College, London
Alexander RD (1979) Darwinism and human affairs. University of Washington Press, Seattle
Alexander RD (1987) The biology of moral systems. Aldine de Gruyter, New York
Armitage D (2005) Adaptive capacity and community-based natural resources management. Environ Manage 35(6):703–715
Armitage D (2007) Building resilient livelihoods through adaptive co-management: the role of adaptive capacity. In: Armitage D, Berkes F, Doubleday N (eds) Adaptive co-management. UBC Press, Vancouver
Armitage D, Plummer R, Berkes F, Arthur RI, Charles AT, Davidson-Hunt IJ, Diduck AP, Doubleday N, Johnson DS, Marschke M, McConney P, Pinkerton E, Wollenberg E (2009) Adaptive co-management for social–ecological complexity. Front Ecol Environ 7(2):95–102
Ayala FJ (1987) The biological roots of morality. Biol Philos 2:235–252
Barash DP (1982) Sociobiology and behaviour, 2nd edn. Elsevier, New York
Berkes F, Folke C (1998) Linking social and ecological systems: management practices and social mechanisms for building resilience. Cambridge University Press, Cambridge
Berkes F, Colding J, Folke C (2003) Navigating social–ecological systems: building resilience for complexity and change. Cambridge University Press, Cambridge
Bowles S (2006) Group competition, reproductive leveling, and the evolution of human altruism. Science 314:1569–1572
Dawkins R (1999) The selfish gene. Oxford University Press, Oxford
De Waal F (1989) "Chimpanzee politics" power and sex among the apes. Johns Hopkins, Baltimore
Diamond J (2005) Collapse: how societies choose to fail or succeed. Viking, New York
Dunbar R (1980) Determinants and evolutionary consequences of dominance among female gelada baboons. Behav Ecol Sociobiol 7:253–265
Ehrlich PR (2000) Human natures: genes, cultures, and the human prospect. Penguin, New York
Fabricius C, Folke C, Cundill G, Schultz L (2007) Powerless spectators, coping actors, and adaptive co-managers: a synthesis of the role of communities in ecosystem management. Ecol Soc 12((1):29, http://www.ecologyandsociety.org/vol12/iss1/art29/. Accessed 10 January 2009
Fazey I, Fazey JA, Fischer J, Sherren K, Warren J, Noss RF, Dovers SR (2007) Adaptive capacity and learning to learn as leverage for social–ecological resilience. Front Ecol Environ 5(7):375–380

Fazey I, Gamarra JGP, Fischer J, Reed MS, Stringer LC, Christie M (In Press) Adaptation strategies to reduce vulnerability to future environmental change. Front Ecol Environ

Fennell DA (2006) Tourism ethics. Channel View Publications, Clevedon

Fennell DA (2009) The meaning of pleasure in pleasure travel. Tourism Recreation Res 34(2):127–134

Fennell D, Plummer R, Marschke M (2008) Is adaptive co-management ethical. J Environ Manage 88(1):62–75

Folke C (2007) Social–ecological systems and adaptive governance of the commons. Ecol Res 22:14–15

Folke C, Carpenter S, Elmqvist T, Gunderson L, Holling CS, Walker B, Bengtsson J, Berkes F, Colding J, Danell K, Falkenmark M, Moberg M, Gordon L, Kaspersson R, Kautsky N, Kinzig A, Levin SA, Mäler K-G, Ohlsson L, Olsson P, Ostrom E, Reid W, Rockstöm J, Savenije S, Svedin U (2002) Resilience and sustainable development: building adaptive capacity in a world of transformations. The Environmental Advisory Council to the Swedish Government Scientific Background Paper

Folke C, Colding J, Berkes F (2003) Synthesis: building resilience and adaptive capacity in social–ecological systems. In: Berkes F, Colding J, Folke C (eds) Navigating social–ecological systems. Cambridge University Press, Cambridge

Folke C, Hahn T, Olsson P, Norberg J (2005) Adaptive governance of social–ecological systems. Annu Rev Environ Resour 30:441–473

Fox R (1997) Conjectures and confrontations: science, evolution, social concern. Transaction, London

Frank R (1988) Passion within reason: the strategic role of emotions. W.W. Norton & Company, New York

Gallopín GC (1991) Human dimensions of global change: linking the global and the local processes. Int Soc Sci J 130:707–718

Gallopín GC (2006) Linkages between vulnerability, resilience, and adaptive capacity. Glob Environ Change 16:293–303

Griffin J (1997) Value judgment: improving our ethical beliefs. Clarendon, Oxford

Grothmann T, Patt A (2005) Adaptive capacity and human cognition: the process of individual adaptation to climate change. Glob Environ Change 15:199–213

Gunderson LH (2000) Ecological resilience – in theory and application. Annu Rev Ecol Syst 31:425–439

Gunderson LH (2003) Adaptive dancing: interactions between social resilience and ecological crises. In: Berkes F, Colding J, Folke C (eds) Navigating social–ecological systems. Cambridge University Press, Cambridge

Hamilton WD (1964) The genetical evolution of social behavior (I and II). J Theor Biol 7:1–52

Hamilton WD (1971) Geometry of the selfish herd. J Theor Biol 31:295–311

Holling CS (1973) Resilience and stability of ecological systems. Annu Rev Ecol Syst 4:1–23

Holling CS, Gunderson LH (2002) Resilience and adaptive cycles. In: Gunderson LH, Holling CS (eds) Panarchy. Island, Washington

Holling CS, Gunderson LH, Peterson GD (2002) Sustainability and panarchies. In: Gunderson LH, Holling CS (eds) Panarchy. Island, Washington

Hutchinson GE (1965) The ecological theater and the evolutionary play. Yale University Press, New Haven

Janssen MA, Kohler TA, Scheffer M (2003) Sunk-cost effects and vulnerability to collapse in ancient societies. Curr Anthropol 44:722–728

Kagan J (1998) Three seductive ideas. Harvard University Press, Cambridge

Keiffer GH (1979) Bioethics: a textbook of issues. Addison-Wesley, New York

Lebel L, Anderies JM, Campbell B, Folke C, Hatfield-Dodds S, Hughes TP, Wilson J (2006) Governance and the capacity to manage resilience in regional social–ecological systems. Ecol Soc 11(1):19, http://www.ecologyandsociety.org/vol11/iss1/art19/. Accessed 3 February 2009

Levin SA (1999) Fragile dominion: complexity and the commons. Perseus Books, Reading, MA

Mayr E (1988) Toward a new philosophy of biology: observations of an evolutionist. Belknap, Cambridge
Nelson DR, Adger WN, Brown K (2007) Adaptation to environmental change: contributions of a resilience framework. Annu Rev Environ Resour 32:395–419
O'Brien MJ, Holland TD (1992) The role of adaptation in archeological explanation. Am Antiq 57(1):36–59
Olsson P, Folke C, Hahn T (2004) Socio-ecological transformation for ecosystem management: The development of adaptive co-management of a wetland landscape in southern Sweden. Ecol Soc 9(4):2, http://www.ecologyandsociety.org/vol9/iss4/art2. Accessed 3 February 2009
Ostrom E (1990) Governing the commons: the evolution of institutions for collective action. Cambridge University Press, Cambridge
Ostrom E, Ahn TK (2003) Introduction. In: Ostrom E, Ahn TK (eds) Foundations of social capital. Edward Elgar, Cheltenham
Pelling M, High C, Dearing J, Smith D (2008) Shadow spaces for social learning: a relational understanding of adaptive capacity to climate change within organizations. Environ Plan A 40:867–884
Peterson G, Allen CR, Holling CS (1998) Ecological resilience, biodiversity, and scale. Ecosystems 1:6–18
Pinker S (2002) The blank slate: the modern denial of human nature. Viking, New York
Plummer R, Fennell DA (2007) Exploring co-management theory: prospects for sociobiology and reciprocal altruism. J Environ Manage 85:944–955
Plummer R, FitzGibbon J (2007) Connecting adaptive co-management, social learning, and social capital through theory and practice. In: Armitage D, Berkes F, Doubleday N (eds) Adaptive co-management. UBC Press, Vancouver
Putnam R (1995) Bowling alone: America's declining social capital. J Democr 6(1):65–78
Richerson PJ, Boyd R (2005) Not by genes alone: how culture transformed human evolution. University of Chicago Press, Chicago
Ridley M (2003) Nature via nurture. HarperCollins, Toronto
Rudd MA (2000) Live long and prosper: collective action, social capital and social vision. Ecol Econ 34:131–144
Ruitenbeek J, Cartier C (2001) The invisible wand: adaptive co-management as an emergent strategy in complex bio-economic systems. Center for International Forestry Research Occasional Paper 34. http://www.cifor.cigar.org. Accessed 18 January 2009
Ruse M (1988) Philosophy of biology today. State University of New York, New York
Scheffer M, Westley FR (2007) The evolutionary basis of rigidity: locks in cells, minds, and society. Ecol Soc 12((2):36, http://www.ecologyandsociety.org/vol12/iss2/art36/. Accessed 15 January 2009
Shanahan T (2004) The evolution of Darwinism: selection, adaptation, and progress in evolutionary biology. Cambridge University Press, Cambridge
Smit B, Wandel J (2006) Adaptation, adaptive capacity and vulnerability. Glob Environ Change 16:282–292
Smithers J, Smit B (1997) Human adaptation to climatic variability and change. Glob Environ Change 7(2):129–146
Trivers R (1971) The evolution of reciprocal altruism. Q Rev Biol 46:35–57
Walker BH, Gunderson LH, Kinzig AP, Folke C, Carpenter SR, Schultz L (2006) A handful of heuristics and some propositions for understanding resilience in social–ecological systems. Ecol Soc 11(1):13, http://www.ecologyandsociety.org/vol11/iss1/art13. Accessed 23 January 2009
Wallace RA, King JL, Sanders GP (1981) Biology: the science of life. Scot, Foresman and Co., Dallas
Wilkinson GS (1984) Reciprocal food sharing in the vampire bat. Nature 308:181–184
Williams SE, Shoo LP, Isaac JL, Hoffmann AA, Langham G (2008) Towards an integrated framework for assessing the vulnerability of species to climate change. PLoS Biol 6(12): e325. doi:10.1371/journal.pbio.0060325
Wilson JQ (1993) The moral sense. Free, New York

Wilson EO (2000) Sociobiology: the new synthesis. Belknap, Cambridge
Withgott J, Brennan S (2008) Environment: the science behind the stories, 3rd edn. Pearson, Toronto
Wright S (1932) The roles of mutation, inbreeding, crossbreeding and selection in evolution. In: Proceedings of the 6th international congress on genetics, vol 1, pp 356–366
Wright RT (2008) Environmental science, 10th edn. Pearson, Upper Saddle River
Yohe G, Tol RSJ (2002) Indicators for social and economic coping capacity – moving toward a working definition of adaptive capacity. Glob Environ Change 12:25–40

Chapter 13
Building Transformative Capacity for Ecosystem Stewardship in Social–Ecological Systems

Per Olsson, Örjan Bodin, and Carl Folke

13.1 Introduction

Current approaches for managing natural resources often fail to match social and ecological structures and processes operating at different spatial and temporal scales (Folke et al. 2007; Carpenter and Gunderson 2001; Cumming et al. 2006; Galaz et al. 2008; Walker et al. 2009a; Rockström et al. 2009). The reasons behind this governance failure lie not only in weak environmental legislation, lack of enforcement power, or poor monitoring and evaluation systems (United Nations Environment Programme 2007), but also in ignorance of ecosystem dynamics and simplistic attempts to control and optimize delivery of specific natural resources (Holling and Meffe 1996). Stabilizing a set of desirable natural resources can create mismatches between institutions and ecosystem dynamics, leading to undesirable regime shifts in the capacity of landscapes and seascapes to generate essential ecosystem services (Scheffer et al. 2001; Galaz et al. 2008). The likelihood of sudden shifts between social-ecological systems, states has profound implications for ecosystem stewardship of essential ecosystem services in a world of rapid and directional change (Chapin et al. 2010). Shifts to more holistic, integrated forms of natural resource management and multilevel governance systems that support ecosystem-based management are urgently needed (Gunderson et al. 1995; Folke et al. 2005). These approaches have the potential to deal with the complexity of interdependent social–ecological systems (SES) and enhance the fit between ecosystem dynamics and governance systems (Berkes and Folke 1998; Berkes et al. 2003).

P. Olsson and Ö. Bodin
Stockholm Resilience Centre, Stockholm University, SE-106 91, Stockholm, Sweden
e-mail: per.olsson@stockholmresilience.su.se

C. Folke
Stockholm Resilience Centre, Stockholm University, SE-106 91, Stockholm, Sweden
The Beijer Institute, The Royal Swedish Academy of Sciences, SE-104 05, Stockholm, Sweden

The search for better approaches towards sustainable outcomes has helped to develop important design principles and protocols for alternative management and governance approaches (Ostrom 1990; Costanza et al. 1998; National Research Council 1999a). These approaches acknowledge ecosystems as complex, dynamic systems and address the mismatch between social systems and ecosystem dynamics (Norberg and Cummings 2008). Although the literature on environmental management and governance recognizes the need for transitions and transformations (National Research Council 1999b; Raskin et al. 2002; Babcock and Pikitch 2004; Walker et al. 2009b), it offers few empirically based insights into social–ecological innovations and strategies that make the shift to new ecosystem stewardship approaches possible. Different disciplines have studied pieces of the puzzle, such as organizational and institutional aspects (Danter et al. 2000; Imperial 1999) and the role of learning (Pahl-Wostl et al. 2008; Armitage et al. 2008) but have rarely captured interdependent social–ecological dynamics. There is still lack of understanding on how to transform SES into new, improved trajectories that sustain and enhance ecosystem services and human well-being. In this chapter, we aim to contribute to the understanding about the transformative capacity required to shift governance from disintegrated resource and environmental management to ecosystem stewardship in SES.

In the first part of this chapter we use a "resilience lens" to identify gaps in the understanding of transformative capacity and highlight some challenges that needs to be addressed. The second part draws on the organizational evolution literature in combination with the latest insights on SES transformations to give a more detailed understanding of what constitutes transformative capacity. This point is further emphasized in the third part where we use findings on transformative capacity from two empirical studies, the Kristianstads Vattenrike, Sweden, and the Great Barrier Reef, Australia, to exemplify ecological changes and social dynamics that lead to shifts to new flexible management and governance approaches. Lastly, we elaborate the key criteria and future needs for developing a framework for analyzing transformations towards ecosystem stewardship and assessing transformative capacity in SES.

13.2 The Problem of Fit and Lock-in Traps in SES

The mismatch between ecosystems and governance systems is often referred to as "the problem of fit" (Young 2002; Folke et al. 2007; Galaz et al. 2008). Resource management institutions that perform in a socially and economically resilient manner, with well-developed collective action and economic incentive structures, may in ignorance degrade the capacity of ecosystems to provide ecosystem services. Such behavior may cause a shift to a degraded ecosystem state (Scheffer et al. 2001), which, in turn, feeds back into the social and economic domains, with the risk of causing unpleasant surprises and undesirable social–ecological regime shifts (Folke et al. 2003). Hence, the interactions between societies and ecosystems can create dynamic feedback loops in which humans both influence and are influenced by ecosystem processes. For example, Gordon et al. (2008) show how agricultural modifications of hydrological flows can produce a variety of ecological regime shifts that operate across a range of spatial and temporal scales ranging from soil structure to salinization and vegetation patchiness. These shifts can have severe implications for food production, the quality and quantity of freshwater resources, and other ecosystem services such as climate regulation and downstream coastal ecosystems (see also Resilience Alliance and Santa Fe Institute 2004). Allison and Hobbs (2004) describe how adaptive behavior that fails to respond to environmental feedback in agricultural systems can result in a "lock-in" trap. A social–ecological resilience approach would treat agriculture as an embedded part of larger landscapes and pay special attention to tipping points and the internal and external dynamics that drive such change in interlinked agricultural, hydrological, and ecological processes (Gordon et al. 2008).

Gunderson and Holling (2002) refer to rigidity traps where people and institutions try to resist change and persist with their current management and governance system despite a clear recognition that change is essential. The tendency to lock into such a pattern comes at the cost of the capacity to respond to new problems and opportunities. In rigidity traps, a high degree of connectivity and the suppression of innovation prolong an increasingly rigid state, which could result in an undesired regime shift in the system. For example, archeological studies shows that people of the Hohokam region, U.S. Southwest, created a way of life that offered few alternatives, which led to a societal collapse (Hegmon et al. 2008). Although conditions worsened, people stayed despite poor health conditions for generations, until the social and physical infrastructure ultimately fell apart. Hence, the misfit between social and ecological systems and the inability to respond to feedbacks can push interconnected SES into undesirable pathways from which it is hard to escape, and may lead to societal collapse and major human suffering.

Scholars from the social sciences and humanities refer to this problem as *path dependence*. A system is path dependent if initial moves in one direction elicit further moves in the same direction; in other words, there are self-reinforcing feedback mechanisms (Kay 2003). Historical institutionalists see institutions as one of the key factors for pushing development along a set of paths (Hall and Taylor 2006) and have focused on explaining how institutions produce such paths. This includes studies of how institutions structure a nation's response to new challenges. Due to stabilizing feedback mechanisms, shifting into new pathways might be very difficult. For example, marine zoning and shifts to ecosystem-based management in the United States have been severely constrained by inflexible institutions, lack of public support, and difficulties in developing acceptable legislation (Crowder et al. 2006). This means that attempts and initiatives to move towards place-based ecosystem management might fail because there are mechanisms, such as peoples' opinions and worldviews, incentives, power relations, and institutions, operating at different scales that do not support such shifts. For example, Berkes et al. (2006) show how trade flows of marine resources at the global level and the lack of legislations to deal with "roving bandits" fishery might stifle attempts to move towards ecosystem-based management at the local/regional level. Understanding traps and path dependence and identifying barriers to change (and how they, for example, are associated at a specific institutional level) are important for developing strategies for SES transformations.

The resilience lens (capacity to deal with change, also abrupt and surprising, and continue to develop) that we use here to discuss social–ecological transformation emphasizes three facets of resilience: (1) persistence (buffer capacity, robustness), which is the most common interpretation of resilience in the literature; (2) adaptability, which is the capacity to reconfigure or reorganize within the same social–ecological regime in the face of disturbance; and (3) transformability, which is the capacity to create a fundamentally new system when ecological, economic, or social conditions make the existing system untenable. From a resilience perspective, one can argue that unsustainable social–ecological regimes can be very persistent to change because there is path dependence and inertia in the system. This has sparked the notion that resilience as persistence is not necessarily a good thing and that building resilience is not an end in itself, especially if you are in a trap or on an unsustainable path (Walker et al. 2009b). The question is how persistence of the undesired SES regime can be reduced in order to enable shifts to a new regime.

Resilience research in transformability focuses on how to "unlock" a locked-in regime and the ability to escape or get out of traps. Gunderson et al. (2009) argues that there are at least two ways to unlock the system. One is the role of crises (external variation that overwhelms system resilience). This means that a crisis, like the current climate change crisis, food crisis, and financial crisis, can potentially be used productively to stimulate experimentation, innovation, novelty, and learning within society (Cumming GS, Olsson P, Chapin FS III, Holling CS (manuscript) "Coping with climate change: the urgent need for a learning agenda", in preparation). The other way is the more "quiet revolution" where internal processes, sometimes eroded by broader scale processes and drivers, reduce the resilience of the system and the resistance to change. As mentioned earlier, lock-in mechanisms operate at different levels and scales and in different parts of the system (social, economical, ecological) and strategies need to be developed to understand such mechanisms and find ways to unlock them. Although Gunderson and Holling (2002) and Gunderson et al. (2009) offers insights on some of the key features for reducing the resilience of undesired regimes, the capacity to unlock social–ecological regimes needs to be explored further.

13.3 Enhancing the Fit and Unlocking SES

Researchers in social sciences and humanities have long recognized that rigidity, lock-in traps, and path dependence are common characteristics of institutional development and public policymaking. They have also focused on understanding sudden change and "punctuated equilibrium," where long periods of stability and incremental change interact with abrupt, nonincremental, large-scale change (e.g.,

Baumgartner and Jones 1991; True et al. 1999). For example, in 1970 the United States experienced an abrupt burst of environmental policy innovations and a number of environmental laws were passed in rapid succession (Repetto 2006). This period lasted for about 5 years and was exceptional in terms of public concern over environmental protection, political mobilization, and legislative consensus. The literature on punctuated equilibrium recognizes that there are *critical junctures* and *branching points* from which historical development moves onto a new path. Understanding the sequence of events that leads to such junctures is of crucial importance for understanding transformative capacity.

Work on environmental and resource regimes has generally focused on institutional arrangements. While historical institutionalists argue for the need to change institutions in order to move into new pathways, other scholars go beyond institutions and argue that whole societal regimes have to be shifted (White 2000; Pahl-Wostl 2009; Fischer-Kowalski and Rotmans 2009). Holtz et al. (2008:629) define a societal regime as follows: "a regime comprises a coherent configuration of technological, institutional, economic, social, cognitive and physical elements and actors with individual goals, values and beliefs." Hence institutions are only one of the components that constitute a regime. Similarly, in the socio-technical literature, a societal regime is defined as "a conglomerate of structure (institutional setting), culture (prevailing perspective) and practices (rules, routines and habits)" (Fischer-Kowalski and Rotmans 2009:12). A regime's institutions provide stability and structure of societal systems, but they can also limit innovation and stifle attempts to deal with new challenges (Geels 2005). This literature recognizes that transformations are more than mere institutional change but rather systemic regime shifts. It points to a broader set of issues that need to be addressed as part of transformative capacity, such as power and social relations, political and economic dynamics, worldviews, and cultural differences.

In the same line, Pahl-Wostl (2009) and Armitage et al. (2008) offer a learning perspective on systemic regime shifts. Learning strategies are particularly pertinent for dealing with complex adaptive social–ecological system where uncertainty is high. Learning strategies involve monitoring, evaluating, and responding to signals of environmental and social change. This literature describes transformations as stepwise social learning processes that involves single-, double-, and triple-loop learning. According to Pahl-Wostl (2009), *single-loop learning* refers to a refinement of actions to improve performance without changing guiding assumptions and question established routines. *Double-loop learning* refers to changing the frame of reference and questioning of guiding assumptions. Reframing implies a reflection on management goals and how problems are framed (define priorities, include new aspects, change boundaries of system analysis) and assumptions on how goals can be achieved. *Triple-loop learning* refers to a change in management paradigms and also in the underlying norms and values that determine the frame of reference. It builds on the recognition that paradigms and structural constraints can impede innovations and effective reframing of resource management. Triple-loop learning most likely needs to be involved in transformation from one regime to another.

Although this literature provides insights on the magnitude of change that needs to happen in order to avoid or escape unsustainable pathways and lock-in traps, the question still remains, what constitutes the capacity to initiate and navigate such transformations? The literature on socio-technological transitions offers some important insights. Geels and Schot (2007) for example, present a typology of transitions in social–technological systems. This literature recognizes transformations as multilevel, multiphase processes that move along in fits and starts. The challenge in SES research is to also capture the ecological dimension of transformations. Addressing only the social dimension of resource management without an understanding of resource and ecosystem dynamics will not be sufficient to guide society toward sustainable outcomes. Societies may go through major regime shifts without improving the capacity to learn from, respond to, and manage environmental feedback from dynamic ecosystems, which in turn can lead to further ecological degradation, SES regime shifts, and deep traps difficult to get out of. For example, the mobilization of Belizian coastal fishermen into cooperatives, which was socially desirable and economically successful, led ultimately to excessive harvesting of stocks of lobster and conch (Huitric 2005). Similarly, focusing only on the ecological aspects as a basis for decision making for sustainability leads to conclusions that are too narrow.

For the reasons outlined above, studies on transformative capacity need to focus on the interconnected SES and more specifically on changes of the feedback loops in these systems (Walker et al. 2004; Chapin et al. 2010). Transformations fundamentally change the structures and processes that alternate feedback loops in SES (Gunderson and Holling 2002), and transformability means defining and creating novel system configurations by introducing new components and ways of governing SES, thereby changing the state variables, and often the scales of key cycles, that define the system (Carpenter and Gunderson 2001).

SES transformations are not scale independent and require an understanding of cross-scale interactions. Changes at smaller scales can trigger changes at larger scales and changes at larger scales can open up and provide windows of opportunity for transformations at regional to local scales, like those in focus here. Transformations from one social–ecological regime will often require sources of resilience, for example, a memory of experiences for creating novelty and innovation, drawn from other scales or other systems (Gunderson and Holling 2002). For example, Gelcich et al. (Gelcich S, Olsson P, Castilla J, Hughes T, Folke C (in preparation) "Governance transformation for the sustainable management of marine coastal SES") describe how, in the late 1980s in Chile, a new governance approach for marine resources emerged at a time of marine resource crisis and political turbulence. The resource crisis triggered a few collaboration initiatives between fishers and scientists who for different reasons started to solve problems together. The political turbulence in the late 1980s provided a window of opportunity for fishers to organize and influence the new national fishery legislation. Hence, transformations at one scale do not take place in a vacuum but in a cross-scale context.

13.4 Initiating and Navigating Purposeful Transformations

In initiating and navigating transformation in SES, developing new ideas and alternatives to existing governance structures is of fundamental importance (Westley 1995, 2002; Gunderson et al. 1995; Gunderson and Holling 2002). Olsson et al. (2004a) describe a sequence of events for the development of new ecosystem management approaches in response to ecological change:

- An ecosystem management approach expands from individual actors, to group of actors, to multiple actor processes.
- Organizational and institutional structures are developed as a response to deal with the broader set of environmental issues.
- Knowledge of social–ecological dynamics develops as a collaborative effort and becomes part of the organizational and institutional structures.
- Social networks are developed to connect institutions and organizations across levels and scales. Social networks facilitate information flows, identify knowledge gaps, and create nodes of expertise of significance for ecosystem management.
- Knowledge for ecosystem management (including local knowledge and scientific knowledge) is mobilized through social networks and complements and refines local practice for ecosystem management.

In the sequence of events, the ability to deal with uncertainty and surprise can be improved, which increases the capacity to deal with future change. Olsson et al. (2004b, 2006) also point out that these transformations are often multilevel and multiphase processes that involve incremental as well as abrupt change. They have identified three phases of SES transformation (Fig. 13.1): (1) preparing for transformation, (2) navigating the transition, and (3) building resilience of the new governance regime. Phases (1) and (2) are linked by a window of opportunity. Important factors for accomplishing transformations include the role of innovation,

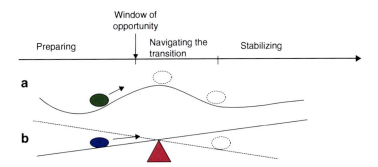

Fig. 13.1 Three identified phases of a social–ecological transformation: preparing for transformation, navigating the transition, and building resilience of the new direction. A window of opportunity links the first and second phases. The transformation is illustrated in two ways: (**a**) as a regime shift between multiple stable states, passing a threshold or (**b**) as a tipping point

transition strategies, enabling legislations, dynamic networks, entrepreneurship, and leadership.

The challenges of initiating and navigating purposeful transformations in SESs have some similarities with developing new commercial organizations. Organizational science has paid much attention to the role of networks, agency, and innovation in transformations. Therefore, we suggest that much can be learnt about transformative capacity in SES by theoretically linking emerging insights from resilience research with theories in the organizational sciences. We focus on the literature on how organizations are created and how these may evolve from early startups to established firms. This line of organizational research is labeled "evolutionary" (Fig. 13.2), and we draw on the seminal work on evolving organizations by Aldrich (1999). We recognize that the amount of literature on organizational change is huge and we do not intend to cover it all in this chapter.

Establishing a new commercial organization in a complex environment involves carrying out many different activities and being able to respond to various challenges in a timely and responsive way. Hence, the nascent entrepreneurs need to be able to cope with uncertainties and to conduct businesses in an adaptive, ad hoc approach (Aldrich 1999). Consequently, few startup attempts are able to survive beyond this very initial phase. Organizational research, as well as research on transformations of SES, is therefore occupied with identifying factors that enable organizations to emerge and develop.

13.4.1 Agency and Dynamic Network

The importance of informal social networks is strongly emphasized in the evolutionary perspective of evolving organizations (Aldrich 1999). Traditionally, much

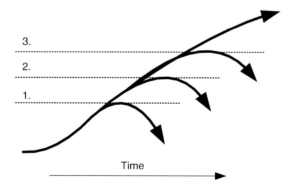

Fig. 13.2 A conceptual model of emerging organizations (adapted from Aldrich 1999). Through time, the organization goes through different phases and challenges (1, 2, and 3 in the figure). At these stages, the organizations will either adaptively embrace the challenges or continue to evolve and develop, or they may not and therefore fail in sustaining themselves. This multiphase model of emerging organizations is here suggested to also capture the vulnerable processes characterizing transformations of social–ecological systems

attention has been directed to the personal characteristics of the commercial entrepreneur, but more recent studies have not been able to identify general characteristics of these entrepreneurs that explain successful take-off of organizations (e.g., Garner et al. 1992). Instead, research has highlighted the role of the social network around entrepreneurs and how these networks can provide for example social, emotional, financial, and material support, which is of crucial importance for successful start-ups (Aldrich 1999). Two specific aspects of the social network is emphasized, namely *diversity* and *strength*.

Firstly, diversity refers to the heterogeneity of the entrepreneurs' contacts in terms of age, sex, occupations, industry affiliations, etc. A high diversity implies access to a wider circle of information, knowledge, capital, and markets, whereas a low diversity means a homogeneity of contacts that only provide access to resources and competencies similar to what the entrepreneur already possesses (see, for example, Granovetter 1973; Carlsson and Sandström 2008). Of particular interests are the relational ties that exclusively bridge different groups of actors. A broker is an actor possessing bridging ties connecting different actors not being directly connected themselves. Such nonredundant ties gives the broker not only access to a diverse sets of actors, but it also enables him/her to strongly influence the flow of information going between the different groups; an ability that by itself provides the broker with a social advantage that can be utilized for various objectives (Burt 1992).

Secondly, the strength of the relational ties is of importance. A strong tie implies a close and durable relationship characterized by a high level of trust and reciprocity, and is typically used for giving/receiving advice, assistance, and support spanning many different areas in life. Strong ties often form the core of a founding team of a new organization (Aldrich 1999). Weaker ties, on the other hand, are typically associated with ties to actors of a different kind, and thus provide the entrepreneur with the diversity discussed above (Granovetter 1973). Conclusively, a common characteristic among many successful commercial entrepreneurs is a diverse and far-reaching social network composed of a mix of strong and weak relational ties (Aldrich 1999).

Emerging insights from empirical research on SES transformations support these findings by ascribing similar characteristics to successful change agents' social networks (e.g., Olsson et al. 2007). For example, Gunderson (1999) and Olsson et al. (2006) identify the role of informal social networks (or shadow networks) for initiating and navigating transformations. These networks emphasize political independence outside the fray of regulation and implementation in places in which formal networks and many planning processes fail. They also emphasize the role of these shadow networks as incubators for new ideas and approaches to governing SES. In this context, a successful change agent often needs to devote considerable amounts of time and energy to expand and reorganize its social network for dealing with new problems and to keep the momentum of the transformational processes. Understanding network dynamics and the ability of key actors to maneuver networks is an important part of the transformative capacity.

A new commercial organization needs to define itself as a bounded entity in order to (1) become a visible actor that can cooperate and compete with other actors and (2) to create a sheltered space where new ideas, concepts, competences, and processes can be initially nurtured and developed internally (Aldrich 1999). From a change agent's perspective, these processes involves defining different roles within the organization, developing work processes, recruiting personnel and external supporters, and making sure that competence and knowledge are developed, maintained, and reproduced within the organization. Also, sooner or later the organization needs to define itself in an *organizational network* of other actors/organizations in order to establish legitimacy and acceptance. However, the organizational networks (where the basic social unit is an organization and not a person and the relations are ties among organizations) referred to here are conceptually different from the previously discussed personal social network of the entrepreneur, even though personal social networks are often used as vehicles to create ties between organizations.

Comparing this finding with insights from research on adaptive governance of SES, we argue that since such systems includes juridical, social, economic, and ecological features that interact across multiple levels, a single actor/organization will have a hard time to accomplish transformations if not firmly connected (in partnerships for example) to other organizations and actors within these spheres. In fact, a key success factor in transforming SES from sectoral natural resource management to ecosystem stewardship of dynamic landscapes and seascapes is the existence of organizations being able to develop and utilize ties with various different actors on different scales. Boundary and bridging organizations are examples of such organizations (Cash et al. 2006; Hahn et al. 2006). The ability among actors to form those connections and shape change depends on the opportunity context and the ability of actors to take advantage of such context.

13.4.2 Making New Approaches Stick

If an organization is successful in moving through the three phases in Fig. 13.1 and develops along a new pathway, it might still be vulnerable to change in its early stages of existence. As for emerging management and governance systems in SES, the question is how to make them "stick". Aldrich (1999) argues that the firm establishment of an organization that is introducing something new is achieved first when several related organizations, together as a group, are able to establish their own common niche in the organizational landscape. The evolving population of organizations can create legitimacy in society, which may involve shifts in existing legislation and cultural norms. Therefore, an important part of transformative capacity is the ability of a single organization to establish itself in an organizational network and collaborate with relevant organization in developing the common niche. Such collaboration is also crucial in facilitating collective learning among organizations, which itself a factor that helps to build resilience of the emerging population of organizations (Aldrich 1999 and references therein).

Again, these insights harmonize with pending insights from recent SES research. For example, Ernstson et al.'s (in press) study on social movements and ecosystem management has shown how a network of volunteer organizations concerned with the protection of green areas in an intensively developed urban region has established itself as an important player with a specific role to play in land-use planning. The network is part of the organizational landscape and is acknowledged as an informal but still legitimate entity by various planning authorities. As we will illustrate in the following sections, the development of new networks, like the shadow networks described earlier, as well as their ability to challenge and change well-established decision-making processes, is an important part of transformative capacity.

13.5 The Kristianstads Vattenrike Biosphere Reserve

The following information is drawn from Olsson et al. (2004b); Hahn et al. (2006) and Schultz et al. (2007). The flooded meadows of Kristianstads Vattenrike Biosphere Reserve have been shaped over several millennia by agricultural practices in combination with the annual flooding of the Helgeå River. Continuous livestock grazing and mowing for haymaking have resulted in a landscape with unique values – biological as well as cultural–historical.

Over the last three centuries, draining, dredging and building of embankments have altered the hydrology and the wetlands have shrunk accordingly. During the 1900s, wetlands were largely seen as wastelands and were used as a dumping ground in the 1960s. However, in 1975, the 35-km stretch of wetlands along the lower Helgeå River came under the protection of the Convention on Wetlands of International Importance. The county administrative board became responsible for managing the Ramsar Convention Site (RCS), which included privately owned land as well as municipal and state-owned. They suggested that almost the whole area (49 km^2) should eventually become a nature reserve.

The Ramsar designation resulted in several conservation plans, policy documents, and protection efforts. However, the natural values continued to disappear even in nature reserves on state-owned land (covering 3% of the RCS in 1989). Inventories conducted by the Bird Society of Northeastern Scania (BSNES) since the 1950s recorded declining waterfowl populations, eutrophication, and overgrowth of lakes subsequently linked to the decreasing use of flooded meadows for grazing and haymaking. Farmers abandoned marginal lands across the country, as agricultural technology enabled intensification of other types of land.

The links between nature and culture were simultaneously explored and illustrated in a series of exhibitions at the Kristianstads County Museum starting in the late 1970s by the curator, Sven-Erik Magnusson (SEM). During the 1980s, the museum established several outdoor museums to give visitors on-site information, to help them interpret the landscape, and to increase their interest and commitment to its associated values.

In 1986, the Municipality of Kristianstad initiated a cultural heritage program and funded an inventory of meadows and pastures focused on the link between biodiversity and agricultural practices in the flooded meadows. It was designed by SEM in collaboration with a member of BSNES, Hans Cronert (HC), and resulted in a detailed land-use map, which helped define and prioritize areas for improving land-use practices, estimate funding needed to maintain and develop these practices, and identify habitats for unique flora and fauna. At this time, BSNES approached officials at the national and county levels responsible for managing the nature reserves and convinced them to improve management practices in these areas. BSNES also proposed that parts of the RCS be made into a national park, but this idea was never realized.

Encouraged by the inventories and inspired by ecomuseums in Europe, SEM created an ecomuseum on the lower Helgeå River. To coordinate ongoing activities under one concept, he coined the term "*Kristianstads Vattenrike*." The name *Vattenrike* roughly translates as "water realm" and "water riches" and underscored the notion of the wetlands as having great value, rather than as water-logged, unhealthy swamps. In early 1988, the term *Ecomuseum Kristianstads Vattenrike* (EKV) had become not just an outdoor museum, but also an organization working to initiate, improve, and build upon ecosystem management of the catchment of the lower Helgeå River.

Once the flooded meadows become overgrown, it is difficult to restore them and the ecosystem assumes an alternative stable state, which is arguably less desirable to users. Several individuals perceived this threat and conducted inventories, produced maps, and worked to increase public awareness of the values and their disappearance. They also initiated restoration projects and analyses of the underlying processes that sustained the flooded meadows. There were parallel processes of sense-making, knowledge generation, and vision building during this stage, the release phase of flooded meadows, and their management. Efforts were largely uncoordinated and took the form of informal meetings between SEM and the BSNES, and between the BSNES and the official managers of the nature reserves.

To garner support for the EKV project, SEM established close relationships and trust with change agents in key organizations. These early contacts resulted in support from five individuals: a researcher at Lund university interested in linking a research project on nutrient loads from agriculture to the EKV; an official at World Wildlife Fund (WWF) Sweden, interested in the project's nature conservation aspects; the rector at Kristianstads university interested research, education, and pedagogy; a hotel director and former president of the Tourism Board intrigued by the EKV's potential to attract tourists, and the director of the National Museum of Natural History. With their support, SEM prepared the first proposal to charter the EKV in late 1988.

An important meeting was held in October 1988 between SEM and a senior municipal politician, who subsequently convinced the chair of the Municipal Executive Board to support the project. By early 1989, SEM had assembled a broad base of support for the EKV from key individuals within various local groups as well as the municipality, the county administrative board, and national organizations. A window of opportunity to establish the EKV and flexible and collaborative management of the ecosystems of KV opened as local politicians were keen to find a profile for the municipality with good potential for recreation and tourism. The municipal board also knew SEM and trusted his capacity, and environmental issues were high on the political agenda in Sweden at the time.

In March 1989, a small team including SEM was funded by the Municipality to develop the EKV idea further. Other funding backers were WWF Sweden and Sweden's National Cultural Advisory Board. In September 1989, SEM and a colleague from the County Museum began working full time with the EKV project at the Municipality. SEM became the Director of the EKV.

The recognition of the value of the flooded meadows and the threats to their future was growing in Kristianstad. Simultaneously, SEM built support for the EKV proposal as a solution to this management failure. He developed a network of potential partners and contacts, leveraged resources to develop the idea further, and made deals with a range of funding sources.

The Ecomuseum has maintained a firm direction and vision since its inception: to enhance the ecosystem services in the area while using them sustainably through strategic collaboration with a diversity of actors. The work relies on voluntary participation by local stewards and actors at the municipal, county, national, and international levels, and depends on trust-building, skilled communication, and identifying benefits for both nature conservation and other public and private goals. Over the past 15 years, the organization has maintained its flexibility, opening and closing projects depending on needs and available resources. Grazed and mowed areas have increased, and so have water fowl populations. The area protected as nature reserves has also increased. The flooded meadows would appear to have a new image among the people of Kristianstad; they are now seen to offer unique esthetic, cultural–historical, and ecological values, as well as services like flood control and recreational opportunities. In 2005, Kristianstads Vattenrike was given Biosphere Reserve status by UNESCO, partly in recognition of these achievements.

13.6 The Great Barrier Reef Marine Park

The following information is drawn from Olsson et al. (2008). The Great Barrier Reef Marine Park (GBRMP) covers 344,000 km^2, an area almost the size of California. Like many other coral reefs, the Barrier Reef generates a multitude of essential ecosystem services. The Park contributes AU$6.9 billion annually to the Australian economy, 85% of which is from tourism. The Australian government enacted The Great Barrier Reef Marine Park Act in 1975 in response to threats to the reef from oil drilling, mining, and unexplained outbreaks of coral-eating starfish. In 1981, the Great Barrier Reef region was also declared a World Heritage

Area. The marine park allows a range of uses based on spatial zoning. The Great Barrier Reef Marine Park Act also established the marine park authority (GBRMPA) in 1976 and required the new agency to initiate zoning plans for the marine park. Between 1983 and 1988, each of the four sections of the park (Far Northern, Cairns, Central, and southern Mackay/Capricorn sections) were zoned for the first time. No-take areas together accounted for 5% of the marine park, mainly in the remote Far Northern area and predominantly covering coral reefs (reefs actually make up only 6% of the entire GBRMP).

Gradually, it became clear that the initial level of protection established for the park did not ensure that the entire ecosystem remained healthy, productive, and resilient. Despite the Act and associated management efforts, studies showed that the Great Barrier Reef was still showing signs of degradation caused by sediment runoff from land, overharvesting, and, more recently, from global warming. Demographic and economic data gathered in the 1980s and 1990s showed rapid growth in human population, land clearing, coastal development, tourist visits, and fishing pressure.

In the late 1990s, there was growing awareness among scientists and reef managers that inshore and deeper habitats were poorly represented in existing no-take zones and that connectivity of larvae and other poorly understood interactions between reef and nonreef habitats were important to the resilience of the entire ecosystem. Unprecedented regional bleaching occurred in the summer of 1997/1998, affecting large parts of the GBR and other reefs in the Western Pacific and most of the tropical Indian Ocean. The initial zoning network did not adequately protect the range of biodiversity of the reefs and hence could not maintain the GBR's resilience in the face of recurrent ecological disturbances. Combined with increased human pressures on the GBR, including the challenges of climate change, individual actors within the GBRMPA were triggered to search for more holistic approaches to governance and management of this large marine ecosystem.

In 1998, the GBRMPA initiated a major rezoning of the marine park called the Representative Areas Program (RAP) protecting representative examples of each type of habitat within a network of no-take areas. Focus was on protecting biodiversity and maintaining ecosystem function and services rather than on maximizing the yield of commercially important fisheries. The idea to rezone the entire reef in one push was controversial and the RAP process required skilful leadership by GBRMPA and its executive team (the Chair and two Executive Directors to whom senior managers report). Five important areas had to be addressed: (1) internal organizational changes, (2) bridging science and policy, (3) changing people's perceptions, (4) facilitating public consultation and participation, and (5) gaining political support.

The emerging concept of rezoning the entire marine park initially occupied a small group, understaffed and underfinanced. By the early 2000s, however, almost all of GBRMPA was involved in the RAP process. The executive team established a Senior Managers' Forum to coordinate activities and advise the organization's Chair. The Senior Management Forum unified internal management and communicated a common vision throughout the organization. The Forum established and nurtured an environment where creativity was encouraged and innovative solutions to problems could emerge. Importantly, this process was achieved without any additional funding and relied entirely on a flexible internal redeployment of staff. The Senior Managers Forum established four regional teams responsible for the comprehensive public consultations associated with the RAP. Using teams helped avoid competition between sectors, increase internal collaboration, and pool experiences and resources.

The RAP process relied heavily on scientific expertise and a new synthesis of data on species and habitats of the Great Barrier Reef. Therefore, GBRMPA created new opportunities for interaction, dialog, and information sharing with researchers. This included establishing committees and panels, facilitating workshops, and communicating GBRMPA's overall vision and goals for the RAP and rezoning. For example, two independent advisory committees (the Scientific Steering Committee and the Social, Economic, and Cultural Steering Committee) were convened to develop operating principles to guide the RAP process. Beginning in 1998–1999, experts compiled more than 40 datasets to characterize the biological and physical diversity of the GBRMP. GIS-based tools and analytical methods identified and mapped 70 bioregions, of which 30 were reef bioregions. Scientists were encouraged to think beyond their individual sample sites or specialized expertise. This dialog was facilitated by a longstanding relationship between GBRMPA and researchers at universities, the Australian Institute of Marine Sciences, and the Cooperative Research Centre for the Great Barrier Reef World Heritage Area (CRC Reef).

Because of its iconic status, there was overwhelming support both nationally and locally for conserving the Great Barrier Reef, which gave the RAP process political leverage. However, not everyone was aware of the threats to the reef or agreed with the proposed management changes. Some local recreational fishers were vocal in their opposition to no-take zones. Many still perceived the Great Barrier Reef as pristine, protected from human impacts by its sheer size and relative isolation. To address this issue, GBRMPA produced a "reef under pressure" information campaign showing threats to the reef from coastal development, land use, shipping, tourism, and fishing. The campaign included Web sites, posters, pamphlets, and television advertisements with celebrity spokespeople. The campaign was followed up by continuous polling to monitor changes in public perceptions.

Public consultation for the RAP greatly exceeded the requirements of the Act and was by far the most extensive in the history of the marine park. GBRMPA attended every public meeting they were invited to. Instead of organizing large public meetings that could be dominated by a few people, they held several hundred community information sessions in regional and local community centers. Periodic updates on the RAP process were posted online. The public response was overwhelming: over 31,000 submissions. A "factory" was set up for handling them, quickly allocating human and financial resources within the organization without additional external funding.

GBRMPA reports to the Australian Federal Minister for Environment and Heritage, whose support was crucial for the RAP process. The rezoning legislation had to pass the two federal houses of Parliament. A new minister for environmental issues was appointed in 2002. The timing of submitting the plan to the Senate was crucial; the new zoning plan needed to be submitted in December 2003 in order to become operational before the upcoming federal election. This provided a narrow political window of opportunity that set the time frame for GBRMPA and the minister to prepare a smooth passage of the plan through both houses of Parliament. Throughout the planning process, senior staff from GBRMPA made frequent trips to inform critical players such as governmental departments and agencies responsible for fisheries and the environment, Members of Parliament and senators (especially those representing constituencies along the Queensland coast), shipping interests, port authorities, and the Defense Department. Senior scientists, conservation nongovernmental organizations, and lobbyists for the tourism and fishing industries also played a role in convincing politicians of the need to pass the reef legislation.

The new, more sophisticated approach that emerged addressed both ecosystem dynamics and the intricate web of interactions in SES. A small team working within the GBRMPA planned the rezoning of the entire marine park, which subsequently led to critical support from the Authority's executive team for the major rezoning effort and the allocation of internal resources for developing the RAP. This happened in three stages, from (1) a relatively minor project within GBRMPA to (2) incorporation across all parts of the Authority and status as an agency priority, to (3) changing national legislation and influencing other areas in Australia (such as the Ningaloo Reef in Western Australia) and becoming a role model for policy development elsewhere.

13.7 Discussion

Drawing on the literature and our case studies, we can identify some key features of transformative capacity. There are at least three distinct dimensions of SES transformations that require capacity building: (1) understanding where you are, (2) figuring out where to go, and (3) developing strategies for how to get there. SES transformations involve agency for changing management paradigms, power distributions, regulatory frameworks, underlying norms and values, knowledge production, and network configurations and interactions among actors. These changes are all important for altering and creating new feedback loops in SES.

Understanding where you are includes identifying the current regime and key feedback mechanisms that keep the social–ecological system in its regime, engaging key actors to recognize dysfunctional states, and raising awareness of the problem. In the Kristianstads Vattenrike case, individual actors were involved in sense-making processes, interpreting ecosystem changes and creating a meaningful order captured in the project proposals as a call for action. The sense-making process helped link ecosystem changes and degradation to social factors including values and perspectives, organizational structures, and institutions at multiple levels.

In the Great Barrier Reef case, collaborative efforts were made to compile a 25-year strategy that made sense of data and information that had been collected over decades. Global, national, and local assessments often provide an understanding of the causes of the trends in environmental, ecological, and social conditions that lead to unsustainable directions for SES at the local to planetary scales. However, this kind of information is often fragmented and requires a capacity to access, compile, and make sense of it in a specific context. In both our case studies, as part of enhancing the fit between ecosystem and governance system, strategies were developed and orchestrated by key agency in bridging organizations to overcome organizational structural problems that produced fragmented responses to change.

Figuring out where to go includes identifying plausible alternative pathways for the social–ecological system. In the Kristianstad case, there was a process of seeking a collective vision for the future and communicating and building support for this vision. Key factors included dialog, conflict resolution, trust-building, and sense-making. The shared vision of ecosystem management helped define the arena for collaboration, connect and coordinate ongoing activities, and develop social networks. In the GBR, a collective vision for the management of the Great Barrier Reef was developed in the *25-Year Strategic Plan for the Great Barrier Reef World Heritage Area* (1994). In addition, a vision was also developed for GBRMPA to guide the interorganizational changes that were instigated in the late 1990s. Scenario building is another key tool for collectively identifying possible futures (Peterson 2007; Carpenter and Folke 2006; Enfors et al. 2008). Building visions and scenarios requires a capacity to coordinate and draw on a number of sources of knowledge including scientific knowledge and local knowledge.

Developing strategies for how to get there includes strategies for identifying and overcoming barriers, keeping the momentum of the transformation and building stability of the new regime. In both cases, knowledge gaps were identified and new knowledge was generated in order to fill these gaps, which were key processes for enhancing the fit in the SES. For example, GBRMPA played a key role in facilitating and coordinating scientists to produce the bioregional map, which was necessary for establishing a network of reserves and managing the GBR at the seascape level. New scientific insights on marine connectivity and spatial ecological resilience were also incorporated in the new approach. Similarly in KV, as part of developing the new approach, the shadow network provided agency and coordinated the synthesis of existing knowledge and initiated a number of inventories that provided new insights for managing social–ecological processes at the landscape level. As an important part of this knowledge production, both case studies show how experiments were initiated to generate innovations that could support the new approaches. In Kristianstad, collaborative experiments were set up to reduce nutrient loads to the rivers, and the GBR experiments showed that the biomass of coral trout was up to six times lower on heavily fished near-shore reefs compared to adjacent no-take areas. Such social–ecological system innovations were important to enhance the fit between the ecosystems and governance systems. This is in line with the findings by scholars in transition management (e.g., Loorbach 2007) who argue that the ability to coordinate experiments in such a manner that they contribute to system innovation is of crucial importance in order to unlock the current lock-in and enable shifts towards new trajectories.

In both case studies, strategies were developed to change peoples' values and perspectives, which were identified as major obstacles to change. The capacity to develop strategies to change public perceptions and attitudes were important to unlock and change existing regimes. The case of Kristianstads Vattenrike Biosphere Reserve also show that it is important within which groups such shifts occur in order to reduce the resilience of the unsustainable regime. Hence, the change in attitudes among a few local politicians was a critical tipping point for moving into a new SES trajectory. Similar linkages between microagency and macrostructures have been described for shifting to more integrated forms of water management in the Netherlands (van den Brink and Meijerink 2005): a shift that was also preceded by a change in people's mental models, from "fighting the water" to "living with the water." A key issue for future research on SES transformations is to go beyond attitudes and mental models and address the cultural dimensions of transformations, like changes in identity.

Transformations in SES require skills that go beyond the capabilities of individual actors. Therefore, networking strategies are needed for connecting nodes of expertise and developing networks of motivated actors. In the Kristianstad case, the development of a shadow network of individuals representing a diversity of skills and backgrounds was crucial for building moral, financial, and political support for the new approach. Both case studies show that, in order to move from an idea that existed in a small network of engaged actors to the institutionalization of the new approach, linking into the political arena was of crucial importance. This means that the entrepreneurial network and their new ideas had to be connected to a strong political leadership in order to create a regime shift. It often happens that the entrepreneurs who are good at developing new ideas and innovations lack the leadership capabilities required to change the social–ecological regime.

The ability of actors to form network connections and shape change depends on the opportunity context and the ability of change agents to take advantage of such context. Both case studies highlight the role of individual actors to scan for and use windows of opportunity to develop and utilize ties with various different actors on different scales and launch new initiatives and innovations. Transformational change is most likely to occur at times of crisis, when enough stakeholders agree that the current system is dysfunctional (Chapin et al. 2010). As a crisis deepens, stakeholders are more likely to negotiate a transformation. However, our case studies also show a capacity to respond to early warning (smaller scale crisis) and steer away from a pending large-scale social–ecological crisis. Using crises occurring at other times or places, change agents developed strategies that enable people to move beyond a state of denial and accept that the system cannot (or should not) continue on its current trajectory. Our social–ecological case studies put the finger on the importance of incorporating understanding and capacity to respond to ecosystem dynamics as an essential part in order to initiate transformations.

13.8 Conclusion

In this chapter, we have developed criteria for a framework to analyze and assess transformations and transformative capacity in SES, or more specifically the capacity to transform SES trajectories toward ecosystem stewardship. These include experimentation and innovation, agency and social networks, opportunity context, diversity, boundaries, and collaboration. We have only started to explore these issues and researchers need to continue to develop such frameworks. This would involve developing criteria for monitoring and evaluating transformations that could provide orientation of "where we are" and the status of transformations at any particular time.

Important questions for future research on SES transformations are *what* needs to be transformed and *how* transformations happen. The *what* question involves research on regime shifts and systemic changes, on trajectories and interconnected SES, and on how to alter and create new feedback loops in SES that sustain flows of essential ecosystem services to society. The *how* question involves understanding the multiphase and multilevel and cross-scale interaction aspects of SES transformations as well as links between microagency and macrostructures. More specifically, it involves research on how to reduce resilience in terms of persistence in undesirable regimes and build transformative capacity; on path dependence and lock-in traps; on unlocking mechanisms and transition strategies; on how to make new approaches stick; on innovations and opportunity contexts; on triggers for transformations and the role of crisis; on dynamic interactions between individuals (including entrepreneurs), organizations, and institutions at multiple levels, and on the aspects of social learning.

Transformations are needed to overcome the mismatch between ecosystems and governance systems where the institutional capacities to manage the earth's ecosystems are evolving more slowly than man's overuse of the same systems. The problem is that transformations of the magnitude that we discuss in this chapter, and that are needed to deal with the global problems that humanity is facing, might take a long time. We argue that if we can increase our understanding of SES transformations and provide strategies and guidelines for initiating and navigating SES' transformations, we could better prepare for and potentially speed up the responses to the rapid changes in the capacity of the earth's ecosystems to sustain our own development and civilization. The issue is pressing, considering the windows of opportunity for transformations towards sustainability that are currently wide open due rapid, pervasive global changes in many dimensions.

References

Aldrich H (1999) Organizations evolving. Sage, London

Allison HE, Hobbs RJ (2004) Resilience, adaptive capacity, and the "lock-in trap" of the Western Australian agricultural region. Ecol Soc 9(1):3, [online] URL: http://www.ecologyandsociety.org/vol9/iss1/art3/

Armitage D, Marschke M, Plummer R (2008) Adaptive co-management and the paradox of learning. Glob Environ Change 18(1):86–98
Babcock EA, Pikitch EK (2004) Can we reach agreement on a standardized approach to ecosystem-based fishery management? Bull Mar Sci 74:685–692
Baumgartner FR, Jones BD (1991) Agenda dynamics and policy subsystems. J Polit 53 (4):1044–1074
Berkes F, Folke C (eds) (1998) Linking social and ecological systems: management practices and social mechanisms for building resilience. Cambridge University Press, Cambridge
Berkes F, Colding J, Folke C (eds) (2003) Navigating social–ecological systems: building resilience for complexity and change. Cambridge University Press, Cambridge
Berkes F, Hughes TP, Steneck RS, Wilson JA, Bellwood DR, Crona B, Folke C, Gunderson LH, Leslie HM, Norberg J, Nyström M, Olsson P, Österblom H, Scheffer M, Worm B (2006) Globalization, roving bandits, and marine resources. Science 311:1557–1558
Burt R (1992) Structural holes: the social structure of competition. Harvard University Press, Cambridge
Carlsson L, Sandström A (2008) Network governance of the commons. Int J Commons 2:33–54
Carpenter SR, Folke C (2006) Ecology for transformation. Trends Ecol Evol 21:309–315
Carpenter SR, Gunderson LH (2001) Coping with collapse: ecological and social dynamics in ecosystem management. Bioscience 51(6):451–457
Cash DW, Adger W, Berkes F, Garden P, Lebel L, Olsson P, Pritchard L, Young O (2006) Scale and cross-scale dynamics: governance and information in a multilevel world. Ecol Soc 11(2):8, (online) URL: http://www.ecologyandsociety.org/vol11/iss2/art8/
Chapin FS III, Carpenter SR, Kofinas GP, Folke C, Abel N, Clark WC, Olsson P, Stafford Smith DM, Walker B, Young OR, Berkes F, Biggs R, Grove JM, Naylor RL, Pinkerton E, Steffen W, Swanson FJ (2010) Ecosystem Stewardship: Sustainability Strategies for a Rapidly Changing Planet. Trends Ecol Evol 25:241–249
Costanza R et al (1998) Principles for sustainable governance of the oceans. Science 281:198–199
Crowder LB et al (2006) Resolving mismatches in U.S. ocean governance. Science 313:617–618
Cumming GS, Cumming DHM, Redman CL (2006) 'Scale mismatches in social–ecological systems: causes, consequences, and solutions'. Ecol Soc 11(1):14, (online) URL: http://www.ecologyandsociety.org/vol11/iss1/art14/
Cumming GS, Olsson P, Chapin FS III, Holling CS (in preparation) Coping with climate change: the urgent need for a learning agenda
Danter KJ, Griest DL, Mullins GW, Norland E (2000) Organizational change as a component of ecosystem management. Soc Nat Resour 13:537–547
Enfors EI, Gordon LJ, Peterson GD, Bossio D (2008) Making investments in dryland development work: participatory scenario planning in the Makanya catchment, Tanzania. Ecol Soc 13(2):42, [online] URL: http://www.ecologyandsociety.org/vol13/iss2/art42/
Ernstson H, Sörlin S, Elmqvist T. Social movements and ecosystem services – the role of social network structure in protecting and managing urban green areas in Stockholm. Ecol Soc 13 (2):39, [online] URL: http://www.ecologyandsociety.org/vol13/iss2/art39/)
Fischer-Kowalski, Rotmans (2009) Conceptualizing, observing and influencing social ecological transitions. Ecol Soc 14(2):3, [online] URL: http://www.ecologyandsociety.org/vol14/iss2/art3/
Folke C, Colding J, Berkes F (2003) Synthesis: building resilience and adaptive capacity in social–ecological systems. In: Berkes F, Colding J, Folke C (eds) Navigating social–ecological systems: building resilience for complexity and change. Cambridge University Press, Cambridge, pp 352–387
Folke C, Hahn T, Olsson P, Norberg J (2005) Adaptive governance of social–ecological knowledge. Annu Rev Environ Resour 30:441–473
Folke C, Pritchard L, Berkes F, Colding J, Svedin U (2007) 'The problem of fit between ecosystems and institutions: ten years later'. Ecol Soc 12(1):30, (online) URL: http://www.ecologyandsociety.org/vol12/iss1/art30/

Galaz V, Olsson P, Hahn T, Folke C, Svedin U (2008) The problem of fit among biophysical systems, environmental and resource regimes, and broader governance systems: insights and emerging challenges. In: Young OR, King LA, Schröder H (eds) Institutions and environmental change – principal findings, applications, and research frontiers. The MIT, Cambridge, pp 147–182

Garner WB, Bird B, Starr J (1992) Act as if: differentiating entrepreneurial from organizational behavior. Entrepren Theor Pract 16(3):13–32

Geels FW (2005) The dynamics of transitions in socio-technical systems: a multi-level analysis of the transition pathway from horse-drawn carriages to automobiles (1860–1930). Tech Anal Strat Manag 17(4):445–476

Geels FW, Schot J (2007) Typology of socio-technical transition pathways. Res Policy 36 (3):399–417

Gelcich S, Olsson P, Castilla J, Hughes T, Folke C (in preparation) Governance transformation for the sustainable management of marine coastal social–ecological systems

Gordon LJ, Peterson GD, Bennett EM (2008) Agricultural modifications of hydrological flows create ecological surprises. Trends Ecol Evol 23(4):211–219

Granovetter M (1973) The strength of weak ties. Am J Soc 76:1360–1380

Gunderson LH (1999) Resilience, flexibility and adaptive management: antidotes for spurious certitude? Conserv Ecol 3(1):7, (online) URL: http://www.consecol.org/vol3/iss1/art7

Gunderson LH, Holling CS (eds) (2002) Panarchy: understanding transformations in social–ecological systems. Island, London

Gunderson LH, Holling CS, Light SS (eds) (1995) Barriers and bridges to renewal of ecosystems and institutions. Columbia University Press, New York

Gunderson L, Allen C, Holling CS (2009) Fundamentals of ecological resilience. Island, Washington

Hahn T, Olsson P, Folke C, Johansson K (2006) Trust-building, knowledge generation and organizational innovations: the role of a bridging organization for adaptive comanagement of a wetland landscape around Kristianstad, Sweden. Human Ecol 34:573–592

Hall PA, Taylor RCR (2006) Political science and the three new institutionalisms. Polit Stud 44:936–957

Hegmon M, Peeples MA, Kinzig A, Kulow S, Meegan CM, Nelson MC (2008) Social transformation and its human costs in the Prehispanic U.S. Southwest. Am Anthropol 110:313–324

Holling CS, Meffe GK (1996) Command and control and the pathology of natural resource management. Conserv Biol 10(2):328–337

Holtz G, Brugnach M, Pahl-Wostl C (2008) Specifying "regime" – a framework for defining and describing regimes in transition research. Technol Forecasting Soc Change 75:623–643

Huitric M (2005) Lobster and conch fisheries of Belize: a history of sequential exploitation. Ecol Soc 10(1):21, http://www.ecologyandsociety.org/vol10/iss1/art21/

Imperial MT (1999) Institutional analysis and ecosystem-based management: the institutional analysis and development framework. Environ Manag 24:449–465

Kay A (2003) Path dependency and the CAP. J Eur Public Pol 10:405–420

Loorbach DA (2007) Transition management: new mode of governance for sustainable development. International Books, Utrecht

National Research Council (1999a) Sustaining marine fisheries. National Academy, Washington

National Research Council (1999b) Our common journey. National Academy, Washington

Norberg J, Cumming G (eds) (2008) Complexity theory for a sustainable future. Columbia University Press, New York

Olsson P, Folke C, Berkes F (2004a) Adaptive co-management for building social–ecological resilience. Environ Manag 34(1):75–90

Olsson P, Folke C, Hahn T (2004b) 'Social–ecological transformation for ecosystem management: the development of adaptive co-management of a wetland landscape in southern Sweden'. Ecol Soc 9(4):2, (online) URL: http://www.ecologyandsociety.org/vol9/iss4/art2

Olsson P, Gunderson LH, Carpenter SR, Ryan P, Lebel L, Folke C, Holling CS (2006) 'Shooting the rapids: navigating transitions to adaptive governance of social–ecological systems'. Ecol Soc 11(1):18, (online) URL: http://www.ecologyandsociety.org/vol11/iss1/art18/

Olsson P, Folke C, Galaz V, Hahn T, Schultz L (2007) Enhancing the fit through adaptive comanagement: creating and maintaining bridging functions for matching scales in the Kristianstads Vattenrike Biosphere Reserve Sweden. Ecol Soc 12(1):28, [online] URL: http://www.ecologyandsociety.org/vol12/iss1/art28/

Olsson P, Folke C, Hughes TP (2008) Navigating the transition to ecosystem-based management of the Great Barrier Reef, Australia. Proc Nat Acad Sci USA 105:9489–9494

Ostrom E (1990) Governing the commons: the evolution of institutions for collective action. Cambridge University Press, Cambridge

Pahl-Wostl C (2009) A conceptual framework for analysing adaptive capacity and multi-level learning processes in resource governance regimes. Glob Environ Change 19:354–365

Pahl-Wostl C, Mostert E, Tàbara D (2008) 'The growing importance of social learning in water resources management and sustainability science'. Ecol Soc 13(1):24, online) URL: http://www.ecologyandsociety.org/vol13/iss1/art24/(last visit November 2008

Peterson GD (2007) Using scenario planning to enable an adaptive co-management process in the northern highlands lake district of Wisconsin. In: Berkes F, Armitage D, Doubleday N (eds) Adaptive co-management: collaboration, learning, and multi-level governance. UBC, Vancouver, pp 289–307

Raskin P, Banuri T, Gallopín G, Gutman P, Hammond A, Kates R, Swart R (2002) Great transition – the promise and lure of the times ahead. Stockholm Environment Institute, Boston

Repetto R (ed) (2006) Punctuated equilibrium and the dynamics of U.S. environmental policy. Yale University Press, New Haven

Resilience Alliance and Santa Fe Institute (2004) Thresholds and alternate states in ecological and social–ecological systems, Resilience Alliance (online) URL: http://www.resalliance.org/index.php?id=183

Rockström J, Steffen W, Noone K, Persson A, Chapin FS 3rd, Lambin EF, Lenton TM, Scheffer M, Folke C, Schellnhuber HJ, Nykvist B, de Wit CA, Hughes T, van der Leeuw S, Rodhe H, Sörlin S, Snyder PK, Costanza R, Svedin U, Falkenmark M, Karlberg L, Corell RW, Fabry VJ, Hansen J, Walker B, Liverman D, Richardson K, Crutzen P, Foley JA (2009) A safe operating space for humanity. Nature 461, 472–475

Scheffer M, Carpenter SR, Foley JA, Folke C, Walker B (2001) Catastrophic shifts in ecosystems. Nature 413(6856):591–696

Schultz L, Folke C, Olsson P (2007) Enhancing ecosystem management through social–ecological inventories: lessons from Kristianstads Vattenrike, Sweden. Environ Conserv 34(2):140–152

True JL, Baumgartner FR, Bryan DJ (1999) Explaining stability and change in American policy-making: the punctuated equilibrium model. In: Sabatier P (ed) Theories of the policy process. Westview, Boulder, pp 97–115

United Nations Environment Programme (2007) Global environment outlook: environment for development, vol 4. Progress Press Ltd, Valletta

van den Brink M, Meijerink S (2005) Implementing policy innovations: resource dependence, struggle for discursive hegemony and institutional inertia in the Dutch river policy domain. Paper prepared for the ERSA Congress, Amsterdam 23–27, August 2005

Walker B, Barrett S, Polasky S, Galaz V, Folke C, Engström G, Ackerman F, Arrow K, Carpenter SR, Chopra K, Daily G, Ehrlich P, Hughes T, Kautsky N, Levin S, Mäler K-G, Shogren J, Vincent J, Xepapadeas T, de Zeeuw A (2009) Looming global-scale failures and missing institutions. Science 325:1345–1346

Walker BH, Holling CS, Carpenter SR, Kinzig A (2004) Resilience, adaptability and transformability in social–ecological systems. Ecol Soc 9(2):5, [online] URL: http://www.ecologyandsociety.org/vol9/iss2/art5/

Walker BH, Abel N, Anderies JM, Ryan P (2009b) Resilience, adaptability, and transformability in the Goulburn-broken catchment, Australia. Ecol Soc 14(1):12, [online] URL: http://www.ecologyandsociety.org/vol14/iss1/art12/

Westley F (1995) Governing design: the management of social systems and ecosystems management. In: Gunderson LH, Holling CS, Light S (eds) Barriers and bridges to the renewal of ecosystems and institutions. Columbia University Press, New York, pp 391–427

Westley F (2002) The devil in the dynamics: adaptive management on the front lines. In: Gunderson LH, Holling CS (eds) Panarchy: understanding transformations in human and natural systems. Island, Washington, pp 333–360

White L (2000) Changing the "whole system" in the public sector. J Org Change Manag 13:162–177

Young OR (2002) Institutional interplay: the environmental consequences of cross-scale interactions. In: Ostrom E, Dietz T, Dolsak N, Stern P, Stonich S, Weber EU (eds) The drama of the commons. National Academy, Washington, pp 265–291

Chapter 14
Adapting and Transforming: Governance for Navigating Change

Derek Armitage and Ryan Plummer

14.1 Introduction

A transformation of earth's systems is under way. Many of the provisioning, regulating, and cultural services upon which national and global economies depend are on the decline (MA 2005; Carpenter et al. 2009), with uncertain consequences for human well-being. To navigate these changes, individuals and societies must develop the capacity to adapt and transform their interactions with ecosystems and ecosystem services (Berkes et al. 2003; Carpenter et al. 2009). Adapting and transforming are linked, but reflect some important differences. Adaptations can be reactive or anticipatory, autonomous, or part of a suite of responses to change. In the context of lock-in traps or unsustainable path dependencies, however, such responses may be inadequate. In systems where the ecological, social, and economic conditions are untenable (Walker et al. 2004), there will be limits to adaptation and a need for more fundamental shifts in strategy that require new ideas and practices (Olsson et al. this volume).

Institutions and multilevel governance arrangements are particularly important in this regard because they can support knowledge building, learning, and conflict resolution, which help to reduce vulnerability, build resilience, and increase adaptive capacity (Berkes 2009). The chapters in this volume highlight these linkages and improve our understanding of their connections. Insights are offered from diverse settings using a variety of analytical approaches (Plummer and Armitage

D. Armitage
Department of Geography and Environmental Studies, Wilfrid Laurier University, Waterloo, ON, Canada N2L 3C5
e-mail: darmitag@wlu.ca

R. Plummer
Department of Tourism and Environment, Brock University, 500 Glenridge Avenue, St. Catharines, ON, Canada L2S 3A1
Stockholm Resilience Centre, Stockholm University, Stockholm, SE-106 91, Sweden
e-mail: rplummer@brocku.ca

this volume). Attributes and practices that can confer greater capacity to adapt and transform are emphasized, and we begin to see in the various chapters how governance can steer societies along pathways that sustain ecosystem services and human well-being.

This volume represents an interdisciplinary perspective on adaptive capacity in the context of environmental governance. In combination, the chapters contribute to (1) a synthesis of current knowledge and understanding of adaptive capacity in a wide range of environment and natural resource contexts, (2) theory development based on the synthesis of experiences from a variety of perspectives, and (3) better understanding of the implications of theory and experience for policy and governance for navigating change. In summarizing these contributions – both conceptual and applied – we make the argument that adaptive capacity and effective governance for navigating change should be seen as closely related. Adaptive capacity is central to effective governance, but building multilevel governance can contribute increased capacity to respond to change. To explore this relationship, we reflect upon the lessons learned from this volume and frame an agenda for practice and research (Fig. 14.1).

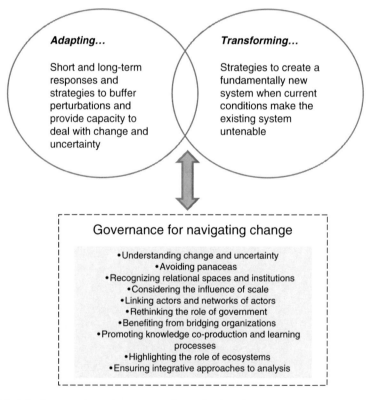

Fig. 14.1 Adaptive capacity and governance for navigating change

14.2 Governance for Navigating Change

Increased vulnerability of people and ecosystems to change, societal lock-in traps, and unsustainable path dependencies present a difficult milieu for environmental governance. Institutions and governance processes are required that match complex social–ecological systems (SESs), adapt as these systems change over time, and help steer these systems along sustainable pathways (Galaz et al. 2008; Pahl-Wostl 2009). Such arrangements must further support strategies to restore, maintain, and enhance the capacity of ecosystems to generate critical ecosystem services (Carpenter et al. 2009; Scheffer 2009). A growing body of environmental governance scholarship is pointing to the importance of multilevel arrangements, interactive networks, and partnerships among state and nonstate actors, in which diverse perspectives and knowledge are shared and social learning processes provide opportunities for adaptability (Folke et al. 2005; Brondizio et al. 2009; Newig and Fritsch 2009).

Folke et al. (2005) outlined four general principles to enhance adaptive capacity and governance: (1) build knowledge of ecosystem dynamics; (2) integrate different forms of knowledge with a focus on learning through adaptive management; (3) support flexible institutions and multilevel governance; and (4) deal with the unpredictable. Such capacity to cope and adapt derives from dynamic and ongoing processes of social learning, facilitated by knowledge coproduction and the formation of vertical and horizontal linkages (Folke et al. 2005; Berkes 2009). Governance arrangements to accommodate these principles and processes typically involve multiple centers of decision making and require many mechanisms for coordinated action (Dietz et al. 2003; Duit and Galaz 2008). Such networked arrangements are presumed to confer stability and adaptability because of enhanced capacity to diffuse the negative effects of change and distribute the benefits (Lebel et al. 2006; Bodin and Crona 2009).

The contributions in this volume reflect on these themes and highlight the manner in which adaptive capacity and governance are linked. Each of the lessons identified below extends in some way the principles identified by Folke et al. (2005) and the interdisciplinary concepts outlined in Chap. 1.

14.3 Lessons Learned

Lessons from this volume are many. As the chapters reveal, identification of attributes and dynamics of adaptive capacity as they relate to governance are multifaceted and complex. It is challenging to isolate a specific social level at which to focus capacity-building efforts, identify a specific asset or process that guarantees the emergence of adaptive capacity for effective governance, or point to a single type of barrier (e.g., institution, political dynamic, resource). Rather, policy makers and practitioners are better served by recognizing the *bundles* of attributes, processes, and practices that support and link capacity and governance. From the perspectives of both research and practice, greater attention is required on how those bundles interact in particular places and at particular times to support or constrain governance for navigating change. We offer below a summary of those insights and the main lessons for practice and research, and thus point to a need for careful assessment of the relationship between adaptive capacity and governance. We make no claim that this list is exhaustive, and recognize that others summarizing the wealth of insights from the chapters in this volume would likely generate a somewhat different list. Nevertheless, we do highlight here what we believe are a set of key lessons with which researchers and practitioners of adaptive capacity and governance can move forward.

14.3.1 Change and Uncertainty

Why are we concerned about building adaptive capacity? The short answer is to deal with change and uncertainty in SESs. Governance systems are required to deal with, individually or in some combination, change of different magnitudes and levels of predictability. For example, directional changes in key drivers of social–ecological processes (e.g., declining fish stocks; see Kalikoski and Allison this volume) or increasing interannual variation and frequency of extreme events (see Wandel and Marchildon this volume) produce a wide range of outcomes of different magnitudes and degrees of uncertainty. The emergence of critical transitions or regime shifts (see Steffen et al. 2004) draws attention to processes of change with outcomes that are becoming increasingly difficult to understand given that the drivers and effects of these changes are cross-scale. A growing body of evidence on critical transitions (Scheffer 2009) thus highlights the challenges for building adaptive capacity to deal with surprise as well as cascade and threshold effects (see Galaz et al. 2008). Cascade and threshold effects present situations in which institutions and governance systems are unable to avoid irreversible shifts in SESs, or where there is an inability to buffer the flow of crises across scales and systems.

The chapters in this volume illuminate these dilemmas as they relate to building adaptive capacity. For example, institutions and governance arrangements in certain resource sectors (e.g., water, wildlife) have been, and still largely are, organized around the assumption that natural systems fluctuate within a predictable envelope of variability (de Loë and Plummer this volume). However, the predictability of this envelope is increasingly less clear. Building adaptive capacity for the former situation is not the same as for the latter. de Loë and Plummer (this volume) raise the problem confronting water managers of how to cope and adapt in this new environment. This problem, as well as the solutions being proposed, extends well beyond the domain of water into other resource sectors. While some, such as Milly et al. (2008), argue that more sophisticated modeling capabilities are required to adapt to climate change impacts, others (e.g., Ivey et al. 2004; de Loë and Plummer this volume) argue that it is necessary to strengthen the institutional capacity of communities, organizations, and societies to address increasing variability.

In contrast, a number of chapters highlight how intersecting system changes in one region can create significant barriers to adaptive capacity and governance. Bohensky et al. (this volume), for example, document the challenges of building adaptive capacity in the Great Barrier Reef system battered by interconnected changes in adjacent terrestrial watersheds, agricultural systems, climate, and the reef itself. Fabricius and Cundill (this volume) raise the more problematic scenario of how to build adaptive capacity in a system that has possibly already crossed a threshold. The highly impoverished and ecologically degraded Macubeni system creates far different adaptive capacity and governance needs, for example, than most other cases in this volume. The implications of ongoing change for adaptive capacity and governance are profound. Because the structure and conditions of complex, multivariate, nonlinear, and multiscale SESs are always in flux (Gunderson and Holling 2002), the task of building adaptive capacity and governance arrangements is forever incomplete.

14.3.2 Blueprints and Panaceas

Ostrom (2007: 15181) has highlighted problems with the "perverse and extensive uses of policy panaceas in misguided efforts to make SESs...sustainable over time." As the chapters in this volume illustrate, efforts to build adaptive capacity and multilevel governance arrangements for complexity are no different. As one example, Diduck (this volume) identifies the recent proliferation of "learning" in relation to adaptive capacity and resource and environmental governance. He argues that it is necessary to untangle the definitions and explanations of learning, and uses five social units of analysis (individual, action group, organizational, network, and societal) to delve into the theoretical nuances and the implications for applied research and practice. de Loë and Plummer (this volume) provide an applied example of the need to "go beyond panaceas." In exploring the challenge of adapting to the impacts of climate change on water supply, they highlight the importance of considering the context of the water system, the anticipated outcomes of climate impacts, and the capability of the system's management to address current and future impacts. There is no simple or linear strategy to build adaptive capacity given the pace and scale of social–ecological change (Fabricius and Cundill this volume), the myriad constraints on the emergence of institutions that match biophysical systems (Matthews and Sydneysmith this volume), and the messy influences of power, ethics, culture, and even genes (Fennell and Plummer this volume).

A key lesson that emerges from the combined insights of this volume is that actors (e.g., resource users, department officials) engaged in collaborative governance processes in "messy social–ecological systems" (see Alessa et al. 2009) must build issue- and place-specific strategies. There is a need for further clarity on how specific experiences and attributes that build adaptive capacity can be transferred across institutions and levels (Keskitalo this volume). Yet it is also clear that there are no panaceas when it comes to building adaptive capacity for governance.

14.3.3 Relational Spaces and Institutions

Pelling et al. (2008) suggested that relational spaces – the places and sites where actors engage in understanding and resolving challenges – are a result of the interplay of institutions and social learning. Such spaces may lead to a process of social transformation and are associated with adaptive capacity. Kalikoski and Allison's (this volume) review of cases from Africa and South America shows how learning and adaptive capacity are built within fishing communities through comanagement. As actors engage in these relational spaces, we can anticipate multilevel interactions. Advances are being made to trace relational spaces by concentrating on the interactions between individual learning, organizational contexts, and the pathways by which adaptive capacity and action are negotiated (Pelling and High 2005; Pelling et al. 2008; see also Bodin and Crona 2009). For example, Diduck (this volume) links learning outcomes at several levels to illuminate adaptive capacity in multilevel governance, and supports the suggestion that relational spaces enhance adaptive capacity. He reveals the importance of accounting for network and societal interactions, the institutions that link individuals and groups, and the power asymmetries that influence relational spaces. The challenge for understanding these spaces, however, stems from their inevitable diversity, and the very different institutional (formal, informal), organizational, and social variables with which they are associated (see Vatn 2005).

The notion of relational spaces highlights the role of institutions in shaping adaptive capacity. Matthews and Sydneysmith (this volume) in fact argue for an institutional perspective that links culture, organizations, and the actions of individual actors, highlighting the limitation of a perspective of institutions as simply normative constraints (see also Ostrom 2005). Emerging from these institutional relationships may be what are referred to as shadow networks (often informal linkages and connections among individuals and groups) because they remain largely hidden, difficult to delineate, and even more difficult to control (Stacey 1996; Pelling et al. 2008). As Matthews and Sydneysmith (this volume) point out, new institutionalists in sociology can offer a unique perspective that is fundamentally concerned with questions of power and authority. Institutions and adaptive capacity are recognized as embedded within a broader context of social structures and a result of particular social relationships that lend them legitimacy. The chapters in this volume illustrate the breath of this institutional perspective as it relates to adaptive capacity and governance. For example, Wesche and Armitage (this volume) emphasize the importance of culture and land-based traditions for adaptive capacity of the Dene in Fort Resolution, Northwest Territories. They also adopt a perspective "from the inside out" and focus on features of social organization in relation to the wider Northern institutional context, highlighting the largely informal, multiscale shadow networks that play an important role in constraining or enabling adaptive capacity. Employing a similar lens, Bohensky et al. and de Loë and Plummer (this volume) provide commentaries on the formal "rules of the game" but also discuss how "the game" is actually being played out in the context of networks and social relationships shaped by asymmetries of power, authority, and legitimacy. Moreover, Kalikoski and Allison, as well as Fabricius and Cundill (both this volume), discuss institutional incentives, such as the importance of participation in comanagement of fisheries, or the role of property rights and economic benefits in building capacity to respond to systems beyond the threshold. As Young et al. (2008) have argued, institutional analysis is cutting-edge social science, and they pointed to the critical role of institutions in shaping "real world" environmental governance systems and policy advances. Understanding adaptive capacity, therefore, demands a careful tracing of formal and informal institutional connections or pathways across spatial and temporal domains.

14.3.4 Scaling Up, Scaling Down, Scaling Out

Scale is a central concern in most of the chapters in this volume. A useful definition as it pertains to adaptive capacity and governance is the notion of scale as the "... spatial, temporal, quantitative, or analytical dimensions used by scientists to measure and study objects and processes" (Gibson et al. 2000). In this regard, scale and level are distinct, with level the units of analysis that are located at different positions on a scale (Cash et al. 2006). Cross-scale dynamics and multilevel interactions are recognized as central to understanding and assessing the attributes and conditions of adaptive capacity in a governance context, and thus, there is inevitably a diversity of variables and processes to consider. For example, the review of experiences with the CAVIAR project (Sydneysmith et al. this volume) emphasizes how exposure sensitivities and adaptive capacity of communities in the Arctic are dynamic across both temporal and spatial scales. They point to an interplay among local leadership and knowledge, social networks, economic resources (e.g., human, financial, equipment), and local institutions with policy decisions and market fluctuations that originate from afar. In the context of global drivers of change, they illustrate how adaptive capacity at the community scale intersects in multiple ways with external events, resources, and higher order governance institutions.

Several insights for adaptive capacity and governance emerge from scale-sensitive analysis. For instance, it is clear that adaptive capacity is scale-dependent, or specific to place and context (Bohensky et al. this volume). This means that it is difficult to scale up from local experience or transfer (scale down) the adaptive capacity of regional bodies to the local level. A clear example of this are the constraints on building adaptive capacity in aboriginal drinking water systems where cross-scale issues like climate change, economic dislocation, and externally driven socio-cultural pressures make coping and adapting particularly difficult. There are, moreover, inevitable trade-offs across scales when seeking to build adaptive capacity, as is the case in the examples from the Northwest Territories and northern Sweden (Wesche and Armitage this volume; Keskitalo this volume). Wesche and Armitage note how building capacity of one group (the Dene) may in fact lead to, or result from, conflict with other groups (the Metis), or how the claims of emerging regional groups through the Akaitcho Territory may paradoxically undermine local capacity-building efforts in the Deninu K'ue Traditional Territory. In contrast, several chapters illustrate how adaptive capacity of certain groups in one part of a system may compensate for a lack of capacity to adapt elsewhere in the system. This is an important relationship for governance indicated by Diduck (this volume), who focuses on the role of different levels and the need to document the relationships among different units of analysis (individual, organizational, etc.) with regard to adaptive capacity. Elsewhere, experiences in the Great Barrier Reef context (Bohensky et al. this volume), the Macubeni region (Fabricius and Cundill this volume), and Brazil fisheries context (Kalikoski and Allison this volume), all provide examples of certain actors or groups operating at different scales that have enabled adaptive capacity in a governance context by compensating for resource, economic, and or political capacity constraints of others (e.g., small-scale producers). As scaling adaptive capacity up, down, or out is not straightforward, strategies to build adaptive capacity in a multilevel governance setting must be carefully targeted to connect with the needs of a diverse range of governance actors.

14.3.5 Actors and Networks, Networks of Actors

Several chapters in this volume (i.e., de Loë and Plummer; Fabricius and Cundill; Olsson et al.; Wesche and Armitage) draw attention to the diverse roles and requirements of state and nonstate actors in building adaptive capacity and creating conditions for adaptive governance. As these contributions illustrate, the "State" is no longer alone in claiming the authority (although see below) or resources to deal with increasingly complex and multiscale problems (Stoker 1998; Sonnenfeld and Mol 2002; Pahl-Wostl 2009). Moreover, it is evident that no one actor can possess the knowledge, experience, resources, or legitimacy necessary to address complex and connected global environmental problems. These insights point to the increasing importance of networked arrangements in which a range of public and private actors coordinate to resolve environmental challenges by sharing information, speeding up the transfer of knowledge about ecosystem dynamics and feedback, and distributing equitably the costs and benefits of action (Bodin and Crona 2009; Berkes 2009; Schultz 2009; Plummer 2009). Networks may thus serve to increase adaptive capacity and build resilience (Tompkins and Adger 2004).

Kalikoski and Allison (this volume) highlight how networks for governance have formed out of comanagement arrangements to support efforts to deal with cross-scale effects. However, their analysis also points out how networks have self-organized in one context but not in an adjacent area. They point to the role of power and legitimacy as key factors in the self-organization of networks prepared to build institutions and respond and adapt to change. Keskitalo's analysis, in contrast, shows what happens when networks are slow to form, creating the potential for conflict in sites of interaction and reducing the capacity to cope and adapt. These findings emphasize the importance of linkages and flexible networks of actors, or partnerships, in building adaptive capacity. However, legitimacy of the network is required to confer stability *and* adaptability, and to create capacity for diffusing the negative effects of change and distributing the benefits (Lebel et al. 2006; Bodin and Crona 2009). In the context of networked governance arrangements, responsibility and accountability are dispersed among a variety of actors (local, regional, private, public). The cases in this volume illustrate that legitimacy may be less a matter of formalized mandates than of the strength of bonds that build trust.

14.3.6 From Governance to Government?

In most environmental governance scholarship, government is no longer considered the sole or even main source of environmental decision-making authority. This shift in thinking is characterized as a transition from government to governance, and reflects an understanding that multiple actors from corporations to nongovernment organizations, public–private partnerships, and quasigovernmental boards are central players in environmental decision making.

The need to consider the adaptive capacity of a broad range of stakeholders and governance actors is well recognized in the literature (e.g., Folke et al. 2005; Boyd 2008; Himley 2008; Armitage et al. 2009). Paradoxically, many case examples in this volume point to a central role for *government* in fostering adaptive capacity. There appear to be a number of reasons for this. For example, state policies and regulations create a very real external limit to the ability of subnational or local actors to adapt and transform (Keskitalo this volume; Wandel and Marchildon this volume). Where states lack legitimacy or functional capacity, prospects for building adaptive capacity at other levels appear to be eroded or at least undermined. This is illustrated to a certain degree by Keskitalo (this volume) as local-level adaptive capacity in the Swedish case is constrained by broader policy interactions between reindeer herders and forestry sectors, and in the case from South Africa (Fabricius and Cundill this volume). In contrast, the case from the Northwest Territories in Canada highlights how government support programs to local harvesters and the emerging land claims agreements (federal-First Nation agreements) may in fact enhance the ability of local communities in the North to proactively respond to change. However, in the case from Alberta Wandel and Marchildon (this volume), note that the "state" is not monolithic, and as a result, it is essential to determine which levels and branches of government are critical to build adaptive capacity in the broader governance process. That chapter related how federal policies, historically, undermined adaptive governance, while provincial policies created opportunities for novel responses to difficult biophysical circumstances.

Recognizing the critical role of the state raises some obvious questions. Are the conventional policy measures of the state – legislation, taxes, and subsidies – up to the task of promoting adaptation and transformation in the face of profound social–ecological change? If not, what types of new and innovative public policy tools do we need to build adaptive capacity and facilitate systemic transformation? How might such tools interact with private-sector-led and community-based governance initiatives? Answering these and other public policy questions will shed further light on suitable roles for the state in adaptive governance. However, a lesson from this volume is that government still has an essential role to play as a leader in seeing that experiences from one setting can be transmitted across scales in an effort to build adaptive capacity.

14.3.7 Organizations on the Edge

Government clearly has a central role to play in providing leadership and creating the enabling conditions for adaptive capacity. However, one interesting theme that has emerged in several chapters is the influence of bridging organizations in helping to build adaptive capacity. Bridging organizations by definition operate on the edges of different domains of practice and in so doing create arenas for social processes and trust-building that support shared capacity for adapting, coping, and transforming. Bridging organizations can thus foster and sustain the relational spaces and governance networks which engender adaptive capacity. Evidence of this appears in several chapters in the volume (notably in the cases from Brazil, South Africa, and Sweden) where the strength of specific organizations that bridge local and higher level interests plays a significant role in the adaptiveness and effectiveness of governance processes. Schultz (2009) has documented the role of bridging organizations specifically in terms of their contributions to knowledge generation, collaborative learning, preference formation, sense making, and conflict resolution among actors in relation to specific environmental issues. Also identified is the role bridging organizations play in supporting the vertical and horizontal linkages that improve information and resource flows and support adaptive capacity (see also Berkes 2009).

Bridging organizations may be particularly important where government capacity is constrained. This appears to be the case in the example from South Africa (Fabricius and Cundill this volume) where the Macubeni Project Advisory and Steering Committee (MPASC) provided critical linkages among community representatives, donor organizations, and government actors. This coordination, enabled at least initially by the MPASC, was necessary to deal with the fragile linkages that existed among these groups in a very challenging context. In the Swedish case examined by Keskitalo (this volume), the lack of an effective bridging organization may have constrained efforts to bring together different groups (reindeer herders, forestry interests, tourism operators) to collaborate more formally, and to build shared capacity to deal with vulnerability and adapt to change. However, bridging organizations also appear to have an important role where the scope of government has been purposefully constrained through decentralization processes. Many of the more recent governance processes in Canada's North provide just such an example as comanagement bodies are emerging as key sites of interaction for different kinds of knowledge sharing (local, scientific), the coordination of tasks enabling cooperation and networking, and learning that promotes adaptive capacity (Armitage et al. 2009; Berkes 2009). Kalikoski and Allison (this volume) document a similar process and set of conditions in the cases they assess, although with variable outcomes. One clear example they cite is the formation of the Forum of Patos Lagoon in Brazil by nongovernmental (i.e., church), community, and government actors. As the Forum itself evolves and learns through experience, it may yet become an important mechanism to bring groups together to deal with the crisis in small-scale fisheries in the lagoon system.

14.3.8 Knowledge Coproduction and Learning Processes

Knowledge is a key component of learning. A diversity of knowledge types and sources about complex SESs, explored through collaborative processes and efforts at sense making, is fundamental to building adaptive capacity for governance (Diduck this volume). Much effort to foster adaptive capacity thus depends on knowledge mobilization and knowledge coproduction. We use knowledge coproduction here to refer to the collaborative process where a plurality of knowledge sources and types are brought together to address a defined problem and to build a systems-oriented understanding of that problem. A willingness to integrate knowledge sources as a basis for testing assumptions and modifying worldviews is the basis for learning processes that enable adaptive capacity (Diduck this volume). With regard to climate change adaptation, for example, bridging science and indigenous knowledge has been shown to produce complementarities in temporal and spatial scales, and thus help understand impacts and adaptations and identify monitoring needs (see Riedlinger and Berkes 2001). Knowledge coproduction thus depends on the openness of actors to share and draw upon a plurality of knowledge in the search for solutions to pressing governance challenges (Berkes 2009).

In the cases examined by Olsson et al. (this volume), governance successes were linked to a large degree to the incorporation of improved understanding of ecosystem dynamics and the capacity to respond to ecosystem feedbacks more effectively as a result. This knowledge was not held by one group or actor, but coproduced. There is good evidence to suggest (Kalikoski and Allison this volume) that knowledge to facilitate adaptive governance and purposeful transformation emerges when a sufficient number of stakeholders agree that system conditions are untenable

(Olsson et al. this volume). Often, this recognition is precipitated by a crisis, and leads to a reevaluation of knowledge about the system and the assumptions and values framing that knowledge. Building adaptive capacity for governance means creating an arena for the coproduction of knowledge, with the assumption that knowledge itself is a dynamic process and contingent upon being formed, validated, and adapted to changing circumstances (Davidson-Hunt and O'Flaherty 2007: 293). At the same time, it is important to remember that changing perspectives, attitudes, and behaviors do not necessarily follow the accumulation of information or the increasing knowledge of individuals (Kollmuss and Agyeman 2002). Mindfulness is therefore required concerning basic questions about the who, what, how, and why of learning in such processes (Diduck this volume; Lundholm and Plummer in press).

14.3.9 The Importance of Ecosystems

A particular strength of this volume is the depth of insight into the institutional arrangements, social attributes and processes that influence adaptive capacity, and implications for flexible, networked governance. In summarizing the chapters, two insights emerge relating to ecosystems. First, it is clear that in the various sites and places where capacity has emerged to support adaptation, that capacity is not developed specifically for environmental change. The cases highlighted from the CAVAIR project (Sydneysmith et al. this volume), for example, show how vulnerability and adaptation emerge in response to the combined effects of environmental change and socioeconomic development. Untangling the relationship between vulnerabilities associated with large-scale resource development versus those associated with climate change is difficult. In fact, many of the cases in this volume point to how communities and other governance actors are dealing with a wide range of stresses and crises cast in terms of their social and economic consequences. Second, several contributions to this volume point out that connections to underlying ecosystem conditions and services which ultimately frame adaptive capacity are not always well articulated. Olsson et al. (this volume) are the most explicit in this regard. They note that significant attention to adapting and transforming can occur in the absence of improved capacity to learn from, respond to, and manage environmental feedback from dynamic ecosystems. This can lead in turn to further ecological degradation, a heightened need to cope, and reduced capacity for adapting and transforming in the longer term. As Newig and Fritsch (2009: 209) point out, "...the relation between public participation on the one hand and multilevel governance and rescaling on the other hand as well as its environmental effects have barely been treated." Ultimately, a focus on building adaptive capacity without adequate ecological knowledge or literacy serves to limit governance arrangements for navigating change.

14.3.10 It Takes Two to Tango!

Pomeroy and Berkes (1997) used the expression, "It takes two to tango", to describe the important interplay between community resources users and government actors in comanagement arrangements. We use the saying here in a different way – to emphasize the union of social–institutional and biophysical systems in relation to adaptive capacity. Complex systems thinking is integrative, and requires us to embrace an SESs perspective (Berkes et al. 2003). Correspondingly, this volume takes an interdisciplinary approach and integrative perspective to adaptive capacity. In so doing, the contributors weave together the bodies of literature and themes that inform our understanding of adaptive capacity (see Plummer and Armitage this volume). For example, Fennell and Plummer (this volume) highlight the connection between the natural and social sciences in regards to adaptive capacity. They unpack adaptive capacity as it has been interpreted by both ecological and social–institutional perspectives. In taking an integrative "nature-in-humans" view, they argue that evolutionary biology provides a metatheoretical basis for adaptive capacity. Both ecological and socio-institutional contexts are requisites for effective environmental governance, and environmental governance in turn shapes SESs. It takes two to tango!

Conducting research and/or engaging in practice at the nexus of social and natural sciences presents considerable challenges for understanding and building adaptive capacity into governance systems, and creating opportunities for transformation (Olsson et al. this volume). However, as Snow (1964:16) highlighted, "…the clashing points of two subjects, two disciplines, two cultures – of two galaxies, so far as that goes – ought to produce creative changes. In the history of mental activity that has been where some of the breakthroughs came." Efforts towards a common or general framework for analyzing SESs, such as that advanced by Ostrom (2009), are critical, as in their absence such knowledge does not cumulate. Persistent effort is required to ensure that (1) bridges are maintained between social and biophysical systems, (2) meaningful integration of the various perspectives occurs when examining adaptive capacity, and (3) insight gained from such endeavors accumulates to advance knowledge.

14.4 Conclusions

The lessons and insights summarized above reflect a growing body of knowledge about adaptive capacity and the emergence of governance systems for navigating change. The chapters have identified, in particular places and contexts, the specific attributes of institutions that enhance learning opportunities and build adaptive capacity. They highlight the different roles participants in governance processes have in fostering the capacity to adapt and transform. And the chapters illustrate how the bundles of attributes and experiences that build capacity in one site can be linked through networks to enhance opportunities for adapting and transforming in other places. In this regard, many of the chapters highlight the importance of vertical and horizontal linkages and interactions that build social capital and trust (Cash et al. 2006; Brondizio et al. 2009). We are provided with empirical examples of how flows of information, resources, and knowledge in emerging multilevel governance processes are linked to the capacity to adapt, and to transform SESs into alternative, sustainable trajectories. At the same time, many of the chapters illustrate how adaptive capacity and governance intersect with the political economy of particular sites and the inherent conflicts over resource access, control, and value systems. As reflected throughout the volume, no one group alone will have the power, resources, or skills to deal with increasing variability, nor provide the portfolio of tools to support adaptive capacity and environmental governance.

Still, further research is required. How institutional arrangements can best create the conditions for adaptive capacity is not yet fully articulated. More detailed studies are needed to examine how linkages in multilevel governance (vertical and horizontal) actually work. Greater clarity is urgently needed on how the attributes and experiences that build capacity in one area (e.g., wildlife management) can be transferred most effectively across institutions and scales to other areas, some of which, such as climate change, are exceedingly complex. Thus, in the context of global environmental change, how to enhance adaptiveness, foster learning, and improve fit between institutions and ecosystems is an area requiring significantly more research (Young et al. 2008; Biermann et al. 2009). If and how adaptive, multilevel governance generates outcomes that are not just socially positive but which also sustain the ecosystem services upon which we depend is one area that demands much more attention and better evaluative tools (Plummer and Armitage 2007; Olsson et al. this volume). Lastly, more effort is required to translate analytical understanding of the links between adaptive capacity and governance into accessible diagnostic tools enabling resource users, managers, and policy makers to identify key drivers of change, the strategies required to foster positive transformation (e.g., institutional mechanisms, leadership requirements), and opportunities to overcome path dependencies. The contributions of this volume are one step in that process, and help us find those windows of opportunity for positive change.

Acknowledgments We thank Alan Diduck, Carina Keskitalo, and Lisen Schultz for their valuable comments and feedback on an earlier version of this chapter. We thank as well the contributors to the volume for the insights and perspectives they have brought to the challenge of adaptive capacity and governance.

References

Alessa L, Kliskey A, Altaweel M (2009) Toward a typology for social–ecological systems. Sustainability Sci Pract Policy 5(1):31–41

Armitage D, Plummer R, Berkes F, Arthur R, Charles A, Davidson-Hunt I, Diduck A, Doubleday N, Johnson D, Marschke M, McConney P, Pinkerton E, Wollenberg L (2009) Adaptive co-management for social–ecological complexity. Front Ecol Environ 7(2):95–102

Berkes F (2009) Evolution of co-management: role of knowledge generation, bridging organizations and social learning. J Environ Manage 90(5):1692–1702
Berkes F, Colding J, Folke C (2003) Navigating social–ecological systems: building resilience for complexity and change. Cambridge University Press, Cambridge
Biermann F, Betsill MM, Gupta J, Kanie N, Lebel L, Liverman D, Schroeder H, Siebenhüner B, Conca K, da Costa Ferreira L, Desai B, Tay S, Zondervan R (2009) Earth system governance: people, places and the planet. Science and implementation plan of the earth system governance project, ESG report no. 1, Bonn, IHDP, The Earth System Governance Project
Bodin O, Crona B (2009) The role of social networks in natural resource governance: what relational patterns make a difference? Glob Environ Change 19(3):366–374
Boyd E (2008) Navigating Amazonia under uncertainty: past, present and future environmental governance. Philos Trans R Soc Lond B 363(1498):1911–1916
Brondizio ES, Ostrom E, Young O (2009) Connectivity and the governance of multilevel socio-ecological systems: the role of social capital. Annu Rev Environ Resour 34:253–278
Carpenter S, Mooney H, Agard J, Capistrano D et al (2009) Science for managing ecosystem services: beyond the millennium ecosystem assessment. Proc Natl Acad Sci USA 106(5):1305–1312
Cash DW, Adger WN, Berkes F, Garden P, Lebel L, Olsson P, Pritchard L, Young O (2006) Scale and cross-scale dynamics: governance and information in a multilevel world. Ecol Soc 11(2):8
Davidson-Hunt I, O'Flaherty RM (2007) Researchers, indigenous peoples and place-based learning communities. Soc Nat Resour 20:291–305
Dietz T, Ostrom E, Stern PC (2003) The struggle to govern the commons. Science 302:1907–1912
Duit A, Galaz V (2008) Governing complexity: insights and emerging challenges. Governance 21:311–335
Folke C, Hahn T, Olsson P, Norberg J (2005) Adaptive governance of social–ecological knowledge. Annu Rev Environ Resour 30:441–473
Galaz V, Olsson P, Hahn T, Folke C, Svedin U (2008) The problem of fit among biophysical systems, environmental and resource regimes, and broader governance systems: insights and emerging challenges. In: Young OR, King LA, Schroeder H (eds) Institutions and environmental change. MIT, Cambridge, pp 147–186
Gibson CC, Ostrom E, Ahn TK (2000) The concept of scale and the human dimensions of global change: a survey. Ecol Econ 32:217–239
Gunderson LH, Holling CS (eds) (2002) Panarchy: understanding transformations in human and natural systems. Island, Washington
Himley M (2008) Geographies of environmental governance: the nexus of nature and neoliberalism. Geogr Compass 2(2):433–451
Ivey JL, Smithers J, de Loë RC, Kreutzwiser RD (2004) Community capacity for adaptation to climate-induced water shortages: linking institutional complexity and local actors. Environ Manage 33(1):36–47
Kollmuss A, Agyeman J (2002) Mind the gap: why do people act environmentally and what are the barriers to pro-environmental behaviour? Environ Educ Res 8(3):239–260
Lebel L, Anderies JM, Campbell B, Folke C, Hartfield-Dodds S, Hughes TP, Wilson J (2006) Governance and the capacity to manage resilience in regional social-ecological systems. Ecol Soc 11(1):19. (online). URL: http://www.ecologyandsociety.org/vol11/iss1/art19/
Lundholm C, Plummer R Learning and resilience: a conspectus for environmental education. Environ Educ Res (in press)
MA (Millenium Ecosystem Assessment) (2005) Ecosystems and human well-being: current state and trends. Island, Washington
Milly PCD, Betancourt J, Falkenmark M, Lettenmaier D, Stouffer RJ (2008) Stationarity is dead: whither water management. Science 319(5863):573–574
Newig J, Fritsch O (2009) Environmental governance: participatory, multi-level – and effective? Environ Policy Governance 19(3):197–214
Ostrom E (2005) Understanding institutional diversity. Princeton University Press, Princeton

Ostrom E (2007) A diagnostic approach for going beyond panaceas. Proc Natl Acad Sci USA 104(39):15181–15187

Ostrom E (2009) A general framework for analyzing sustainability of social–ecological systems. Science 325(24):419–422

Pahl-Wostl C (2009) A conceptual framework for analysing adaptive capacity and multi-level learning processes in resource governance regimes. Glob Environ Change 19(3):354–365

Pelling M, High C (2005) Understanding adaptation: what can social capital offer assessments of adaptive capacity? Glob Environ Change 15(4):308–319

Pelling M, High C, Dearing J, Smith D (2008) Shadow spaces for social learning: a relational understanding of adaptive capacity to climate change within organisations. Environ Plan A 40:867–884

Plummer R (2009) The adaptive co-management process: an initial synthesis of representative models and influential variables. Ecol Soc 14(2):24. (online). URL: http://www.ecologyandsociety.org/vol14/iss2/art24/

Plummer R, Armitage D (2007) A resilience-based framework for evaluating adaptive co-management: linking ecology, economy and society in a complex world. Ecol Econ 61(1):62–74

Pomeroy RS, Berkes F (1997) Two to tango: the role of government in fisheries co-management. Mar Policy 21:465–480

Riedlinger D, Berkes F (2001) Contributions of traditional knowledge to understanding climate change in the Canadian Arctic. Polar Rec 37:315–328

Scheffer M (2009) Critical transitions in nature and society. Princeton University Press, Princeton

Schultz L (2009) Nurturing resilience in social–ecological systems: lessons learned from bridging organizations. Stockholm University Press, Stockholm

Snow CP (1964) The two cultures. Cambridge University Press, London

Sonnenfeld DA, Mol APJ (2002) Globalization and the transformation of environmental governance. Am Behav Sci 45(9):1318–1339

Stacey R (1996) Complexity and creativity in organizations. Berrett-Koehler, San Francisco

Steffen W et al (2004) Global change and the earth system: a planet under pressure. Springer, Berlin

Stoker G (1998) Governance as theory: five propositions. Int Soc Sci J 50(155):17–28

Tompkins EL, Adger WN (2004) Does adaptive management of natural resources enhance resilience to climatic change? Ecol Soc 9(2). (online). URL: http://www.ecologyandsociety.org/vol9/iss2/art10/

Vatn A (2005) Institutions and the environment. Edward Elgar, Cheltenham

Walker BH, Holling CS, Carpenter SR, Kinzig A (2004) Resilience, adaptability and transformability in social–ecological systems. Ecol Soc 9(2): 5. (online). URL: http://www.ecologyandsociety.org/vol9/iss2/art5/

Young OR, King LA, Schroeder H (eds) (2008) Institutions and environmental change: principal findings, applications, and research frontiers. MIT, Cambridge

Index

A

Action group, 204, 212
Actors and networks, role of actors, 294
Adaptation, 26, 287
   forestry, 101
   reindeer herding, 99
   winter tourism, 100
Adaptive capacity, 1, 2, 5, 6, 10, 15, 25, 27, 70, 112, 147, 216, 217, 223–225, 227, 229, 230, 233–235, 237, 239, 244, 245, 287–300
   aboriginal communities, 173
   aboriginal-state relations, 118
   as an enabling factor in multi-level governance, 10
   bottom-up approach to understanding, 229
   and bridging organizations, 296
   for building environmental governance, 15
   climate change studies perspective, 8
   community, 255
   and complex adaptive systems, 11
   to deal with change and uncertainty, 290
   deductive approaches, 28
   determinants, 227, 228
   economic transition, 117
   and environmental governance, 6, 10, 24
   in environmental governance discourse, 2
   evolution, 255
   Fort Resolution, 113
   fostering in social-ecological systems, four factors, 2
   four general principles to enhance adaptive capacity and governance, 289
   geographic scale, 228
   government support programs, 116
   Great Barrier Reef sub-regions, 35
   inductive approaches, 28
   information and knowledge, 37
   and institutions, 11, 147
   local conditions, 228
   natural science perspective, 7
   policy response, 127
   political ecology perspective, 7
   resilience thinking and social-ecological systems perspective, 9
   risks and hazards perspective, 8
   role of government, 296
   role of networks of actors, 295
   and scale, 35, 36, 293
   self-governance, 120
   and social capital and networks, 12
   social-ecological systems, 26
   social sciences perspective, 7
   social systems, 26
   socio-institutional, 246
   temporal scale, 228
   tourism, 35
   vulnerability and exposure, relationship to, 227
Adaptive capacity and institutional dynamics, 231
Adaptive capacity of institutions, 236
Adaptive co-management, 43, 61, 213
   theory, 43
Adaptive learning, 79
Adaptive management, 54, 58
Adaptive strategies, 252
Akaitcho Territory Government (ATG), 119

Index

Alberta
　drought, 191
　open-range cattle ranching, 183
　social-ecological system (*see* Special Areas)
Altruism, 249

B
Bantu Authorities, 47
Biological evolution, 243
Bridging organizations, 296, 297
Brundtland report, 224
Bundles of attributes, 289, 299
　supporting and linking capacity and governance, 290

C
Canada, drinking water supply, 162
Canadian Great Plains. *See* Great Plains, The
Capacity and capacity building, 11
　functionalist perspective, 11
　relational perspective, 11
Capital, financial, human, natural, physical, 53
CAVIAR project, 134, 152
　adaptive strategies, 136
　vulnerability, 135
Change and uncertainty, 290
　cascade and threshold effects, 290
　transitions and regime shifts, 290
Climate change
　aboriginal communities, 169
　adaptation, 159
　adaptive capacity, 159
　drinking water treatment, 163
Cognition, 204
　collective cognition, 205
　multiple cognition, 205
　mutual cognition, 215
Collaboration, 5, 11, 12
Collective action dilemmas, 12
Co-management, 62, 78, 81
　devolution, 85
　fisheries, 69, 72
　power imbalance, 74
Commercial fishing industry, 29
Communicative processes. *See* Deliberative processes
Community-based monitoring, 58
Community-based social marketing, 214
Complex adaptive systems, 10
　and complex systems thinking, 11
Complex systems thinking, 299
Concerted action, 202, 204, 211, 217
Contemporary challenges of global-local environmental change, 3

D
Deliberative processes, 200, 203, 208, 217
Dewey, J., 208, 211
Drinking water
　aboriginal communities, 167
　governance, 166, 167
Driving forces-pressure-state-impact-response (DPSIR), 33

E
eCacadu River, 46
Environmental governance, 4, 5, 15, 245
　crucial aspects, 5
　governance *vs.* government, 4
　interplay, 5
　limitations of command and control, 4
　making environmental governance operational, 5
　models, 5
Exposure-sensitivities, 137, 141, 142, 144

F
Fisheries management, 73
Fishers, artisanal, 75
Fit, 180
　functional, 181, 184, 189
　institutional, 184, 192
　spatial, 180, 184, 188
　temporal, 180, 188
Flexibility, 144, 150
Forestry, 96
　environmental protection, 98
　fertilisation, 95
Fort Resolution, 108
　cultural development, 124
　social change, 111
Forum of the Patos Lagoon, 76
Freire, Paulo, 217
From governance to government, 295

G
Gällivare, 91
　land use, 103
Gene-culture
　co-evolution, 251
　evolution, 255
Governance, 289, 299
Government, 295
Great Barrier Reef, 23
　social-ecological system, 29

Index 305

water quality, 30
Great Barrier Reef Marine Park Authority
 (GBRMPA), 24
Great Plains, The, 179, 181

H
Habermas, Jurgen, 203, 208

I
Inclusive fitness, 249
Institutional analysis and development (IAD)
 framework, action arenas, 235
Institutional Human Dimensions Program
 (IHDP), 235
 IDGEC and institutional capacity, 236
 Institutional Dimensions of Global
  Environmental Change (IDGEC), 235
Institutional interplay, 181
Institutional perspective, 292
Institutions, 5, 11, 201, 206, 208, 209, 224, 230,
 231, 233, 237, 287, 289, 299
 and adaptive capacity, 224
 constraints, 231
 fit, interplay, and scale, 11
 institutional analysis, 224, 293
 institutional arrangements, 230, 300
 institutional capacity, 291
 institutional incentives, 293
 and power, 292
 and social relationships, 293
Institutions and behaviour, 233
 and adaptive capacity, 232
 calculus approach, 232
 cultural approach, 231
Institutions and organizations
 distinction between, 231, 234
 relationship between, 231, 234
Intergovernmental Panel on Climate Change
 (IPCC), 3, 225
 Fourth Assessment Report (AR4), 225
Interim Measures Agreement (IMA), 122
International Human Dimensions
 Programme (IHDP), 225
Interplay, 184
 horizontal, 189
 institutional, 192
 vertical, 189

K
Knowledge co-production and learning
 processes, 297
Knowledge transfer, 126
Kristianstads Vattenrike, 274

L
Lake Malombe, 76
Lake Malombe Participatory Fisheries
 Management Program (PFMP), 72
Learning, 12, 291, 297, 299
 action group learning, 204, 206, 207, 211,
  212, 215
 double-loop or transformative, 12
 as a feature of innovative governance, 12
 individual learning, 202, 209, 211
 multi-level learning, 199, 213, 216, 217
 multi-level learning connections, 201, 203,
  209, 214
 network learning, 207, 216
 non-formal learning, 212–214
 organizational learning, 206, 207, 211
 single-loop, 12
 social learning, 200, 201, 204, 208, 215
 societal learning, 208, 213, 214, 216
 transformative learning, 203, 211
Legitimacy, 77
Lock-in traps, 265, 289

M
Macubeni
 adaptive co-management, 48
 fine resolution approach, 64
 governance, 56
 hysteresis effect, 64
 land management, 46
 management plans, 60
 social-ecological resilience, 45
 social vulnerability, 47
Macubeni Dam, 47
Macubeni Project Advisory and Steering
 Committee (MPASC), 52
 model, 58
Macubeni Technical Committee, 52
 capacity, 57
Marine extractive reserve (MER), 73, 75, 82
Meme, 255
Millennium ecosystem assessment (MA), 2, 3
 drivers, 3
 findings, 3
 millenium development goals, 3
Multi-level governance, 287, 288, 291, 300
 contributions to capacity, 288
Multiple cognition, 211

N
Natural resource management, 52
Networks, 294, 300
 co-management arrangements, 294

Networks (*cont.*)
conflicts, 295
legitimacy, 295
shadow networks, 292
New institutional analysis (NIA), 225, 231, 232, 234–236
nkacha, 71, 78
Non-formal education, 211, 217
critical non-formal education, 217

O
Organizational frames, 212
Organizational memory, 206, 212
Organizational networks, 207, 272
Ostrom, E., 225

P
Panaceas, 291
Panarchy, 223
Path dependent, 265
Platforms for learning, 205, 211
Power, 232–234, 238, 292, 295
and adaptive capacity, 233
and institutions, 233
Power, ethics, culture, 291
Power relations in learning, 207, 208, 214, 217
Praxis, 204
Principal components analysis, 30
Problem of fit, 264
Punctuated equilibrium, 266

R
Reciprocal altruism, 250, 253
Reef plan, 30
Reindeer husbandry, 93
forest management, 95
Relational spaces, 217, 292
and adaptive capacity, 292
interplay of institutions and social learning, 292
and multi-level interactions, 292
Resilience, 6, 26, 223, 224, 234, 245, 287
arctic communities, 171
transformability, 266
Resource and environmental governance, 215
multi-level governance, 216
Rigidity trap, 265

S
Scale, 51, 293
cross-scale dynamics, 293
trade-offs and constraints for building adaptive capacity, 294

Six Nations of the Grand River, 170
Slave River Delta, 108
Social capacity, 123
Social capital and networks, 12
bonding, 12
bridging, 12
and collaboration, 12
vertical linkages, 12
Social-cognitive filters, 209, 212
Social-ecological systems (SES), 1, 10, 26, 37
Social-ecological systems (SES) transformations, 725
Social network
diversity, 271
strength, 271
Social resilience, 30
Social vulnerability, 90
Societal regime, 267
Sociobiology, 249
Socio-cultural attributes, 31
Socio-ecological system, 247
South Africa, Macubeni, 44
Special Areas, 181
community pastures, 188
drought, 185, 191
institutional reorganization, 190
societal reorganiztion, 190
socio-ecological stability, 191
Stationarity, 157
Subsistence harvesting, 133
Systems perspective, 223, 224

T
Territorial Use Rights to Fisheries (TURFs), 70, 80
The Great Barrier Reef Marine Park, representative areas program (RAP), 276
The importance of ecosystems, 298
The role of local institutions in adaptation to climate change, 230
Tourism, 143
Traditional ecological knowledge, 139, 148
Traditional livelihoods, 138
Transformative capacity, 268, 278
double-loop learning, 267
single-loop learning, 267
triple-loop learning, 267

U
Unsustainable path dependencies, 289

V

Vulnerability, 25, 90, 287
    adaptive capacity, 25
Vulnerability and livelihoods, 13
    connection to resilience or adaptive capacity, 13
    sustainable livelihoods framework, 13
Vulnerability contexts, 137
    community infrastructure, 145
    local culture and society, 137
    market-related enterprises, 142
    subsistence-related livelihoods, 140

W

Water management, 32, 161
Whitehorse, Yukon, 225
    Arctic Gateway City, 236
Winter tourism, 98

Y

Yukon Territorial Government, 237